T0198236

Time-Resolved Mass Spectrometry

Time-Resolved Mass Spectrometry

From Concept to Applications

Pawel Lukasz Urban

Department of Applied Chemistry,
National Chiao Tung University, Taiwan

Yu-Chie Chen

Department of Applied Chemistry,
National Chiao Tung University, Taiwan

Yi-Sheng Wang

Genomics Research Center, Academia Sinica, Taiwan

WILEY

This edition first published 2016
© 2016 John Wiley & Sons, Ltd

Registered office
John Wiley & Sons Ltd, The Atrium, Southern Gate, Chichester, West Sussex, PO19 8SQ, United Kingdom

For details of our global editorial offices, for customer services and for information about how to apply for permission to reuse the copyright material in this book please see our website at www.wiley.com.

The right of the author to be identified as the author of this work has been asserted in accordance with the Copyright, Designs and Patents Act 1988.

All rights reserved. No part of this publication may be reproduced, stored in a retrieval system, or transmitted, in any form or by any means, electronic, mechanical, photocopying, recording or otherwise, except as permitted by the UK Copyright, Designs and Patents Act 1988, without the prior permission of the publisher.

Wiley also publishes its books in a variety of electronic formats. Some content that appears in print may not be available in electronic books.

Designations used by companies to distinguish their products are often claimed as trademarks. All brand names and product names used in this book are trade names, service marks, trademarks or registered trademarks of their respective owners. The publisher is not associated with any product or vendor mentioned in this book.

Limit of Liability/Disclaimer of Warranty: While the publisher and author have used their best efforts in preparing this book, they make no representations or warranties with respect to the accuracy or completeness of the contents of this book and specifically disclaim any implied warranties of merchantability or fitness for a particular purpose. It is sold on the understanding that the publisher is not engaged in rendering professional services and neither the publisher nor the author shall be liable for damages arising herefrom. If professional advice or other expert assistance is required, the services of a competent professional should be sought

The advice and strategies contained herein may not be suitable for every situation. In view of ongoing research, equipment modifications, changes in governmental regulations, and the constant flow of information relating to the use of experimental reagents, equipment, and devices, the reader is urged to review and evaluate the information provided in the package insert or instructions for each chemical, piece of equipment, reagent, or device for, among other things, any changes in the instructions or indication of usage and for added warnings and precautions. The fact that an organization or Website is referred to in this work as a citation and/or a potential source of further information does not mean that the author or the publisher endorses the information the organization or Website may provide or recommendations it may make. Further, readers should be aware that Internet Websites listed in this work may have changed or disappeared between when this work was written and when it is read. No warranty may be created or extended by any promotional statements for this work. Neither the publisher nor the author shall be liable for any damages arising herefrom.

Library of Congress Cataloging-in-Publication Data applied for

ISBN: 9781118887325

A catalogue record for this book is available from the British Library.

Typeset in 10/12pt TimesLTStd by SPi Global, Chennai, India

Printed and bound in Singapore by Markono Print Media Pte Ltd

1   2016

Contents

Author Biographies

 **Pawel Lukasz Urban** received his education from the University of Warsaw (MSc) and the University of York (PhD). His research training was supplemented with pre- and post-doctoral stays at the University of Alcala and ETH Zurich. He currently holds an academic position at the National Chiao Tung University. His research and teaching interests encompass biochemical analysis, development of instrumentation, and engineering smart biosystems.

 **Yu-Chie Chen** received her education from the National Sun Yat-Sen University (MSc) and Montana State University (PhD). She is Professor of Chemistry at the National Chiao Tung University. Her research interests include biological mass spectrometry, nanomedicine, and nanotechnology. She is the co-inventor of several ionization techniques for mass spectrometry (SALDI, UASI, C-API, PI-ESI), which are useful in the monitoring of chemical reactions.

 **Yi-Sheng Wang** received his education from the Feng Chia University (BEng) and the National Taiwan University (PhD). He is currently Associate Research Fellow in the Genomics Research Center of Academia Sinica. His research covers in-depth developments in mass spectrometry instrumentation, fundamental research on ionization chemistry, method development and applications of biological mass spectrometry.

Preface

Mass spectrometry has become an important part of teaching curricula in many university programs related to chemistry, biology, physics, and engineering. The field is highly interdisciplinary. This book takes the readers into the field of time-resolved mass spectrometry (TRMS) providing more detailed conceptual background and practical guidance than the specialized review papers published in the past few years. It is intended to serve as a comprehensive monograph on TRMS. Since the definition of TRMS (cf. Chapter 1) is broad, the scope of this book is extensive. In the first three chapters, we outline the main principles of mass spectrometry. We discuss common ion sources and mass analyzers emphasizing the features of these devices that make them useful in time-resolved measurements (Chapters 2 and 3). Subsequently, we introduce and discuss the design of instrumentation for such measurements considering sample delivery and treatment stages (Chapters 4–9). In Chapter 5, we highlight the trade-off between acquisition speed (temporal resolution), mass resolution, and sensitivity. Eventually, we enumerate examples of different detection/analysis schemes in TRMS that can be implemented in real-world applications (Chapters 10–13). Finally, we present the prospects for future developments and applications of TRMS (Chapter 14). The contents of this book are arranged in a logical sequence but individual chapters can also be read separately – not following the order of the table of contents. Some of the chapters provide a general background on mass spectrometry related technology (Chapters 2 and 3) while the others contain reviews of the previous achievements related to TRMS along with numerous references to original papers and specialized reviews (Chapters 4–13).

We hope this monograph can help science and engineering students better understand the concept of TRMS, and recognize the usefulness of dynamic monitoring of chemical and biochemical processes. We believe it will benefit researchers in academia (including research students and assistants) in the fields of chemistry, physics, and biology; students attending courses in analytical chemistry, organic chemistry, biochemistry, biophysics; as well as industrial scientists.

The field of TRMS is growing rapidly, and almost every month one can read about new exciting developments in the chemistry literature which take advantage of mass spectrometry as a tool to follow processes in time or to detect short-living species. Therefore, we suggest readers consider this book as a primer to TRMS but also recommend following the current progress in TRMS by reading recent articles published in peer-reviewed journals related to mass spectrometry and analytical chemistry. The authors welcome feedback from readers. We are keen to read comments and listen to criticisms. We also apologize to those authors of original papers, whose excellent mass spectrometry work has not been cited in this short monograph, or discussed to any great extent, due to space restrictions.

Pawel Lukasz Urban
Yu-Chie Chen
Yi-Sheng Wang

Acknowledgments

We wish to thank our talented co-workers with whom we interact on a daily basis. These interactions have certainly been beneficial to the writing of this book. Special thanks are due to Professor Yen-Peng Ho (National Dong Hwa University) and Dr Kent Gillig (Academia Sinica) for their comments on the book's manuscript. Yi-Sheng Wang would like to thank Professor Sheng Hsien Lin (National Chiao Tung Univeristy) and Dr Sabu Sahadevan (Bruker Taiwan) for useful discussions and comments. Any outstanding errors are the authors' responsibility.

List of Acronyms

2D	two-dimensional
3D	three-dimensional
ABS	acrylonitrile butadiene styrene
AC	alternating current
ADP	adenosine diphosphate
AMS	aerosol mass spectrometry
APCI	atmospheric pressure chemical ionization
API	atmospheric pressure ionization
APPI	atmospheric pressure photoionization
ASAP	atmospheric solids analysis probe
ATP	adenosine triphosphate
BIRD	blackbody infrared radiative dissociation
CAD	computer-aided design
C-API	contactless atmospheric pressure ionization
CCS	collision cross-section
CE	capillary electrophoresis
CEC	capillary electrochromatography
CFA	continuous flow analysis
CGE	capillary gel electrophoresis
CI	chemical ionization
CID	collision-induced dissociation
cITP	capillary isotachophoresis
CSI	cold spray ionization
CTI	charge-transfer ionization
CZE	capillary zone electrophoresis
DAPCI	desorption atmospheric pressure chemical ionization
DART	direct analysis in real time
DBDI	dielectric barrier discharge ionization
DC	direct current
DESI	desorption electrospray ionization
DHB	2,5-dihydroxybenzoic acid
DIOS	desorption/ionization on silicon
DMF	digital microfluidics
EASI	easy ambient sonic-spray ionization

ECD	electron capture dissociation
EESI	extractive electrospray ionization
EI	electron ionization
EIC	extracted-ion current
EKC	electrokinetic chromatography
ELDI	electrospray-assisted laser desorption/ionization
ELISA	electrostatic ion storage ring, Aarhus
EM	electron multiplier
EOF	electroosmotic flow
ESI	electrospray ionization
ESSI	electrosonic spray ionization
ETD	electron transfer dissociation
EWOD	electrowetting-on-dielectric
FAB	fast atom bombardment
FAPA	flowing atmospheric-pressure afterglow
FIA	flow-injection analysis
FID	flame ionization detection
FPOP	fast photochemical oxidation of proteins
FT	Fourier transform
FWHM	full width at half maximum
GC	gas chromatography
HDX	hydrogen/deuterium exchange
HETP	height equivalent to a theoretical plate
HILIC	hydrophilic interaction chromatography
HPLC	high-performance liquid chromatography
HV	high voltage
ICAT	isotope-coded affinity tag
ICP	inductively coupled plasma
ICR	ion cyclotron resonance
IM	ion mobility
IMAC	immobilized metal ion affinity chromatography
IMS	ion-mobility spectrometry
IR	infrared
IRMPD	infrared multiphoton dissociation
ISD	in-source decay
IT	ion trap
LAESI	laser ablation electrospray ionization
LC	liquid chromatography
LDI	laser desorption/ionization
LIT	linear ion trap
LMJ	liquid microjunction
LOD	limit of detection
LTP	low-temperature plasma
MALDI	matrix-assisted laser desorption/ionization

MAMS	micro-arrays for mass spectrometry
MCP	microchannel plate
MEKC	micellar electrokinetic chromatography
MIMS	membrane inlet mass spectrometry
MPI	multiphoton ionization
MRM	multiple reaction monitoring
MS	mass spectrometry
MudPIT	multidimensional protein identification technology
nanoDESI	nanospray desorption electrospray ionization
nanoESI	nanospray electrospray ionization
NBDPZ	4-nitro-7-piperazino-2,1,3-benzoxadiazole
NIMS	nanostructure-initiator mass spectrometry
NMR	nuclear magnetic resonance
NP-LC	normal-phase liquid chromatography
OIT	orbital ion trap
PA	proton affinity
PAT	process analytical technology
PDMS	polydimethylsiloxane
PE	potential energy
PEG	polyethylene glycol
PESI	probe electrospray ionization
PET	polyethylene terephthalate
PFTBA	perfluorotri-*n*-butylamine
PICT	photoionization charge transfer
PI-ESI	polarization induced electrospray ionization
PMMA	poly(methyl methacrylate)
PSD	post-source decay
PTR	proton-transfer reaction
Q	quadrupole
QMF	quadrupole mass filter
QqQ	triple quadrupole
RF	radio frequency
RMS	root-mean-square
RP	reversed phase
RP-LC	reversed-phase liquid chromatography
S	sector
SALDI	surface-assisted laser desorption/ionization
SAT	spectrum acquisition time
SDS	sodium dodecyl sulfate
SEC	size-exclusion chromatography
SID	surface-induced dissociation
SIFT	selected ion flow tube
SILAC	stable isotope labeling by amino acids in cell culture
SIM	selected ion monitoring

SIMS	secondary ion mass spectrometry
SPE	solid-phase extraction
SPME	solid-phase microextraction
S/N	signal-to-noise
SPI	single photon ionization
SRM	selected reaction monitoring
SSP	surface sampling
TE	translational energy
TIC	total-ion current
TIMS	trapped ion mobility spectrometry
TLC	thin-layer chromatography
TOF	time-of-flight
TRMS	time-resolved mass spectrometry
TWIG	traveling wave ion guide
UASI	ultrasonication-assisted spray ionization
UHPLC	ultra-high-performance liquid chromatography
UV	ultraviolet
UVPD	ultraviolet photodissociation
V-EASI	Venturi easy ambient sonic-spray ionization
VIS	visible
VUV	vacuum ultraviolet
μPESI	micropillar array electrospray ionization
μTAS	microscale total analysis system

1

Introduction

Time flies over us, but leaves its shadow behind.

Nathaniel Hawthorne (1804–1864)

1.1 Time in Chemistry

According to one definition, time is "the indefinite continued progress of existence and events in the past, present, and future regarded as a whole" [1]. For millennia, time has intrigued philosophers and artists. The inevitability of time flow has triggered frustration and hope. While basic chemical knowledge was gathered in antiquity, modern chemistry has been developed since the 17th century. During the first ~200 years, the notion of time in chemistry was obscure and elusive. It appeared in the spotlight when Peter Waage and Cato Guldberg began to develop the concept of chemical kinetics. Since then, the time dimension was instantly promoted to become an important factor in chemical reactions. In modern chemistry textbooks, potential energy diagrams often represent changes in the energy of reactants along an axis labeled as the *reaction path*. However, time is the variable that describes the progress of every chemical transition and physical process. Thus, time has always been among the key factors studied in chemical science [2]. Investigating chemical phenomena in relation to time has turned out to be vital for the understanding of fundamental concepts – in particular, reaction kinetics.

Temporal resolution is the ability of a method to discern consecutive transitions in the studied dynamic systems. In the field of analytical chemistry, there exist numerous methods that allow one to measure concentrations of substances in solutions or gaseous mixtures at given time points. However, many conventional methods possess limited temporal resolution. In some cases, samples are obtained from reaction mixtures at specific time points. As a result the temporal characteristics of the studied process can only be described considering the limited frequency of sampling points. The obtained samples can

Time-Resolved Mass Spectrometry: From Concept to Applications, First Edition.
Pawel Lukasz Urban, Yu-Chie Chen and Yi-Sheng Wang.
© 2016 John Wiley & Sons, Ltd. Published 2016 by John Wiley & Sons, Ltd.

be regarded as *zero-dimensional*. We live in a *four-dimensional* world that is described by three spatial dimensions. Time is the fourth elusive dimension that describes happenings in the other three dimensions (change of position). Various novel analytical methods have been developed to grasp the 3D nature of chemistry. For example, optical methods are irreplaceable when it comes to 2D and 3D spatial analysis (imaging) of chemical processes [3, 4]. Frequently these methods also provide superior temporal resolutions. Due to numerous technical obstacles, advancement of these multidimensional analysis tools could only happen because of the developments in physics, optics, photonics, and physical chemistry.

Some physical or chemical processes are so fast that their existence can only be verified using highly refined analytical approaches. For example, using infrared (IR) spectroscopy and computational methods, it was possible to confirm the existence of the simplest Criegee intermediate (CH_2OO) which has a lifetime counted in microseconds [5]. Other processes (e.g. radioactive decay of uranium-238 with a half-life of $\sim 4.5 \times 10^9$ years) are so slow that their progress cannot easily be observed during a human lifetime. However, most reactions occurring in the biological world are accelerated by biocatalysts (enzymes). Such catalytic processes can be observed on timescales of seconds and minutes. Reaction kinetics encompasses experimental methodology and the associated mathematical treatment aiming to describe the progress of chemical reactions in time. Understanding chemical kinetics can help us to optimize important reactions, so that they can be applied in large-scale synthesis, and used by industry. The kinetic profiles of reactions let us gain fundamental insights on the reaction mechanisms. Similarly, to chemical reactions, there exist other processes which serve chemists every day – distillation and extraction are just two examples. Studying kinetic properties of dynamic processes involving molecules requires the use of appropriate analytical methodology – capable of recording molecular events in the time domain.

Several physical techniques were introduced to chemistry laboratories in order to enable monitoring of chemical and physical processes in time. They include such dissimilar platforms as: fluorescence detection [6], IR spectroscopy [7], diffraction [8, 9], nuclear magnetic resonance (NMR) [10–12] as well as crystallography [13]. Since ultrafast phenomena are relevant to many fundamental studies in physics and chemistry [14], various spectroscopic techniques have been developed which enable investigation of molecular events in the time range from 10^{-9} to 10^{-18} s [15, 16]. Ultrafast IR and Raman spectroscopies enable measurements of phenomena which occur on the pico- and femtosecond timescales (1 fs = 10^{-15} s); corresponding to the elementary steps that affect chemical reactivity, including changes in the electron distribution, molecular structure and translocation of chemical moieties [17]. Pulse fluorometry and phase-modulation fluorometry enable the measurement of fluorescence lifetimes, which typically last over $10^{-8}–10^{-11}$ s [18, 19]. Time-resolved luminescence methods are routinely used in fundamental and applied sciences. For instance, by monitoring the intensity of light emitted following an excitation pulse in the nano- to millisecond range, one can distinguish contributions of de-excitation of individual fluorophores and/or phosphors. This approach enables sensitive detection of labeled molecules (or supramolecular probes) in complex biological samples which possess intrinsic luminescence (e.g., autofluorescence). Lifetime spectroscopy is nowadays almost routinely used in the analysis and imaging of biological specimens [20].

Optical methods have grounded their place in chemistry. They also have intrinsic limitations: the most prominent one is low molecular selectivity. Monitoring unknown substances and identification of unknown analytes, which are frequently present in complex mixtures, often requires powerful analytical strategies. Two particularly significant ones are NMR and mass spectrometry (MS). These two techniques have dissimilar principles: while MS separates and detects ions in the gas phase, NMR recognizes nuclei based on the characteristic electromagnetic radiation emitted by them at specific conditions in a magnetic field. Due to its versatility in structure elucidation, it is certainly worthwhile studying the principles and applications of NMR in chemistry with the aid of recent monographs (e.g., [21, 22]).

1.2   Mass Spectrometry

MS is one of the main techniques used in chemical analysis nowadays [23–28]. Due to its capabilities in molecular identification and structure elucidation, as well as high sensitivity [29], it has already enabled important discoveries in chemistry and the biosciences. A mass spectrometer consists of three main parts: an ion source, a mass analyzer, and a detector (Figure 1.1). They are supplemented with a number of auxiliary components, including interfaces for sample pre-processing and introduction, ion optic elements for manipulating ion beams, and electronic data acquisition systems.

In the gas phase, it is generally easier to handle and separate ions than neutral molecules. Gas-phase ions are produced in the ion source (see Chapter 2), while the mass analyzer subsequently separates them according to their characteristic mass-to-charge ratios (m/z). There are various types of mass analyzers (see Chapter 3). For example, in a quadrupole mass analyzer, ions pass through a zone between two pairs of metal rods spaced radially along the ion propagation axis. The ratio of alternating and direct current electric fields produced in this section enables selection of ions with a specific m/z value, and discrimination of other ions. A detector – positioned "at the end" of the quadrupole (on the side opposite to the ion source) – counts the ions which have passed through the quadrupole zone. In order to prevent collisions of gas-phase ions with gas molecules, analyzers and detectors are held under high vacuum. Some ion sources also operate under vacuum while in others the process of ion generation occurs at a higher pressure up to atmospheric pressure.

MS has evolved in the past few decades. There have been several milestone discoveries and inventions (Table 1.1). They encompass fundamental physical phenomena related to ion formation, as well as technical aspects – including the construction of ion sources and mass analyzers. Since the operation of these two components is crucial for the measurements conducted in the time domain, they will be discussed extensively in the following chapters.

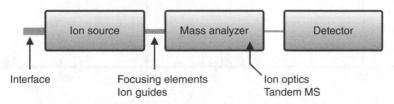

Figure 1.1   *Main components of a mass spectrometer. Reproduced from Kandiah and Urban [29] with permission of The Royal Society of Chemistry*

Table 1.1   *Selected milestones in MS*

Year	Discovery/Invention	Credited inventor	Ref.
1897	Electron	J.J. Thomson	[48]
1918	Electron ionization, first modern mass spectrometer	A. Dempster	[49]
1919	Accurate determination of the masses of individual atoms	F. Aston	[50]
1946	Time-of-flight analyzer	W.E. Stephens	[51]
1953	Ion trap analyzer	W. Paul	[52]
1957	Time-resolved mass spectrometry of flash photolysis	G.B. Kistiakowsky, P.H. Kydd	[36]
1960	Quadrupole analyzer	W. Paul	[53]
1984	Electrospray ionization	J. Fenn	[54]
1985	Matrix-assisted laser desorption/ionization	F. Hillenkamp	[55]
2000	Orbital ion trap	A. Makarov	[56]

A mass spectrum is the primary result of mass spectrometric analysis. The pioneer of MS – Francis Aston – recorded mass spectra of separated ions by exposing a photographic emulsion to the ion stream: the resulting dark bands marked the regions corresponding to the beams with high ion intensities (Figure 1.2a). The positions of these bands along the horizontal axis of such a "radiogram" are related by the m/z of the impinging ions: the darker the band the higher the intensity of the ion beam. Modern mass spectrometers use electronic sensors to measure the intensities of ion fluxes – in space, in time, or in both

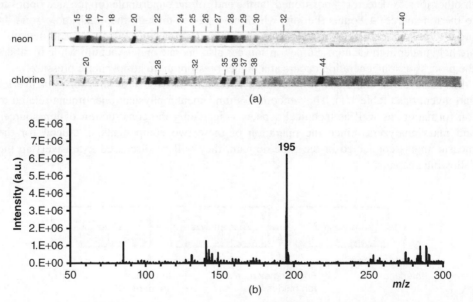

Figure 1.2   *(a) Mass spectra obtained by Aston using the early mass spectrometer. Reproduced from Squires [50] with permission of The Royal Society of Chemistry. (b) Mass spectrum of caffeine obtained using a modern ESI-IT mass spectrometer. Courtesy of E.P. Dutkiewicz*

dimensions (depending on the design of mass analyzer). These intensities are classified into discrete channels. In the course of mass calibration, these channels are related by the m/z. As a result of this operation, a mass spectrum is produced (Figure 1.2b). It can be viewed as a histogram representing ion counts (intensities) in very narrow intervals of the m/z scale.

1.3   Time-resolved Mass Spectrometry

In biochemistry it is often necessary to capture changes in the metabolic composition of samples (e.g., [30, 31]). For instance, animals communicate with each other chemically by releasing signaling molecules called pheromones, which persist in the proximity of an individual for a short period of time [32]. In the domain of exact sciences, it is also important to study temporal changes of chemical systems (reactions) and physical phenomena (e.g., molecular transport). Although many mass spectrometric analyses are often conducted as single-time-point measurements, it is evident from numerous reports published over the past few decades that MS can facilitate the studies of dynamic processes in which concentrations or structures of analyte molecules change over time. In particular, the technique enables structural determination of reactants while preserving temporal resolution [33]. While other techniques – notably, electronic and vibrational spectroscopy – can record time-dependent data, time-resolved mass spectrometry (TRMS) provides orthogonal information (m/z vs. abundance vs. time) to those of other established techniques. Since short-living reaction intermediates can be detected by various MS methods, this technique is ideal for studies on mechanisms of chemical reactions, chemical kinetics, and biochemical dynamics.

Dynamic phenomena have been investigated with MS for half a century now (e.g., [34–37]). Ions cannot move as fast as photons while mass spectrometers typically operate in the microsecond regime. Therefore, MS cannot equalize temporal resolutions achieved with some optical techniques (which are currently limited by the speed of detection systems rather than the speed of photons). Temporal resolution is a critical limitation when it comes to identifying reaction intermediates and measuring reaction kinetics by MS [38]. As it will be evident later on, the speed of MS is currently limited by the speed of sample processing rather than the speed of ions.

Here we define **TRMS** as a **mass spectrometric approach that allows one to differentiate between two chemical states that can be observed sequentially at two points on the timescale**. Such a broad definition encompasses various methods with quite dissimilar temporal resolutions and diverse areas of applications. A large variety of physical and chemical processes can be studied by TRMS: from protein folding, to enzyme kinetics, to mechanisms of organic reaction, extraction, and combustion [33]. Since short-lived reaction intermediates can be detected using several TRMS methods, this platform is ideal for the studies on mechanisms of chemical reactions, chemical kinetics, and biochemical dynamics. Depending on the focus area, temporal resolutions typically range from microseconds to minutes. While many organic reactions occur in hours, they do not require sophisticated setups to follow their kinetics or characterize intermediates (see Chapter 11). On the other hand, some of the studies on protein folding may require temporal resolutions below 1 ms (see Chapter 12).

The need to achieve satisfactory speed of MS analysis encouraged the development of delay-free sampling strategies. Early attempts to demonstrate TRMS date back to the 1950s, when radicals were recorded for thermally decomposing gases [34, 35], and flash photolysis reactions were captured using customized time-of-flight (TOF) instruments [36] (see also Chapter 4). In some cases, obtaining time-resolved data required construction of complicated experimental systems and solving cumbersome technical problems. Importantly, it is not always necessary to take a lot of effort to assemble complex apparatus since many pieces of hardware are already available commercially, and often minor modifications of these standard instruments are sufficient to carry out TRMS measurements. TRMS encompasses an assortment of methods based on different principles. Some TRMS systems are so dissimilar that the only common feature is the use of a mass spectrometer as the detector. Using customized apparatus, a multitude of physical, chemical and biochemical phenomena can be characterized in the time domain.

1.4   Dynamic Matrices

The concept of *sample* is one of the most widely used in analytical chemistry. According to the International Union of Pure and Applied Chemistry, sample is defined as a portion of material selected from a larger quantity of material [39]. Often we talk about *representative samples* which are supposed to reflect the chemical composition of the sampled medium at the time of sampling. However, many chemical media do not represent a constant chemical composition. Therefore, samples collected at different time points may have different components – qualitatively and quantitatively. For example, the chemical composition of river water may change in the course of rainfall or discharge of effluent from a wastewater treatment plant. Contents of a chemical reactor (reaction vessel) change as the reaction moves on, that is the concentrations of reactants decrease while the concentrations of products increase. If we continuously collect aliquots of media from such systems (river, reactor), and perform chemical analysis, we will find that the composition of the obtained fluid varies qualitatively and quantitatively. Therefore, in some cases it would be helpful to use the term *dynamic sample* (or *dynamic matrix*) to describe chemically unstable media that change with time. The concept of *dynamic sample* is in opposition to *static sample* (or *steady sample*) – which does not change its composition over time. However, in the following discussion we will also deal with cases in which spatial gradients of reaction products are formed downstream from micromixing devices (see Chapter 4). Generation of such gradients enables studies of some fast phenomena (e.g., protein folding measurements, see Chapter 12). Nonetheless, probing analytes from spatial concentration gradients does not rely on time-resolving capabilities of MS. Thus, the composition of an aliquot obtained from a fixed position within a spatial gradient can be referred to as *quasi-static* (or *quasi-steady*).

1.5   Real-time *vs.* Single-point Measurements

It is necessary to distinguish between two modes of recording temporal information in TRMS:

*   *Real-time monitoring* provides the means to record data continuously with a specified temporal resolution. Multiple spectra are recorded at a high frequency (typically, several

spectra per second). Here, the spectral patterns are assumed to reflect the chemical compositions of the investigated systems at given time points. The consecutive spectra are expected to represent different patterns reflecting qualitative and/or quantitative changes in the investigated systems. The time interval between two consecutive analyses (data points) should be as short as possible in order to achieve high temporal resolution of real-time monitoring.

* *Single-point measurements* provide spectral "snapshots" of the investigated systems at selected time points. If the system is dynamic or unstable, it may require quenching before introduction to the ion source. In this case, the time of quenching rather than measurement itself puts the time tag on the molecular composition of the dynamic sample. In some cases, one of the ionization steps (e.g., removal of solvent) may be regarded as quenching.

When investigating highly dynamic systems, the *sampling time* or the *incubation time* is an important attribute of MS measurements. Other chemical systems exist in equilibrium, in which case *steady samples* are obtained, and the sampling time becomes a less important descriptor. As mentioned above, in some designs of TRMS, reaction mixtures are generated continuously in the flow of the reactant stream or droplet plume. In those cases, *quasi-steady samples* are produced. Thus, single-point measurements of reaction products or intermediates can be conducted following a very short incubation period.

1.6 Further Reading

The reference lists included in every chapter contain numerous examples of representative reports on TRMS. For more succinct summaries of TRMS, as well as the coverage of specific topics, readers are also encouraged to consult recent review articles and chapters [33, 40–46]. While some of the following chapters cover important parts of MS workflow, readers interested in more general aspects of the technique are encouraged to consult MS textbooks [23–28]. Those who are specifically interested in the application of MS in the studies of reaction intermediates of chemical reactions, and would like to gain a more extensive overview of that subject, are encouraged to read the book edited by Santos [47].

References

1. Oxford Dictionaries – "Time", http://www.oxforddictionaries.com/definition/english/time (accessed September 11, 2015).
2. Benfey, O.T. (1963) Concepts of Time in Chemistry. J. Chem. Educ. 40: 574–577.
3. Bánsági Jr, T., Vanag, V.K., Epstein, I.R. (2011) Three-Dimensional Turing Patterns in a Reaction Diffusion System. Science 331: 1309–1312.
4. Hsieh, K.-T., Urban, P.L. (2014) Spectral Imaging of Chemical Reactions Using Computer Display and Digital Camera. RSC Adv. 4: 31094–31100.
5. Su, Y.-T., Huang, Y.-H., Witek, H.A., Lee, Y.-P. (2013) Infrared Absorption Spectrum of the Simplest Criegee Intermediate CH_2OO. Science 340: 174–176.
6. Damnjanovic, B., Apell, H.-J. (2014) Role of Protons in the Pump Cycle of KdpFABC Investigated by Time-Resolved Kinetic Experiments. Biochemistry 53: 3218–3228.

7. Mitri, E., Kenig, S., Coceano, G., Bedolla, D.E., Tormen, M., Grenci, G., Vaccari, L. (2015) Time-Resolved FT-IR Microspectroscopy of Protein Aggregation Induced by Heat-Shock in Live Cells. Anal. Chem. 87: 3670–3677.
8. Helliwell, J.R., Rentzepis, P.M. (eds) (1997) Time-resolved Diffraction (Oxford Series on Synchroton Radiation, 2). Clarendon Press, Oxford.
9. Kim, T.K., Lee, J.H., Wulff, M., Kong, Q., Ihee, H. (2009) Spatiotemporal Kinetics in Solution Studied by Time-Resolved X-Ray Liquidography (Solution Scattering). ChemPhysChem 10: 1958–1980.
10. Sans, V., Porwol, L., Dragone, V., Cronin, L. (2015) A Self Optimizing Synthetic Organic Reactor System Using Real-Time In-Line NMR Spectroscopy. Chem. Sci. 6: 1258–1264.
11. Boisseau, R., Bussy, U., Giraudeau, P., Boujtita, M. (2015) *In Situ* Ultrafast 2D NMR Spectroelectrochemistry for Real-Time Monitoring of Redox Reactions. Anal. Chem. 87: 372–375.
12. Dass, R., Koźmiński, W., Kazimierczuk, K. (2015) Analysis of Complex Reacting Mixtures by Time-Resolved 2D NMR. Anal. Chem. 87: 1337–1343.
13. Kupitz, C., Basu, S., Grotjohann, I., Fromme, R., Zatsepin, N.A., Rendek, K.N., Hunter, M.S., Shoeman, R.L., White, T.A., Wang, D., James, D., Yang, J.-H., Cobb, D.E., Reeder, B., Sierra, R.G., Liu, H., Barty, A., Aquila, A.L., Deponte, D., Kirian, R.A., Bari, S., Bergkamp, J.J., Beyerlein, K.R., Bogan, M.J., Caleman, C., Chao, T.-C., Conrad, C.E., Davis, K.M., Fleckenstein, H., Galli, L., Hau-Riege, S.P., Kassemeyer, S., Laksmono, H., Liang, M., Lomb, L., Marchesini, S., Martin, A.V., Messerschmidt, M., Milathianaki, D., Nass, K., Ros, A., Roy-Chowdhury, S., Schmidt, K., Seibert, M., Steinbrener, J., Stellato, F., Yan, L., Yoon, C., Moore, T.A., Moore, A.L., Pushkar, Y., Williams, G.J., Boutet, S., Doak, R.B., Weierstall, U., Frank, M., Chapman, H.N., Spence, J.C.H., Fromme, P. (2014) Serial Time-Resolved Crystallography of Photosystem II Using a Femtosecond X-ray Laser. Nature 513: 261–265.
14. de Nalda, R., Bañares, L. (eds) (2013) Ultrafast Phenomena in Molecular Sciences: Femtosecond Physics and Chemistry (Springer Series in Chemical Physics). Springer, Berlin.
15. Weinstein, J.A., Hunt, N.T. (2012) In Search of Molecular Movies. Nature Chem. 4: 157–158.
16. Caldin, E.F. (2001) The Mechanisms of Fast Reactions in Solution. IOS Press, Amsterdam.
17. Fayer, M.D. (ed.) (2001) Ultrafast Infrared and Raman Spectroscopy. Marcel Dekker, New York.
18. Valeur, B. (2002) Molecular Fluorescence. Principles and Applications. John Wiley & Sons, Ltd, Weinheim.
19. Demchenko, A.P. (2009) Introduction to Fluorescence Sensing. Springer, Berlin.
20. Marcu, L., French, P.M.W., Elson, D.S. (eds) (2014) Fluorescence Lifetime Spectroscopy and Imaging: Principles and Applications in Biomedical Diagnostics. CRC Press, Boca Raton.
21. Günther, H. (2013) NMR Spectroscopy: Basic Principles, Concepts and Applications in Chemistry. John Wiley & Sons, Ltd, Weinheim.
22. Benesi, A.J. (2015) A Primer of NMR Theory. John Wiley & Sons, Ltd, Weinheim.

23. McLafferty, F.W., Tureček, F. (1993) Interpretation of Mass Spectra. University Science Books, Sausalito.
24. Gross, J.H. (2004) Mass Spectrometry: A Textbook. Springer-Verlag, Berlin.
25. Dass, C. (2007) Fundamentals of Contemporary Mass Spectrometry. John Wiley & Sons, Inc., Hoboken.
26. de Hoffman, E., Stroobant, V. (2007) Mass Spectrometry: Principles and Applications. John Wiley & Sons, Ltd, Chichester.
27. Watson, J.T., Sparkman, O.D. (2007) Introduction to Mass Spectrometry: Instrumentation, Applications, and Strategies for Data Interpretation. John Wiley & Sons, Ltd, Chichester.
28. Lee, M.S. (2012) Mass Spectrometry Handbook. John Wiley & Sons, Inc., Hoboken.
29. Kandiah, M., Urban, P.L. (2013) Advances in Ultrasensitive Mass Spectrometry of Organic Molecules. Chem. Soc. Rev. 42: 5299–5322.
30. Tseng, T.-W., Wu, J.-T., Chen, Y.-C., Urban, P.L. (2012) Isotope Label-aided Mass Spectrometry Reveals the Influence of Environmental Factors on Metabolism in Single Eggs of Fruit Fly. PLOS ONE 7: e50258.
31. Liang, Z., Schmerberg, C.M., Li, L. (2015) Mass Spectrometric Measurement of Neuropeptide Secretion in the Crab, *Cancer borealis*, by *In Vivo* Microdialysis. Analyst 140: 3803–3813.
32. Agosta, W.C. (1992) Chemical Communication. The Language of Pheromones. Scientific American Library, New York.
33. Chen, Y.-C., Urban, P.L. (2013) Time-Resolved Mass Spectrometry. Trends Anal. Chem. 44: 106–120.
34. Lossing, F.P., Tickner, A.W. (1952) Free Radicals by Mass Spectrometry. I. The Measurement of Methyl Radical Concentrations. J. Chem. Phys. 20: 907.
35. Lossing, F.P., Ingold, K.U., Tickner, A.W. (1953) Free Radicals by Mass Spectrometry. Part II. – The Thermal Decomposition of Ethylene Oxide, Propyline Oxide, Dimethyl Ether, and Dioxane. Discuss. Faraday Soc. 14: 34–44.
36. Kistiakowsky, G.B., Kydd, P.H. (1957) A Mass Spectrometric Study of Flash Photochemical Reactions. I. J. Am. Chem. Soc. 79: 4825–4830.
37. Price, D., Todd, J.F.J. (1976) Dynamic Mass Spectrometry. John Wiley & Sons, Inc., New York.
38. Lee, J.K., Kim, S., Nam, H.G., Zare, R.N. (2015) Microdroplet Fusion Mass Spectrometry for Fast Reaction Kinetics. Proc. Natl. Acad. Sci. USA 112: 3898–3903.
39. IUPAC Gold Book, http://goldbook.iupac.org/S05451.html (accessed February 23, 2015).
40. Santos, L.S., Knaack, L., Metzger, J.O. (2005) Investigation of Chemical Reactions in Solution Using API-MS. Int. J. Mass Spectrom. 246: 84–104.
41. Eberlin, M.N. (2007) Electrospray Ionization Mass Spectrometry: A Major Tool to Investigate Reaction Mechanisms in Both Solution and the Gas Phase. Eur. J. Mass Spectrom. 13: 19–28.
42. Santos, L.S. (2008) Online Mechanistic Investigations of Catalyzed Reactions by Electrospray Ionization Mass Spectrometry: A Tool to Intercept Transient Species in Solution. Eur. J. Org. Chem. 2008: 235–253.

43. Kaltashov, I.A., Eyles, S.J. (2012) Chapter 6, Kinetic Studies by Mass Spectrometry. In: Mass Spectrometry in Structural Biology and Biophysics: Architecture, Dynamics, and Interaction of Biomolecules, 2nd Edition. John Wiley & Sons, Inc., Hoboken.

44. Ma, X., Zhang, S., Zhang, X. (2012) An Instrumentation Perspective on Reaction Monitoring by Ambient Mass Spectrometry. Trends Anal. Chem. 35: 50–66.

45. Rob, T., Wilson, D.J. (2012) Time-Resolved Mass Spectrometry for Monitoring Millisecond Time-Scale Solution-Phase Processes. Eur. J. Mass Spectrom. 18: 205–214.

46. Zhu, W., Yuan, Y., Zhou, P., Zeng, L., Wang, H., Tang, L., Guo, B., Chen, B. (2012) The Expanding Role of Electrospray Ionization Mass Spectrometry for Probing Reactive Intermediates in Solution. Molecules 17: 11507–11537.

47. Santos, L. S. (ed.) (2010) Reactive Intermediates: MS Investigations in Solution. John Wiley & Sons, Ltd, Weinheim.

48. J.J. Thomson – Biographical, http://www.nobelprize.org/nobel_prizes/physics/laureates/1906/thomson-bio.html (accessed September 11, 2015).

49. Dempster, A.J. (1918) A New Method of Positive Ray Analysis. Phys. Rev. 11: 316.

50. Squires, G. (1998) Francis Aston and the Mass Spectrograph. J. Chem. Soc., Dalton Trans. 1998: 3893–3900.

51. Stephens, W.E. (1946) A Pulsed Mass Spectrometer with Time Dispersion. Phys. Rev. 69: 691.

52. Paul, W., Steinwedel, H. (1953) Ein neues Massenspektrometer ohne Magnetfeld. Z. Naturforsch. A 8: 448–450.

53. Paul, W., Steinwedel, H. (1960) patent number US2939952 (A).

54. Yamashita, M., Fenn, J.B. (1984) Electrospray Ion Source. Another Variation on the Free-Jet Theme. J. Phys. Chem. 88: 4451–4459.

55. Karas, M., Bachmann, D., Hillenkamp, F. (1985) Influence of the Wavelength in High-Irradiance Ultraviolet Laser Desorption Mass Spectrometry of Organic Molecules. Anal. Chem. 57: 2935–2939.

56. Makarov, A. (2000) Electrostatic Axially Harmonic Orbital Trapping: A High-Performance Technique of Mass Analysis. Anal. Chem. 72: 1156–1162.

2

Ion Sources for Time-resolved Mass Spectrometry

The ion source is the component of a mass spectrometer where ionization takes place. It can be viewed as a "security guard" because it screens analytes which enter the mass spectrometer, while it also renders them amenable to mass spectrometric analysis. It is the place where analytes become ionized; for example, by gaining/losing an electron or attracting/detaching positively charged species such as protons. In many cases, it also facilitates transfer of analytes from the condensed phase to the gas phase by adjusting the form of the analytes to the operating conditions of the mass analyzer, which can only separate ionic species in the gas phase. The analytes which (i) are transferred to the gas phase, and (ii) acquire/maintain electric charges, can be effectively directed towards further stages of the mass spectrometer for determination of the mass-to-charge (m/z) ratio.

Some ion sources, such as electron ionization (EI) [1, 2] and chemical ionization (CI) [3] impose the requirement that the samples are in the gaseous form even before they are introduced to the ion source chamber. Condensed-phase samples need to be vaporized through heating prior to introduction to such ion sources. In addition to molecular ions, fragment ions dominate EI mass spectra, while fewer fragment ions are observed in CI mass spectra. Thus, separation techniques (such as gas chromatography) are often combined with mass spectrometers equipped with these ion sources when mixtures need to be analyzed. On the other hand, condensed-phase samples can directly be introduced to the ion sources based on electrospray ionization (ESI) [4, 5], atmospheric pressure chemical ionization (APCI) [6], or desorption/ionization (DI) [7–9]. Few fragments are observed when using most of these techniques. This is mainly because a large portion of the energy implanted by the ion source is used for the phase transition protecting fragile molecules from damage. Development of these ion sources has led to considerable advances of the ionization methodology in the first decades of the 21st century [10]. In fact, various ionization techniques – which operate at atmospheric pressure – are derived from the canonical ion sources based on electrospray [4, 5], plasma effects [6], and laser desorption [7–9]. The most common ionization

Time-Resolved Mass Spectrometry: From Concept to Applications, First Edition.
Pawel Lukasz Urban, Yu-Chie Chen and Yi-Sheng Wang.
© 2016 John Wiley & Sons, Ltd. Published 2016 by John Wiley & Sons, Ltd.

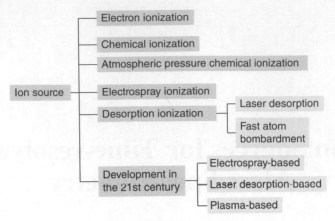

Figure 2.1   *Ion sources which have been utilized in time-resolved mass spectrometry (TRMS) studies (a dichotomy)*

techniques – including EI, CI, APCI, ESI, DI, as well as the recently developed ion sources operating at atmospheric pressure – will be described in this chapter (Figure 2.1).

It should be noted that the time spent on ionization contributes to temporal resolution of TRMS; this intrinsic "dead time" of the analytical procedure will be discussed separately for different ionization technoques. Although the time spent in the ion source is quite short (sub-microseconds to sub-milliseconds), it may in some cases affect the observation of the reaction of interest. Thus, it is worthwhile knowing the time spans characterizing different ionization techniques.

2.1   Electron Ionization

Electron ionization – introduced by Dempster in 1918 [1, 2] – is the oldest ionization technique for mass spectrometry (MS) analysis of gaseous samples. Although EI has already been used for almost a century, it is still widely utilized in the analysis of small organic molecules – especially those that are relatively non-polar, thermally stable, volatile, and highly volatile. Analytes with a vapor pressure of $\sim 10^{-5} - 1$ Pa are heated to 200–250 °C in order to enable vaporization. They are subsequently introduced to the ion source compartment. A high energy electron cloud (70 eV), generated at a heated tungsten filament, is directed towards gaseous analyte molecules. It induces the ejection of electrons from the analyte molecules leading to the formation of molecular ions ($[M^{+\bullet}]$):

$$M + e^- \rightarrow M^{+\bullet} + 2e^-$$

The excess internal energy of the molecular ions results in fragmentations. The fragment ions – appearing in the mass spectra – are characteristic features of the EI technique. However, it is more problematic to record molecular ions in the EI mass spectra, especially in the case of analyte molecules with fragile structures. Figure 2.2 shows a schematic illustration of the EI ion source. An energetic electron beam emanating from a metallic filament is directed into the ion source lumen filled with gas-phase analytes. At the section, the

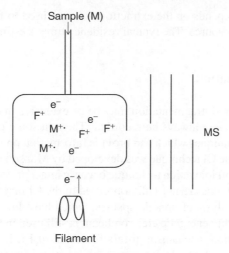

Sample (M)

MS

Filament

Figure 2.2 *Schematic illustration of the EI source. M, analyte; M⁺•, molecular ion; F⁺, fragment ion*

energetic electron beam effectively interacts with the neutral molecules to generate ionic species. On the basis of some studies [11–13], the maximum ionization cross-section is observed when the energy of the electron beam is set to 70 eV. In fact, reproducible mass spectra and fragmentation patterns can be obtained when the energy of the electron beam is set to 70 eV (Figure 2.3). The minimum ionization energy for common organic species is usually in the range of 5–20 eV, which leads to losing one electron (Figure 2.3). During collisions with the high energy electron beam, the analyte species can generally gain over 20 eV. The excess energy planted on the analyte ions results in excessive fragmentation.

It should be noted that the time required to produce ionic species is related to the speed of the energetic electron beam and the size of the analyte molecules. When the electron beam is supplied at 70 eV, it is estimated that the ionization time in EI is within the sub-femtosecond to low femtosecond range [13, 14]. However, the residence time of the

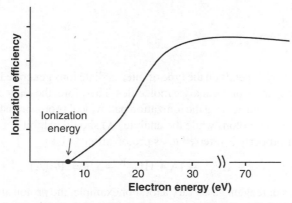

Figure 2.3 *Plot showing the relationship between the ionization efficiency and the energy of the electron beam used in EI*

ions in the ion source depends on the extraction voltages used to focus the ion beam and the dimension of the ion source. The typical residence time is estimated to be ~ 1 μs [15].

2.2   Chemical Ionization

When using the EI method, fragmentation may be so extensive that the molecular weight of intact analyte ions cannot always be recorded. Thus, in some cases, it is necessary to use softer ionization techniques which can provide information on the molecular weight of intact molecular ions. The CI technique was developed by Munson and Field in 1966 [3]. It has been regarded as a soft ionization technique because intact protonated pseudomolecular ions and a few fragment ions are normally observed in the CI mass spectra. Similar to EI, CI is suitable for the analysis of organic species, which have high volatility and thermal stability. Although a highly energetic electron beam is still used in the course of ionization (Figure 2.4), the ionization mechanism differs from that of EI. In CI, a reagent gas such as methane [3] is introduced into the ion source cavity at a pressure of ~ 100 Pa, while the sample pressure is much lower (e.g., $\sim 10^{-3}$ Pa). Because of its relatively high pressure, it is more probable that the high energy electrons (e.g., 200 eV) impinge on the reagent gas molecules (R) than analyte molecules (M) [3]. Electrons in the reagent gas molecules are knocked out by the high energy electron beam, leading to the generation of cationic radicals ($R^{+\bullet}$)

The short-lived cationic radicals then initialize a series of reactions between the generated cations and reagent gas molecules. The resulting product ions (F^{+}), which do not further react with reagent gas, promote chemical reactions with the analyte molecules, leading to formation of gas-phase ions. For example, the intermediate ions, CH_5^+, generated from methane, are capable of transferring protons to the analyte molecules with a higher proton affinity. Ion–molecule reactions in the gas phase are mainly involved in the ionization of analyte molecules in CI. The ionization is *de facto* induced during the collisions with an ionized reagent gas.

In general, there are four main pathways leading to the ionization of analyte molecules in CI [12, 13]:

1. proton transfer;
2. anion abstraction;
3. electrophilic addition;
4. charge exchange.

The ionization pathway depends on the type of intermediate ions generated from the reagent gas as well as the properties of the analyte molecules. Therefore, the choice of reagent gas is particularly important. Considering the ionization pathway of proton transfer, intermediates should be able to donate protons while the analyte (A) should exhibit a tendency to accept a proton. The latter property is referred to as proton affinity (PA).

$$A_{(g)} + H_{(g)}^+ \rightarrow [A + H]_{(g)}^+ \ -\Delta H = \ PA(A)$$

The proton affinities of reagent gases vary a lot. For example, the proton affinity of methane during formation of protonated species (CH_5^+) was found to be 131.6 kcal mol^{-1} [13, 16]. In order to cause ionization, the proton affinity of the analyte must be higher than that of

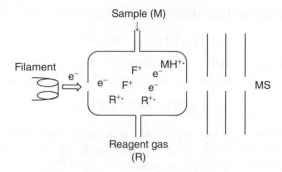

Figure 2.4   *Schematic depiction of the CI source*

the reagent gas (R). Protonation occurs as long as the process is exothermic. Differences in the proton affinities of the analytes and the proton donor determine the degree of fragmentation. The excess energy obtained from the proton transfer process renders the analyte ions unstable, leading to their fragmentation. When the difference between proton affinity of the analyte and the reagent gas [PA(A) > PA(R)] is large, more fragmentations can be observed. Nonetheless, fragmentation in CI is much less pronounced than in the case of EI. Therefore, there is a high chance that the intact pseudomolecular ions (e.g., $[M + H]^+$) are recorded in CI mass spectra. Abundances of the fragment ions observed in CI can also provide hints for structure elucidation. By selecting a proper reagent gas, fragmentation of analyte molecules can be controlled to some extent.

Reagent gases commonly used in CI include methane, isobutane, and ammonia. When using methane as the reagent gas, three types of reactions (listed below) occur simultaneously during the ionization process [3, 12, 13]:

1. Formation of primary ions:

$$CH_4 + e^-_{(200\ eV)} \rightarrow CH_4^{+\bullet} + 2e^-\ \text{(ionization)}$$

$$CH_4^{+\bullet} \rightarrow CH_3^+ + H^\bullet\ \text{(fragmentation)}$$

$$CH_4^{+\bullet} \rightarrow CH_2^{+\bullet} + H_2\ \text{(fragmentation)}$$

2. Generation of secondary reagent ions:

$$CH_4^{+\bullet} + CH_4 \rightarrow CH_5^+ + CH_3^\bullet\ \text{(ion-molecule reaction)}$$

$$CH_3^+ + CH_4 \rightarrow C_2H_5^+ + H_2\ \text{(ion-molecule reaction)}$$

$$CH_2^{+\bullet} + CH_4 \rightarrow C_2H_3^+ + H_2 + H^\bullet\ \text{(ion-molecule reaction)}$$

$$C_2H_3^+ + CH_4 \rightarrow C_3H_5^+ + H_2\ \text{(ion-molecule reaction)}$$

3. Formation of product ions:
   In the case of saturated hydrocarbons (RH), ions are formed via hydride abstraction:

$$RH + CH_5^+ \rightarrow R^+ + CH_4 + H_2$$

In the case of unsaturated hydrocarbons (M), protonation occurs:

$$M + CH_5^+ \rightarrow MH^+ + CH_4$$

In addition, ion–molecule adducts of polar molecules are frequently observed in the CI mass spectra. This type of reaction can be regarded as *gas-phase solvation*:

$$M + CH_3^+ \rightarrow (M + CH_3)^+$$

The temporal characteristics of reagent ion generation have been investigated using methane as substrate [17]. In this case, the abundances of CH_5^+ and $C_2H_5^+$ are much higher than the abundances of other ions such as CH_3^+ and CH_4^+. Furthermore, intensities of CH_5^+ and $C_2H_5^+$ ions remain steady, indicating that CH_5^+ and $C_2H_5^+$ do not react further with neutral methane molecules. CH_5^+ and $C_2H_5^+$ are good proton donors for ionization of neutral analytes when using methane as the reagent gas. In fact, methane is the most common reagent gas because it can assist ionization of many organic molecules studied by CI-MS.

Another common reagent gas used in CI is isobutane (i-C_4H_{10}). It possesses an even higher proton affinity (196 kcal mol^{-1}) than methane [16, 17]. In this case, the ionization process is described by the following reactions:

$$i\text{-}C_4H_{10} + e^- \rightarrow i\text{-}C_4H_{10}^{+\bullet} + 2e^-$$

$$i\text{-}C_4H_{10}^{+\bullet} + i\text{-}C_4H_{10} \rightarrow i\text{-}C_4H_9^+ + C_4H_9^\bullet + H_2$$

When using isobutane in CI, fragmentation is usually low since the proton affinity of the analyte (which can receive protons from the reagent gas ions) is close to that of the reagent gas ions. Furthermore, C_4H_9 adduct ions of the analyte ([M + 57]) are frequently formed in the case of polar analytes during the CI process.

Another type of reagent gas used in CI encompasses nitrogen-containing gases such as ammonia and dimethylamine. In general, ammonia is not suitable for the analysis of most organic compounds unless they also possess nitrogen-containing functional groups. NH_4^+ and $N_2H_7^+$ species are formed in CI through ion–molecule reactions [18]:

$$NH_3 + e^- \rightarrow NH_3^{+\bullet} + 2e^-$$

$$NH_3^{+\bullet} + NH_3 \rightarrow NH_4^+ + NH_2^\bullet$$

$$NH_4^+ + NH_3 \rightarrow N_2H_7^+$$

The proton affinity of NH_3 is 204 kcal mol^{-1} [16], which is higher than that of most organic molecules that do not possess functional groups with nitrogen atoms. The advantage of using ammonia as the reagent gas in CI is its selectivity because only analytes with a higher proton affinity than ammonia can be ionized. Thus, target analytes with nitrogen-containing functional groups in mixtures can be directly analyzed by CI-MS – even without complete separation in the GC column.

As proposed by Griffith and Gellene [19], the residence time of the generated ions in the CI source mainly depends on the pressure of the reagent gas and the analytes. For an estimation, it is also necessary to know the electronic states of all species in the ion source and the kinetics of the system including the diffusion rate of these gas ions toward the wall of the ion source. It is not easy to obtain all the information characterizing such a complex process. Thus, Griffith and Gellene used a mixture of O_2/Ar as the model ion–molecule reaction system [19]. The experimental results showed that the ion residence time is independent of the ratio of O_2/Ar, but it is closely related to the gas pressure in the ion source (Figure 2.5a). Specifically, as the pressure in the ion source is increased, the ion residence time gets longer. For example, the ion residence time in such a system was estimated to be ~300 ns at a pressure of 1 torr (~100 Pa) (Figure 2.5a and 2.5b).

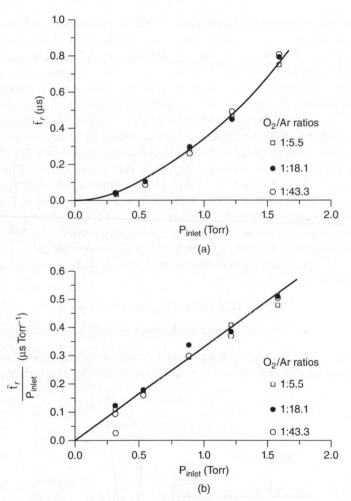

Figure 2.5 *The residence time of the ions in the CI source versus gas pressure. (a) Effective ion source residence time as a function of inlet pressure for three different initial Ar/O$_2$, compositions. (b) The ratio of effective ion source residence time to inlet pressure as a function of inlet pressure [19]. Reprinted with permission from Griffith, K.S., Gellene, G.I. (1993) A Simple Method for Estimating Effective Ion Source Residence Time. J. Am. Soc. Mass Spectrom. 4: 787–791. Copyright (1993) American Chemical Society*

EI and CI are suitable for the analysis of small organic species with high volatility and thermal stability. However, many molecules studied by organic chemists, biochemists, or environmental scientists, do not possess these features. Furthermore, EI and CI need to be operated under high vacuum. On the other hand, atmospheric pressure ionization techniques can be used in the analysis of polar and non-polar organic compounds that are thermally labile with low volatility.

2.3   Atmospheric Pressure Chemical Ionization

Atmospheric pressure chemical ionization (APCI) is commonly used to analyze com pounds with varied polarity. Its early version was disclosed by Horning and co-workers in 1974 [6, 20]. It is amenable to coupling APCI-MS with liquid chromatography (LC) [21–23] because of its tolerance to high flow rates of mobile phases ($0.2–2.0$ ml min^{-1}) which are commonly used in LC. Unlike in CI, the APCI ionization process takes place at atmospheric pressure. Similarly to the vacuum CI, energetic electrons that are generated from a metal needle initiate the ionization process (Figure 2.6), while ion–molecule reactions contribute to the subsequent ionization steps. Electrons are generated in the course of corona discharges [24] or by a ^{63}Ni β-radioactive source [6]. Note that the heated filament cannot be used under atmospheric pressure conditions. Corona discharge is widely used in APCI because it provides a larger range of dynamic response, leading to improved sensitivity [24], while its implementation does not raise radiological safety concerns. It has been estimated that ion densities in the region of the sampling aperture were two orders of magnitude higher in the corona source than in a ^{63}Ni source [24]. In this case, primary ions are generated from air components through corona discharges. Reactions of nitrogen and oxygen molecules present in air with electrons initially give rise to radical cations:

$$N_2 + e^-(\text{corona discharges}) \rightarrow N_2^{+\bullet} + 2e^-$$

$$O_2 + e^-(\text{corona discharges}) \rightarrow O_2^{+\bullet} + 2e^-$$

The sample dissolved in a solvent is introduced directly into the ion source cavity. Solvent vapor (S) is reacted with primary ions to generate secondary ions which act as proton donors

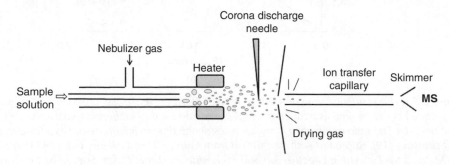

Figure 2.6   *Schematic representation of the APCI source*

during the subsequent ionization of analytes (A) [6]:

$$N_2^{+\bullet} + S \rightarrow N_2 + S^{+\bullet}$$

$$S^{+\bullet} + S \rightarrow [S + H]^+ + [S - H]^\bullet$$

$$[S + H]^+ + A \rightarrow [A + H]^+ + S$$

$$S^{+\bullet} + A \rightarrow [A + H]^+ + [S - H]^\bullet$$

Both positive and negative analyte ions can be obtained as a result of ion–molecule reactions. Protonated and deprotonated pseudomolecular ions usually dominate the positive-ion and negative-ion mode mass spectra, respectively. Overall, ion–molecule reactions involved in the APCI include proton transfer, charge transfer, and hydride abstraction.

In APCI, the sample solution is pushed by either a syringe pump or LC pump through a capillary surrounded by coaxial-flow of nebulizing gas heated up to ~200 °C (Figure 2.6). The combination of nebulizer gas and heat facilitates vaporization of the effluent stream while the hot carrier gas also stabilizes the corona discharge [23]. Fine droplets – resulting from the vaporized effluent stream – encounter primary ions generated by corona discharge at a needle (diameter: ~10 μm) connected to a high voltage power supply (2–5 kV), in which the current generated from corona discharge is several microamperes [24, 25]. The droplets are mainly composed of sample solvent. Thus, there is a high probability that the ionized gas collides with solvent vapor molecules (S) in the gas phase. Consequently, proton donors are formed in the course of ion–molecule reactions to generate protonated analyte ions. Protonated pseudomolecular ions of analytes (A) usually dominate the APCI mass spectra. Based on experimental results, it has been estimated that the residence time of ions in the APCI source, with a distance between the corona needle and the MS orifice of 5–25 mm, is ~10^{-4} s [26–28].

APCI is a relatively soft ionization technique. In fact, only few fragment ions are normally recorded. Nevertheless, analyte decomposition may occur due to heating. It can be used in the analysis of polar and low polar analytes with molecular weights up to ~1500 Da [29] as long as the proton affinity of the analytes is higher than that of the solvents. On the other hand, ionization techniques such as ESI and matrix-assisted laser desorption/ionization (MALDI) are mainly used in the analyses of polar, less volatile, and thermally labile analytes, for example, large biomolecules. These two ionization techniques are discussed in the following (Sections 2.4 and 2.6).

2.4 Electrospray Ionization

Polar and non-volatile molecules can readily be analyzed using ESI in combination with MS. Fenn and co-workers realized the combination of ESI with MS for the analysis of organic and biological molecules in the 1980s [4, 5]. ESI-MS is a very practical analytical tool that provides molecular characteristics of analytes over a wide mass range. Although the ESI-MS technique was developed in the middle of the 1980s [4], researchers had become interested in this technique at the beginning of the 20th century. In 1914, Zeleny generated an electrospray and observed the formation of a cone-jet from the outlet of the electrospray emitter by a microscope in the presence of an electric field [30]. The setup was very simple: a metal needle was used as the sprayer while diluted hydrochloric acid with high electrical conductivity was used as the sprayed solution. The electric field applied to the metal capillary tip leads to partial separation of ions carrying positive and negative charges. In the positive-ion mode, positive ions tend to aggregate at the surface of the liquid meniscus at the capillary tip, while negative ions are attracted to the wall of the capillary. Coulomb repulsion of the positive ions at the liquid surface acts against surface tension, leading to formation of a liquid cone on the liquid surface. This liquid cone is often referred to as the *Taylor cone* [31]. When the cone reaches the Rayleigh limit, its structure disintegrates because of the imbalance between the surface tension and the electrostatic force, and a stream of droplets is formed (Figure 2.7) [32]. Zeleny learned that the disintegration follows hydrodynamic instability. This hydrodynamic stability can be obtained at a certain value of the applied voltage. However, the Taylor cone is formed due to the effect of the electric field even before the hydrodynamic instability occurs [30]. Important studies related to ESI were conducted by Dole *et al.* [33], who demonstrated that multiply charged ions of zein (proteins from corn ~50 kDa) and polystyrene (~51 kDa) could be generated by ESI. The generated ionic species were detected only using a Faraday cup detector. Thus, the information on molecular weights of analytes was not directly obtained although Dole *et al.* tried to estimate the molecular weights of the model analytes based on

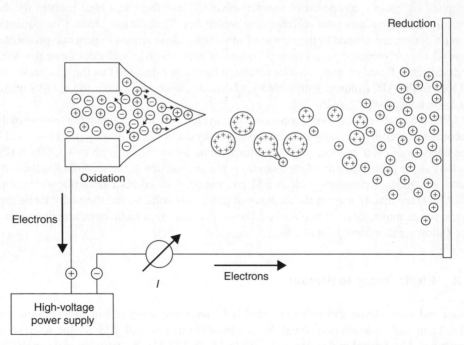

Figure 2.7 *General schematic representation of the processes in ESI-MS [32]. Reprinted with permission from Kebarle, P., Tang, L. (1993) From Ions in Solution to Ions in the Gas Phase. The Mechanism of Electrospray Mass Spectrometry. Anal. Chem. 64: 972A–986A. Copyright (1993) American Chemical Society*

the electrospray ion current by varying the voltage applied to the ESI emitter. The combination of ESI and MS was later realized by Fenn and co-workers using small molecules as the test samples in 1984 [4]. The same research group further demonstrated that multiply charged ions could be generated from polyethylene glycol [34], and soon demonstrated the possibility of using ESI-MS in the analysis of peptides and proteins [5]. Generating multiply charged species from large molecules is advantageous because it is possible to observe large ions using mass analyzers with a narrow mass range such as quadrupole analyzers.

The earlier version of ESI-MS that was proposed in 1989 [5] is still widely used without major modifications. Liquid sample is infused through a needle, which is co-axial with respect to a cylindrical electrode. The voltage of a few kilovolts is applied to the needle while the sample is pumped at a flow rate in the range of typically $1-20 \, \mu l \, min^{-1}$. Fine droplets are generated because of charge accumulation on the surface of liquid, followed by Coulomb explosions. A drying gas (e.g., nitrogen) is pumped through the ionization chamber to facilitate solvent evaporation. Very fine droplets are formed during repeated cycles of solvent evaporation and Coulomb explosions. The generated droplets are directed to the ion transfer capillary because of the applied electric field and pressure difference. The ion transfer capillary is heated to above 100 °C to further assist evaporation of solvent. The pressure is decreased gradually using the primary and secondary pumping systems – from the atmospheric pressure to the high vacuum ($< 1.33 \times 10^{-3}$ Pa). Eventually, gas-phase ions are formed through the ion transfer capillary. The ions are further directed into the mass analyzer through a skimmer, which can eliminate neutral species.

Protonated pseudomolecular ions of low-molecular-weight analytes usually dominate ESI mass spectra, while multiply charged ion peaks, derived from large molecules (differing by one charge from those of adjacent peaks), are observed. Using the peak patterns of the multiply charged ions derived from large molecules, molecular weights can be easily calculated based on the adjacent peaks with their m/z values recorded in the mass spectra [35].

An in-depth overview of the ESI mechanisms has been presented by Kebarle and Tang [32]. It was estimated (based on experimental results and calculations) that the time required for ion formation – from generation of fine micro-droplets to the formation of gas phase ions – is below 1 ms (Figure 2.8a) [32]. However, the ionization time depends on the interface used in the ESI setup and can be extended to several milliseconds [32]. The image in the inset (Figure 2.8a) depicts a charged droplet which experiences fissions. A miniature Taylor cone is created when the droplet distorts and gives rise to a string of offspring droplets [36]. The sample flow rate in ESI-MS is usually in the order of a few microliters per minute. When the flow rate and the diameter of the ESI emitter are particularly small, the method is referred to as *nano electrospray ionization* (nanoESI). It was proposed by Wilm and Mann in the 1990s [37, 38]. It has been demonstrated that nanoESI can provide higher sensitivity and better ionization efficiency than conventional ESI. In nanoESI, a tapered capillary with a diameter of $1-2 \, \mu m$ is used as the emitter. The sample flow rate is in the order of a few tens of nanoliters per minute [37, 38]. Therefore, droplets with much smaller size (compared with the conventional ESI) can be formed. ESI and nanoESI are routinely

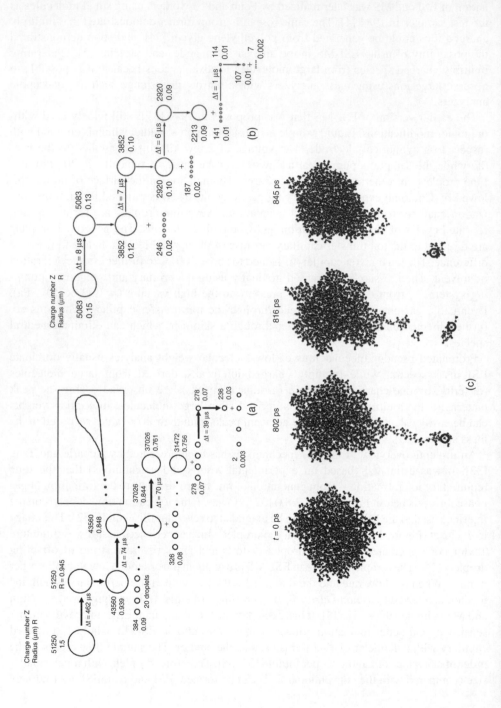

Charge number Z
Radius (μm) R

(a)

(b)

(c)

$t = 0$ ps   802 ps   816 ps   845 ps

Figure 2.8   *Time spent on ion formation in ESI. (a) Schematic representation of time history of parent and offspring droplets. Droplet at the top left is a typical parent droplet created near the ESI capillary tip at low flow rates. Evaporation of solvent at constant charge leads to uneven fission. The number beside the droplets gives radius R (μm) and number of elementary charge N on droplet; Δt corresponds to the time required for evaporative droplet shrinkage to the size where fission occurs. Only the first three successive fissions of a parent droplet are shown. At the bottom right, the uneven fission of an offspring droplet to produce offspring droplet is shown. The timescale is based on $R = R_0 - 1.2 \times 10^{-3}t$ (R and R_0, radius of droplet; t, time), which produces only a rough estimate. Inset: Tracing of photograph by Gomez and Tang [36] of droplet undergoing "uneven" fission. Typical droplet loses 2% of its mass, producing some 20 smaller droplets that carry 15% of the parent charge [32]. Reprinted with permission from Kebarle, P., Tang, L. (1993) From Ions in Solution to Ions in the Gas Phase. The Mechanism of Electrospray Mass Spectrometry. Anal. Chem. 64: 972A–986A. Copyright (1993) American Chemical Society. (b) Droplet histories for charged water droplets produced by nanoESI. The first droplet is one of the droplets produced at the spray tip. This parent droplet is followed for three evaporation and fission events. The first generation droplets are shown as well as the fission of one of these to lead to second generation offspring droplets [40]. Reproduced from Peschke, M. et al. (2004) [40] with permission of Springer. Data shown based on experimental results [41] and calculations [40]. (c) In this molecular dynamic simulation, an unfolded protein chain that was initially placed within a Rayleigh-charged water droplet gets ejected via the chain ejection model. Side chains and backbone moieties are represented as beads [42]. Reproduced with permission from Konermann, L., Ahadi, E., Rodriguez, A.D., Vahidi, S. (2013) Unraveling the Mechanism of Electrospray Ionization. Anal. Chem. 85: 2–9. Copyright (2013) American Chemical Society. See colour plate section for colour figure*

used to couple separation techniques with MS nowadays (see Chapter 6). NanoESI generates very small (nano-sized) droplets. Thus, the time required for ion formation is further reduced (Figure 2.8b) [39–41]. It is supposed to be ~1 order of magnitude shorter than that estimated for conventional ESI. Additionally, molecular dynamics simulations have been used to estimate the time required for the gas phase ions of an unfolded protein molecule to form [42]. It was estimated that ~1 μs is required for protein ion formation from the liquid phase to the gas phase (Figure 2.8c) [42, 43]. These estimations imply the feasibility of using ESI and nanoESI as the ionization techniques for detecting reacting species from fast reactions in a timescale down to microseconds. Thus, nanoESI as well as ESI seem to be suitable ionization techniques for the monitoring of ultrafast processes.

Two mechanisms are generally used to describe the formation of gaseous ions in ESI [33, 39, 44]. The so-called *single ion in droplet theory* (or "charge residue model") is used to explain the formation of multiply charged ions [33]. In a nutshell, the fine droplets generated from electrospray shrink as the solvent evaporates. The increased charge density (due to solvent evaporation) causes large droplets to divide into smaller droplets. Repetition of this process leads to nanodroplets which consist only of single molecules/ions and multiple charges (Figure 2.9a). The above is the most likely mechanism explaining the formation of gaseous macromolecular ions with multiple charges. The other mechanism, the so-called *ion evaporation theory*, was proposed by Iribarne and Thomson [45]. According

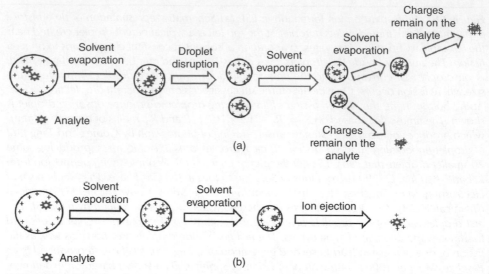

Figure 2.9 *Ionization mechanisms in ESI. (a) Schematic illustration of the single ion in droplet theory and (b) ion evaporation theory*

to this theory, the formation of fine droplets – due to Coulomb explosions (whenever the Rayleigh limit is reached) and solvent evaporation – leads to an increased charged density, and contributes to Coulomb explosions. When the surface tension of the liquid droplets is overcome, ions are released from the droplets. This mechanism can be used to explain the formation of singly charged ions.

2.5 Atmospheric Pressure Photoionization

Atmospheric pressure photoionization (APPI) complements APCI and ESI in its functionality and application scope. It can be used to ionize analytes with relatively low polarity, which are difficult to be ionized using other atmospheric pressure ionization techniques [46, 47]. In the APPI source, a krypton lamp that emits photons at 10.0 eV is normally implemented. The photon energy provided by the lamp is sufficient to ionize a large number of organic compounds. However, it is insufficient to ionize air and solvents such as water, methanol, and acetonitrile (Table 2.1). Thus, the interference from solvent can be minimized. Dopants, which have a lower ionization energy than the energy of photons emitted by the lamp, are usually added to the solvent in APPI. Thus, dopants are easily ionized, and they also facilitate the ionization of target analytes. They act as intermediates between photons and analytes. The dopant–analyte reactions involve electric charge exchange or proton transfer. The charge transfer process is illustrated by the following reactions:

$$D + h\nu \rightarrow D^{+\bullet}$$

$$D^{+\bullet} + M \rightarrow M^{+\bullet} + D$$

Table 2.1   *List of the ionization energies of common gases, organics, and solvents [47]. Energies of photons emitted by different lamps [47, 48]. [Atmospheric Pressure Photoionization Mass Spectrometry. Raffaelli, A., Saba, A. Mass Spectrom. Rev. 22/5. Copyright (2003) John Wiley and Sons]*

Lamp	Compound	Ionization energy (eV)
	Nitrogen	15.58
	Water	12.62
	Acetonitrile	12.20
	Oxygen	12.07
Ar: 11.2 eV		
	Methanol	10.84
	Isopropanol	10.17
	Hexane	10.13
Kr: 10.0 eV		
	Heptane	9.93
	Isooctane	9.80
	Acetone	9.70
	Pyridine	9.26
	Benzene	9.64
	Furan	8.88
	Toluene	8.83
Xe: 8.4 eV		
	Naphthalene	8.14
	Triethylamine	7.53

Dopants (D) are photoionized first, followed by charge exchange with the analytes (M). The ionization of analytes can also be achieved by means of proton transfer processes:

$$D + h\nu \rightarrow D^{+\bullet}$$

$$D^{+\bullet} + nS \rightarrow [D - H]^{\bullet} + S_nH^+$$

$$S_nH^+ + M \rightarrow nS + MH^+$$

Dopants are ionized first, followed by the ion–molecule reaction with solvent (S). Protonated solvent ions are generated, and subsequently they transfer protons to the analytes.

2.6   Desorption/Ionization

In the beginning of the 1970s (i.e., before the development of ESI and APCI), the repertoire of available ion sources was quite narrow. The analysis of polar and thermal labile analytes by MS was problematic. In fact, at that time, sample vapors were normally generated by application of heat prior to ionization. However, this method was not suitable for thermally labile compounds since the heating rate was too low, which – in many cases – led to thermal decomposition. If the sample vaporization was faster than the decomposition rate, intact molecules could be transferred from the condensed phase to the gas phase

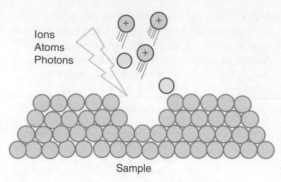

Sample

Figure 2.10 *Schematic representation of an energetic beam (e.g., ions, atoms, and photons) impacting on samples*

efficiently. Thus, finding an efficient way to rapidly convert condensed samples, containing labile analytes, to the gas phase – without causing decomposition – was a major challenge. In the 1970s, an idea emerged that irradiating samples with high energy beams (Figure 2.10) such as laser light could aid bringing the analytes to the gas phase. In the mid-1970s, researchers explored a plasma based desorption/ionization approach. In this technique, a high energy plasma was used to bombard the sample to release the analyte molecules trapped in the condensed phase [49]. Macfarlane and Torgerson employed the radioactive nuclide californium-252 (^{252}Cf, half-life: 2.6 years) which can generate energetic fission fragments to irradiate solid sample [49]. A typical pair of fission fragments is ^{142}Ba^{+18} and ^{106}Tc^{+22} with kinetic energies of ~ 79 and ~ 104 MeV, respectively [49]. The heating process through the bombardment by the fission fragments can be very efficient. For example, it was estimated that the fission fragments (10 MeV) only took ~ 0.1 ps to pass through a 1-μm-thick film made of an organic material [49]. The temperature of the irradiated area (80 Å in diameter) was estimated to be as high as $\sim 10\,000$ K [49]. Because of rapid heating and high energy input, polar and thermally labile analytes such as natural toxins and peptides could be desorbed and maintained their structures prior to MS analysis. This approach showed the possibility of using an energetic beam as the ion source to desorb intact molecules from the condensed phase to the gas phase. However, the ^{252}Cf plasma desorption/ionization required a couple of hours to obtain satisfactory MS results [49].

2.6.1   Fast Atom Bombardment

Secondary ion mass spectrometry (SIMS) uses an energetic ion beam such as Ar$^+$ to desorb condensed-phase samples for MS analysis [50, 51]. This technique evolved in the 1970s. However, because of the high energy impact, fragment ions usually dominate the SIMS results. Furthermore, the primary ion beam may cause surface charging of the non-conductive sample, resulting in poor focusing of samples. In another approach, Barber *et al.* [52] used an energetic beam of atoms (e.g., Ar) to irradiate samples for the analysis of thermally labile and non-volatile samples. The technique has been termed *fast atom bombardment* (FAB). Moreover, liquid matrix, which possesses high viscosity and low vapor pressure, was applied to the primary sample to facilitate the ionization process in FAB [53]. The energy of the fast atoms (Ar or Xe) is quite high (4000–10 000 eV). Glycerol, thioglycerol, 3-nitrobenzyl alcohol, diethanolamine, and triethanolamine were demonstrated to be useful FAB matrices. They can maintain the sample in high vacuum, buffer the high energy impact, and facilitate the ionization process by providing protons or removing protons from analyte molecules [53]. Protonated or deprotonated analyte ions are observed in the FAB mass spectra in the positive- or negative-ion mode, respectively. If an ion beam (e.g., Cs^+) is used in place of an atom beam, the ionization technique is called liquid SIMS [54]. The MS results of the same sample obtained from FAB and liquid SIMS are similar.

2.6.2 Laser Desorption/Ionization

Laser light has been implemented in the ionization of molecules prior to MS analysis since the early 1960s [55]. The analytes are irradiated directly with short but intense pulses of laser light with wavelength in the ultraviolet (UV) range. The laser beam, carrying high energy, directly impinges on the solid sample, thus initiating the desorption and ionization process [55–58]. In the beginning, pulses of laser beam with diameter of 150 μm and duration of 50 μs were used – leading to a broad kinetic energy distribution of the generated ions [56]. This led to poor mass resolution of the resulting spectra. Thus, reducing the laser beam diameter down to 20 μm, and shortening the pulse down to the nanosecond range (1–100 ns) were used to limit the kinetic energy distribution [56], and to lower the probability of thermal decomposition of labile analytes. Hillenkamp *et al.* [57] modified the laser focusing system by implementing microscope optics, and they could focus the laser beam down to a spot with diameter of ~ 0.5 μm. The upper mass limit for biopolymers was ~ 1 kDa. After two decades of using this technique, researchers discovered that adding an organic matrix [7, 59] or inorganic materials [8, 9] can assist the ionization process. These substances act as mediators – absorbing the energy from the laser light, and transferring it to the analyte molecules. In this way, the upper mass range has been extended to enable analysis of large biomolecules such as proteins. Since a large amount of photonic energy is inputted to the sample in a short period of time, laser desorption/ionization (LDI) is used to transfer intact analytes from the condensed phase to the gas phase. During the desorption process, ion–molecule reactions (e.g., proton transfer) may occur, leading to ionization of neutral species in the gas phase. It is generally accepted that desorption and ionization occur almost at the same time. The residence time of ions in the ion source can be calculated using the following equation [60]:

$$t = \frac{vm}{z(eV/d)}$$

where t is residence time of the ionized species at the ion source before entering the mass analyzer; v is velocity of the ions; d is length of the ion source; while V/d is electric field strength inside the ion source. If we assume the applied voltage on the ion source is 20 kV, the flight distance in the ion source is 2 cm, the target ion has m/z 1000 and then the velocity of the ion is 6.2×10^4 m s^{-1}. The time of the ions in the ion source is ~ 0.32 μs. The following sections provide more details on the implementation of organic matrices and inorganic materials used as energy absorbers in LDI-MS.

2.6.2.1 *Matrix-assisted Laser Desorption/Ionization*

MALDI was introduced in the 1980s by Karas and Hillenkamp [7, 59]. It has quickly gained popularity because it enables identification of biomolecules within a broad mass range: from a few hundred daltons to several tens of kilodaltons. In particular, it is applied in the analysis of biomolecules such as peptides, proteins, and nucleic acids. The MALDI "phenomenon" was discovered in the course of extremely careful studies [59]. Karas *et al.* [59] first observed that dipeptides containing aromatic amino acids (e.g. tryptophan) require lower laser power than dipeptides containing a non-aromatic amino acid (e.g., valine or proline) (Figure 2.11). In fact, high ionization efficiency was obtained from dipeptides containing amino acids such as tryptophan. Protonated pseudomolecular ions derived from dipeptides were observed in the mass spectra (Figure 2.11). The experimental results suggested that the molecular structure and the ability to absorb laser light are important factors that inherently affect the analysis process. Thus, it was concluded that adding organic molecules (*matrix*) [59] – capable of absorbing laser energy, in large excess with respect to the amount of analytes – gaseous analyte ions can be formed. Furthermore, it was also proposed that the matrix molecules, which contain acidic functional groups, can work as proton donors and assist the ionization of analytes. The inventors demonstrated that mixing an organic matrix (such as nicotinic acid) with peptides (such as melittin) greatly enhances ionization of these test analytes [61]. Their paper was the first one to report application of LDI-MS in the analysis of compounds with a molecular weight of $\sim 3\,kDa$ [61]. In fact, this mass scale is relevant to many proteomic studies carried out nowadays. Karas and Hillenkamp also demonstrated the suitability of MALDI-MS for the analysis of proteins [7]. The innovation together with ESI-MS has elevated the significance of MS analysis. It solved some long-lasting challenges in the analysis of polar, non-volatile and large organic molecules. The discovery of these ionization techniques has greatly benefited various research fields, including biochemical analysis, biomedicine, and synthetic chemistry.

Due to the significance of the MALDI technique for mass spectrometric analysis of molecules, it is appropriate to spend some time discussing its technical aspects. Efficient and controllable energy transfer to the sample requires that the molecules can absorb the laser light. In order to circumvent thermal decomposition of thermally labile molecules, the energy must be delivered within a very short time. Typically, lasers with pulse widths in the range $1-100\,ns$ are utilized. Given these short durations and the fact that laser beams can easily be focused to spot sizes which are very small compared with the other dimensions of the ion source, the ions are generated essentially in one point (in space and time). This feature makes MALDI highly compatible with the time-of-flight (TOF) mass analyzer (see Chapter 3).

In a nutshell, the matrix absorbs and transfers energy, and facilitates ionization. A large molar excess of the organic matrix with the absorption capacity in the UV region and favorable crystallization properties are required to achieve high ion yields in MALDI. When analyzing proteins, the mass ratios of proteins to matrix can be set between 1.0×10^{-4} and 1.5×10^{-2}, depending on the protein size [59, 60, 62]. Against common sense, a higher ratio of analyte molecules, relative to matrix molecules, contributes to the decrease of ion abundances. MALDI-MS is a particularly sensitive analytical tool when compared with other MS-based platforms [63]. It is routinely used in the analysis of metabolites, peptides, and proteins at attomole levels and beyond. Analysis of metabolites at low attomole levels using special matrices (9-aminoacridine) and arrayed sample supports in the negative-ion mode has been achieved using this technique [63].

In addition to its usefulness in the analysis of a variety of analytes, MALDI-MS also possesses the advantage of rapid and simple sample preparation. Easy sample preparation

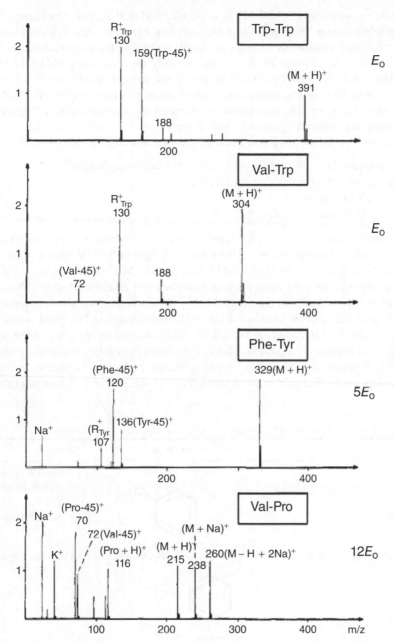

Figure 2.11   *LD mass spectra of dipeptides at their threshold irradiance, $\lambda = 266$ nm [59]. Reproduced with permission from Karas, M., Bachmann, D., Hillenkamp, F. (1985) Influence of the Wavelength in High-Irradiance Ultraviolet Laser Desorption Mass Spectrometry of Organic Molecules. Anal. Chem. 57: 2935–2939. Copyright (1985) American Chemical Society*

is one of the reasons why MALDI-MS is so popular and well-accepted by many scientists. It only requires mixing the analyte and the matrix in a proper ratio. Typically, microliter and sub-microliter volumes of samples are used. Following the evaporation of solvent at atmospheric or reduced pressure, the sample is ready for conducting MALDI-MS analysis [7]. The matrix is usually prepared in an organic solvent or mixture of water and an organic solvent. This sample preparation makes the matrix medium volatile, and easy to remove prior to analysis. The co-crystallized samples are irradiated with UV laser light in the vacuum compartment to generate gaseous ions.

Overall, the MALDI matrix possesses multiple functions [7, 59–62, 64–67]:

1. It has resonant absorption of the molecule at the laser wavelength.
2. It transfers energy to the analyte molecules.
3. It facilitates ionization.
4. It isolates individual biomolecules.

MALDI matrix compounds generally have a molecular structure comprising an aromatic ring with functional groups because lasers emitting light in the UV region (e.g., 337 and 355 nm) are generally used in MALDI-MS. Besides, the MALDI matrix should form fine crystalline deposits since the energy transfer can be more efficient when analytes are well mixed and co-crystallized with the matrix [7, 64]. Two structurally related compounds, that is gentisic acid [62] and cinnamic acid [64, 65], were recognized as good matrices soon after the invention of the MALDI technique. They are commonly used in the MALDI-MS analysis of peptides and proteins (Table 2.2). Note, the energy acquired by the matrix can quickly be transferred to the surrounding analyte molecules to enable desorption of sample components. The acidic functional groups in the matrix structure also provide a

Table 2.2   *Commonly used MALDI matrices [59–62, 64–67]*

Matrix	Structure	Analyte
Gentisic acid (2,5-dihydroxybenzoic acid)		Peptides, proteins, lipids, nucleic acids, carbohydrates
Cyano-4-hydroxycinnamic acid		Proteins, peptides
Sinapinic acid (3,5-dimethoxy-4-hydroxycinnamic acid)		Peptides, polymers, proteins

sufficient number of protons for ionizing (protonating) analytes. The matrix concept in MALDI is very elegant: one molecule possesses multiple functions enabling a complex physical process. Apart from the multiple functions of the MALDI matrix, the solid support (referred to as the "plate" or "target") can also possess multiple functions [68] which greatly facilitate sample preparation and make the MALDI-MS applicable to many challenging problems in chemical science.

Although MALDI-MS enables very sensitive analysis of most polar analytes, the background ions generated from matrices cause serious interferences when analyzing small molecules (< 800 Da). Thus, efforts have been made to reduce the interference in the low mass region [69, 70]. Lin and Chen covalently bonded common MALDI matrix, 2,5-dihydroxybenzoic acid (DHB) into SiO_2 sol-gel film (Figure 2.12) [69]. The resulting film possesses the functions of typical MALDI matrix. It acts as an absorber of laser energy and a proton source. Interferences due to the presence of numerous ions in the low m/z range are greatly reduced. In this case, covalent trapping of the matrix in the network structure of the SiO_2 film contributes to the reduction of these interferences. The formed nanostructured surface of the inorganic material, incorporating 2,5-DHB, can assist loading analyte molecules and facilitate subsequent desorption/ionization processes. Thus, a

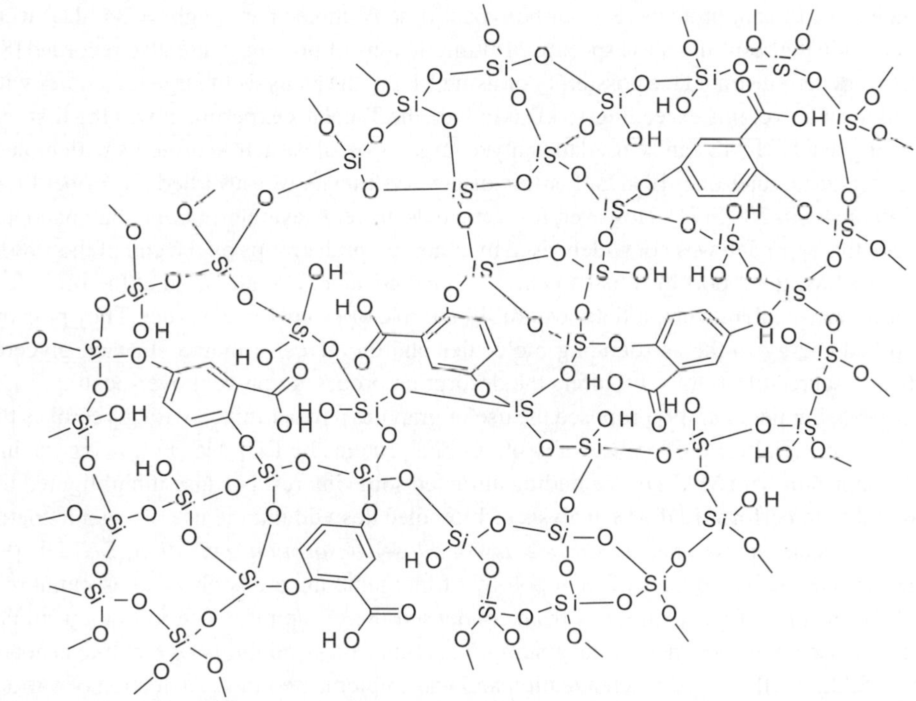

Figure 2.12 *Proposed polymeric structure for entrapping 2,5-dihydroxyl benzoic acid into the sol-gel polymer [69]. Reproduced with permission from Lin, Y.-S., Chen, Y.-C. (2002) Laser Desorption/Ionization Time-of-Flight Mass Spectrometry on Sol-Gel-Derived 2,5-Dihydroxybenzoic Acid Film. Anal. Chem. 74: 5793–5798. Copyright (2002) American Chemical Society*

matrix interference-free of small molecules can be obtained. Irradiation with laser power of 70–110 µJ is not likely to generate any background ions from this sol-gel-matrix derived film. However, a higher laser power >110 µJ was required to desorb small proteins [69]. This method also simplifies sample preparation. The sample solution is directly placed on top of the sol-gel film. The analyte signals are homogeneously distributed on the target surface because crystallization of the matrix is not necessary.

2.6.2.2   *Surface-assisted Laser Desorption/Ionization*

Although MALDI-MS can be very useful in the analysis of biomolecules, it is generally less suitable for the analysis of small molecules <1000 Da. This limitation is due to the high ionic background in the low *m/z* region, derived from the MALDI matrix. Thus, other LDI-MS variants have been to enable analysis of small molecules. For example, surface-assisted laser desorption/ionization (SALDI) [9] uses inert inorganic materials as "assisting materials" to facilitate desorption/ionization of condensed phase samples. It produces relatively low numbers of background ions.

The early work using inorganic substances as the assisting materials for LDI-MS was conducted by Tanaka *et al.* [8]. The authors claimed that the discovery was serendipitous [71]: Tanaka accidentally mixed cobalt nanoparticles ∼30 nm with, and conducted LDI-MS of proteins [8]. Ions of proteins (e.g., carboxypeptidase A, molecular weight = 34 kDa) were observed in the resulting mass spectra. Multimeric ions of proteins were also recorded [8]. This work demonstrated the possibility of using LDI in the analysis of large molecules with the molecular weights exceeding 10 kDa. In fact, the Tanaka's experiment was the first one showing that LDI-MS can be used to analyze large molecules such as proteins with molecular weights greater than 10 kDa. For this discovery, Tanaka was awarded the Nobel Prize in Chemistry in 2002 [71]. However, for some reasons (e.g., availability of cobalt nanoparticles), this approach was not widely used by other research groups by the end of the 1980s. Nevertheless, the report by Tanaka *et al.* attracted considerable attention at the time. It is clear now that cobalt nanoparticles can fulfill the role of an energy absorber. They pass the absorbed energy to the surrounding molecules and the glycerol matrix. In fact, glycerol acts as a source of protons. It buffers the desorption process to prevent over-heating.

In 1995, Sunner *et al.* [9] proposed the use of graphite powder mixed with glycerol as the assisting material for the analysis of peptides and proteins by LDI-MS. Following the initial fascination with MALDI, the finding attracted some interest because it highlighted the possibility to perform LDI-MS analysis of biomolecules without the use of organic matrices. The technique was named *surface-assisted laser desorption/ionization* (SALDI) [9]. The inventors believed the surface topology of inorganic materials plays an essential role in the formation of gas-phase ions from condensed-phase samples. The follow-up studies [72, 73] also confirmed their initial assumptions. Since the availability of graphite is not an issue, SALDI-MS gained much attention and was implemented in the detection of various analytes [74–80]. Moreover, the burgeoning field of nanotechnology promoted the development of SALDI-MS. Inorganic materials – including silicon nanostructures [81], titania nanomaterials [82–85], core–shell nanoparticles [85], iron oxide nanoparticles [86, 87], gold nanoparticles [88], silver nanoparticles [89], and platinum nanoparticles [90] – have been shown to be suitable assisting materials for SALDI-MS. In fact, various types of nanoparticles can be synthesized via many simple routes. Thus, the synthesis of SALDI assisting materials is within the reach of most MS laboratories.

Several reports also emphasized the low ionic background in SALDI-MS, which rendered this technique suitable for analysis of small molecules [91]. Glycerol – initially used as the source of protons [9] – imposed certain limitations, and it was soon replaced by ammonium citrate and citric acid [83–86]. Although the vapor pressure of glycerol is relatively low, and glycerol depositions are not readily vaporized in high vacuum, it is not convenient to introduce numerous samples containing glycerol into the ion source compartment. If the analytes (e.g., saccharides) have predisposition to form adducts with alkali metals, there is no need to include a proton source in the sample preparations. In fact, alkali metals are ubiquitous in the laboratory environment, and their trace quantities are sufficient to promote formation of adducts during the SALDI process. Besides, the suitability of using nanomaterials as nanoprobes for target species has led to wide interest in SALDI-MS. In fact, nanomaterials can participate in affinity interactions with biomolecules. For example, nanoparticles can be used in both affinity-based enrichment of analyte molecules as well as ionization [82–87]. When nanoparticles are used to selectively concentrate target species, the conjugates of nanoparticles and analytes can be directly analyzed by SALDI-MS without conducting additional sample preparation steps. Additionally, because of the relatively low background ions in the low mass region, SALDI-MS is suitable monitoring chemical reactions for TRMS studies (see Chapter 11). Reactants in chemical reactions usually have small molecular sizes; thus, full compatibility can be obtained.

When discussing matrix-free LDI strategies, we should also mention one more related approach that resonated in the specialist literature over the past two decades. In *desorption/ ionization on silicon* (DIOS), a nanostructured silicon chip is utilized as the SALDI-assisting material [81]. After depositing sample solution on such a chip, the sample is ready for LDI-MS analysis. Furthermore, DIOS chips are compatible with microfluidic and microreactor systems [92, 93]. Thus, DIOS-MS has occasionally been implemented in TRMS-related measurements (see also Chapters 7 and 13).

2.7 Innovations in the 21st Century

The ionization techniques mentioned above have mostly been developed in the 20th century. They are used to analyze almost all types of molecules characterized with diverse properties. The time that chemical species spend in these common ion sources is typically in the range of sub-microseconds to sub-milliseconds (Table 2.3). In the first decade of the 21st century, developments in the construction of mass analyzers were paralleled by developments in ionization techniques. Strong emphasis was put on the sensitivity and resolution of MS. It is now straightforward to generate ions from samples present in the condensed phase, and detect trace amounts of analytes [63]. During the past few years, the number of available ionization techniques has grown from ~ 10 to > 50. Nevertheless, many new ionization techniques are derived from the canonical schemes of ESI, LDI, and plasma ionization – developed decades ago. Most ionization techniques developed in the 21st century operate at atmospheric pressure rather than in vacuum to enable analysis without extensive sample preparation and instrument conditioning (e.g., evacuating ion source). Many developments in atmospheric pressure ion sources followed the invention of desorption

Table 2.3   *Residence times of molecules/ions in different ion sources. Typical estimated performance data are provided*

Ion source	Sample phase	Residence time	Ref.
Electron ionization	Gas, liquid, solid	~μs	[15]
Chemical ionization	Gas, liquid, solid	~μs to sub-μs	[19]
Atmospheric pressure chemical ionization	Liquid	Sub-ms	[26–28]
Electrospray ionization	Liquid	Sub-ms	[32]
Nanoelectrospray ionization	Liquid	Several tens of μs	[39–41]
Matrix-assisted laser desorption/ionization	Liquid, solid	Sub-μs[a]	[60]

[a]The m/z of the target ions is 1000.

electrospray ionization (DESI) in 2004 [94]. A number of the new techniques are classified as the so-called *ambient ionization techniques* to emphasize the minimal sample preparation and no need for vacuum operation [95]. This latter feature makes those ionization techniques suitable for temporal monitoring of chemical and biochemical processes taking place at atmospheric conditions. Specialized review papers on this topic have been published [96–101]. Herein, we do not attempt to describe all the published innovations; instead we will introduce some important ionization techniques which – in our view – have the potential to facilitate TRMS analyses.

2.7.1   Ion Sources Derived from Electrospray Ionization

Several ionization techniques that are derived from ESI are described in the following sections.

2.7.1.1   Desorption Electrospray Ionization

DESI [94] is a technique that combines the concepts of ESI and desorption/ionization. An electrospray plume is directed towards the sample in the condensed phase, which is placed a few millimeters away, in order to desorb analyte molecules (Figure 2.13a). The electrospray plume is initially produced from a mixture of an organic solvent and water containing trace acid infused through a capillary (flow rate 3–15 μl min^{-1}) connected to a high voltage power supply (e.g., 4 kV). It contains charged micro-droplets. It is believed that when these micro-droplets impinge on the sample surface, secondary micro-droplets and/or ionic species are desorbed, and directed toward the MS inlet. The DESI-MS approach is suitable for the analysis of large proteins and small organic molecules. Protonated pseudo-molecular ions of small molecules usually dominate DESI mass spectra. Multiply charged ions – derived from large molecules – give rise to similar spectral patterns as in the case of ESI. Various mechanisms have been proposed to explain how ions are formed during the DESI process [101]. Two mechanisms describe the generation of ions from small molecules and large biomolecules, respectively. Unlike LDI operated in a high vacuum, DESI involves low energy processes, and it is operated at atmospheric pressure. When the electrospray plume – containing electrons, protons, and ions – impinges on the sample, the sample

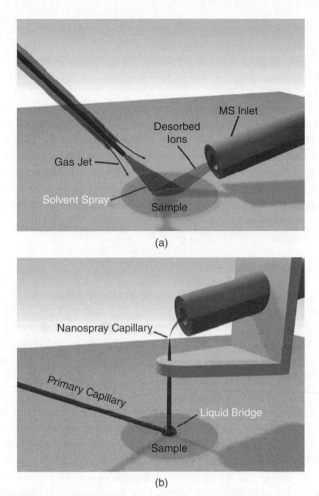

(a)

(b)

Figure 2.13 *Experimental configuration of DESI and nanoDESI. (a) Traditional DESI setup. Solvent is electrosprayed with the aid of a nebulizer gas jet. Solvent and analyte are removed from the surface. (b) NanoDESI setup. A solvent bridge formed between the primary and nanospray capillaries contacts the analyzed surface. Analyte-containing solvent is removed from the surface by self-aspirating nanospray [102]. Reproduced from Roach, P.J. et al. (2010) [102] with permission of the Royal Society of Chemistry*

surface may acquire electric charge. If the primary electrospray plume micro-droplets possess sufficiently high momentum, charged analytes are released from the sample surface. This effect could explain formation of ions from small molecules. On the other hand, the so-called *droplet-pickup mechanism* attempts to explain formation of multiply charged ions of large molecules such as proteins [101]. When the electrospray plume is directed toward the analyte deposited on a solid surface, the charged droplets from the electrospray plume form a thin liquid film on that surface, and solubilize the analytes. Since the electrospray plume continuously impinges on the sample surface, the micro-droplets hit on the newly formed thin liquid film, and fine droplets containing analyte molecules are detached from the surface due to hydrodynamic forces. Similarly to ESI, fine droplets shrink due to evaporation of solvent. The shrinkage of secondary micro-droplets is followed by Coulomb explosions. After a few cycles, gaseous analyte ions are formed through ion ejection or ion charge residue processes, as described in Section 2.4.

In addition, there exists a variant of DESI which is referred to as *nanospray desorption electrospray ionization* (nanoDESI) [102]. The setup of nanoDESI (Figure 2.13b) is quite different from the conventional DESI. A sample (e.g., supported on a solid surface) is placed between two capillaries. The primary capillary supplies solvent to create and maintain the liquid bridge, while the second capillary transports the dissolved analyte from the bridge to the mass spectrometer. A high voltage applied between the inlet of the mass spectrometer and the primary capillary creates a self-aspirating nanospray. Since two capillaries are involved in the setup, the primary capillary can deliver substrates which may react with co-substrates delivered in proximity to the liquid junction. Applications of nanoDESI for on-line reaction monitoring will be described in Chapter 11.

2.7.1.2   Electrosonic Spray

Another technique derived from ESI is *electrosonic spray ionization* (ESSI) [103]. ESSI utilizes a higher pressure of nebulizer gas (e.g., N_2) than ESI (ESI : $\sim 10^5$ Pa; ESSI : $\sim 10^6$ Pa). Therefore, finer droplets can be formed. Similarly to gas-assisted ESI, the sample and the nebulizer gas flows are coaxial. Another ESSI-derived ionization technique is the so-called *Venturi easy ambient sonic-spray ionization* (V-EASI). In this variant, high voltage is not applied to the sample emitter [104]. The sample solution is delivered in proximity to the MS inlet due to the Venturi effect evoked by the flow of the nebulizing gas supplied at a high pressure [~ 10 bar ($\sim 10^6$ Pa); sample flow rate: $3.5 \, 1 \, min^{-1}$] [104]. Very fine droplets are formed due to the assistance of the pressurized nebulizing gas. The ESSI techniques are useful in the monitoring of chemical reactions because it is easy to setup reactions in a sample vial and deliver the reaction solution to the inlet of the mass spectrometer in real time without the use of a syringe pump (Figure 2.14a). Additionally, this setup can also be used to relay solvent/reactant from the sample vial onto solid samples for direct analysis (Figure 2.14b). Unlike in DESI, there is no electric voltage applied to the solvent reservoir. The fine droplets of solvent are formed during the nebulization process assisted by high pressure gas [~ 10 bar ($\sim 10^6$ Pa)]. Applications of these techniques in TRMS are further discussed in Chapters 11 and 13.

2.7.1.3   Fused Droplet Electrospray Ionization and Extractive Electrospray Ionization

Fused droplet electrospray ionization (FDESI) [105], also called *extractive electrospray ionization* (EESI) [106], is a two-step ionization technique. Ultrasonication or the flow of gas assists nebulization of samples in the absence of high voltage. A mist of micro-droplets – formed from sample – further merges with an ESI plume initiating ionization. A methanol solution, containing a trace amount of acetic acid (0.1–1.0 vol%) can be used to generate the electrospray plume. The resulting mass spectra have similar profiles as those obtained using the conventional ESI approach. Multiply charged ions of large molecules dominate fused droplet/EESI mass spectra; however, singly charged ions of small organic molecules can also be recorded.

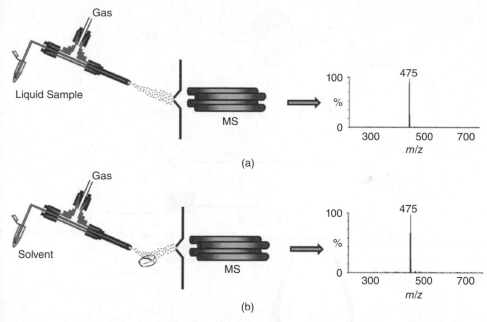

(a)

(b)

Figure 2.14   *Illustration of the V-EASI dual mode of operation: (a) V_L-EASI performed for a 30 ng ml^{-1} sildenaphil solution in acidified methanol; and (b) V_S-EASI performed for a commercial sildenaphil tablet using acidified methanol as the spray solvent [104]. Reproduced with permission from Santos, V.G., Regiani, T., Dias, F.F.G., Romao, W., Jara, J.L.P., Klitzke, C.F., Coelho, F., Eberlins, M.N. (2011) Venturi Easy Ambient Sonic-Spray Ionization. Anal. Chem. 83: 1375–1380. Copyright (2011) American Chemical Society*

2.7.1.4   Polarization-induced ESI

In some models of ESI-MS instruments, the MS inlet is biased against the earth by a few kilovolts while the sample outlet is grounded. This arrangement helps to assure electrical safety while providing stable electric field to support the ESI plume. Building on the same principle, *polarization induced electrospray ionization* (PI-ESI) has recently been developed [107–114]. In this variant, the ESI plume is generated due to the existence of a heterogeneous electric field present in proximity to the MS inlet. Polarization of electric charges on the non-conductor surfaces of the sample emitter gives rise to a virtual potential that is sufficient to support the electrospray. Since there is no electrical contact between the sample emitter and the power supply, this type of ion source is very easy to build. Moreover, the obtained mass spectral profiles look similar to those obtained by conventional ESI-MS. Singly charged ions dominate mass spectra of small organics while multiply charged ions – derived from large biomolecules – are observed in the corresponding mass spectra.

The predecessor of PI-ESI was *ultrasonication-assisted spray ionization* (UASI) [107, 108]. In that early version, one end of a sample capillary was dipped in the sample vial held in an ultrasonicator (frequency: 40 kHz) while the other (tapered) end was placed close (∼3 mm) to the MS inlet. Although no electric potential was applied to the tapered end of the capillary, electrospray could be observed, and mass spectra of analytes with

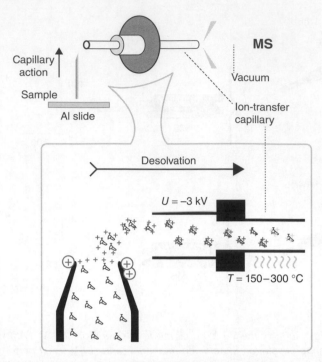

Figure 2.15   *Schematic representation of C-API MS, with sample delivery enabled by capillary action. A short tapered silica capillary [length, 1 cm; base o.d., 363 μm (or 323 μm without polyimide); tip o.d., 10 μm] was positioned vertically above an electrically isolated aluminum slide, with the outlet end placed orthogonal to the inlet of a metal capillary attached to the orifice of an ion trap mass spectrometer. The distance between the outlet of the silica capillary and the inlet of the metal capillary, attached to the MS orifice, was ~1 mm. Before the measurements, the silica capillary was filled with a makeup solution [deionized water/acetonitrile (1 : 1, v/v)] by means of capillary action. The inlet end of the silica capillary was then dipped into a droplet of a sample (10 μl) put onto the surface of the aluminum slide. The inset provides an illustration of the hypothetical mechanism of C-API [109]. Reproduced with permission from Hsieh, C.-H., Chang, C.-H., Urban, P.L., Chen, Y.-C. (2011) Capillary Action-supported Contactless Atmospheric Pressure Ionization for the Combined Sampling and Mass Spectrometric Analysis of Biomolecules. Anal. Chem. 83: 2866–2869. Copyright (2011) American Chemical Society. See colour plate section for colour figure*

diverse molecular weights could be recorded [107]. The purpose of using ultrasound in this ion source was to direct the sample from the inlet to the outlet of the sample capillary in order to promote the formation of fine droplets at the outlet.

Later, it was discovered that a short tapered capillary (~1 cm), placed in proximity (~1 mm) to the MS inlet, could also support electrospray – even in the absence of a power supply, electrical grounding, or ultrasound [109]. Since there was no direct electric contact at the sample capillary, this method was named *contactless atmospheric pressure ionization* (C-API) [109] or *contactless ESI* [110]. Figure 2.15 shows the putative mechanism of the ion formation in C-API. Rearrangement of electric charges at the end of the tapered capillary is induced by the electric field present near the MS inlet.

The charges, with the signs opposite to the sign of the electric potential applied to the MS inlet, accumulate on the liquid meniscus at the capillary outlet. This process is followed by the formation of a Taylor cone and Coulomb explosions which lead to the formation of fine droplets. The mechanism responsible for generation of charged droplets and ions in PI-ESI appears to be similar to that found in the conventional ESI process (Section 2.4).

Recently, an even simpler version of PI-ESI [114] was proposed, in which the use of a capillary as sample emitter was circumvented. A sample droplet (4–10 μl) containing analytes was placed in front of the MS inlet connected to the source of electric potential (~3 kV). The ions corresponding to the analyte molecules present in this droplet were instantly recorded by a mass spectrometer. Importantly, the sample droplet was deposited on a dielectric substrate. Polarization of electric charges on the surface of the dielectric and the sample contributes to detachment of smaller droplets which are directed toward the MS orifice (Figure 2.16). This step is followed by desolvation which may occur in a similar way to that in ESI. Because of the simplicity of the setup, PI-ESI is suitable for use in TRMS studies (see Chapter 11).

2.7.2   New Ion Sources Derived from Laser Desorption/Ionization

Several relatively new ionization techniques are derived from the canonical LDI approach. Some of them do not require the use of a vacuum [115–118]. For example, in the first step, neutral species can be desorbed from the condensed phase by a laser light beam, and in the second step, they can be mixed with the charged droplets present in an ESI plume. Several ionization schemes – which bear distinct names – are based on the same concept. For example, *electrospray-assisted laser desorption/ionization* (ELDI) (Figure 2.17) generally

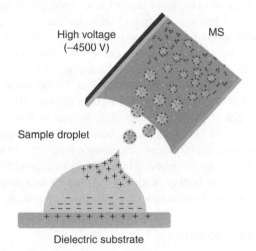

Figure 2.16 *Schematic representation of the putative mechanism of the PI-ESI-MS [114]. [Polarization Induced Electrospray Ionization Mass Spectrometry for the Analysis of Liquid, Viscous, and Solid Samples. Meher, A.K., Chen, Y.-C. J. Mass Spectrom. 50/3. Copyright (2015) John Wiley and Sons]*

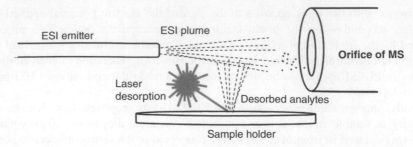

Figure 2.17   Schematic illustration of ELDI. Samples are desorbed by a laser and the desorbed species are fused with the electrospray plume

uses a pulsed UV laser [115] to transfer samples from the condensed phase to the gas phase although a pulsed IR laser is also applicable [117]. ELDI is carried out at atmospheric pressure. Unlike MALDI, it does not normally use a chemical matrix to enhance absorption of laser light.

Biomolecules such as peptides and proteins can survive irradiation by laser and give rise to identifiable ions. Similarly to ESI, multiply charged ions derived from large biomolecules are observed in the ELDI mass spectra [115]. Furthermore, small molecules can be recorded in the course of ELDI-MS analysis. Because no organic matrix is required, spectral interferences are less significant than in MALDI-MS. Another approach takes advantage of a pulsed IR laser for desorption prior to ionization by ESI. It is called *laser ablation electrospray ionization* (LAESI) [117]. Both ELDI and LAESI have been shown to be useful in MS imaging [116, 117] because the laser beams can be focused to very small spots, enabling selective ablation of the sample surface.

In another approach, a diode laser ($\lambda = 808$ nm) was used to irradiate the rear side of a sample loading chip to selectively desorb small and polar species from complex samples in the condensed phase followed by post-ionization in an ESI plume [118]. This approach has been termed *thermal desorption-based ambient mass spectrometry*. Ionization was assisted by a spray [1 : 1 (v/v) water–acetonitrile mixture containing 0.1% acetic acid] generated from a short tapered capillary. A layer-by-layer multilayer of Au nanoparticles created on a glass slide was used as the photonic energy absorber and as the sample holder. This sample chip absorbs light in the near IR region at a wavelength corresponding to the light of the diode laser. The feature of this approach is that large molecules cannot be liberated to the gas phase through a thermal desorption process. That is, when small molecules are the target species, sample pretreatment is not necessary since large biomolecules will not be desorbed. Since a gold nanoparticle-coated chip is used for this setup, the chip can be further modified to function as a microreactor. Therefore, TRMS can be performed on chemical reactions using this approach.

2.7.3   Plasma-based Ion Sources

As mentioned above (Section 2.3), corona discharge is normally used to support ionization in the APCI source. The discharge process produces a large number of charged species which constitute plasma. Ion–molecule reactions are responsible for ionization of neutral analyte species in APCI. Active species generated from different types of reagents (e.g., water, organic solvents, inert gases) in the electric field have also been used for ionization of analytes at atmospheric pressure.

In *desorption atmospheric pressure chemical ionization* (DAPCI) [119–121], charged species – generated by a corona discharge – are used to directly desorb and ionize analyte molecules. In this technique, different dopant vapors are mixed with the nitrogen gas flow, and a high voltage (3–6 kV) is applied to the discharge needle. Reagent ions are generated and directed toward the sample in order to initiate the ionization process. The concept is somewhat similar to DESI; however, the desorption/ionization is due to the presence of corona discharge rather than the electrospray plume. It has been demonstrated that DAPCI can readily be used to analyze samples such as petroleum [120] and explosives [121]. In DAPCI, ion–molecule reactions are involved in the ionization of neutral species.

2.7.3.1 Atmospheric Solids Analysis Probe

The *atmospheric solids analysis probe* (ASAP) [122, 123] is another ionization source, in which the corona discharge is used to facilitate ionization. In one implementation, a "melting-point tube" containing sample is inserted into a stream of hot nitrogen gas, and the species evaporated from the tube are ionized at atmospheric pressure with the aid of corona discharge – similar to the one observed in an APCI source. The hot gas stream (350–500 °C) may either be provided by the APCI sprayer or by an ESI sprayer. The unique feature of this approach is that no solvent is required for the ionization process. Evaporation of sample present in the condensed phase in ASAP resembles – to some extent – the use of the direct probe in EI and CI. The advantage of this approach over EI and CI is that the generation of the evaporated species is conducted at atmospheric pressure. Since the probe does not need to be inserted into a vacuum, it is faster and easier to operate. Furthermore, the extent of fragmentation is smaller than in EI and similar to that in CI.

2.7.3.2 Direct Analysis in Real Time

Direct analysis in real time (DART) [124, 125] is suitable for analysis of gases, liquids, and solids in "open air" conditions. A gas flow (~ 1 l min^{-1}), typically helium or nitrogen, is used to generate electronic excited-state or vibronic excited-state species through discharge at a high voltage (1–5 kV) [124, 125]. The gas temperature may be set from room temperature to 250 °C [124, 125]. The charged species are removed by applying an electric field, and the remaining electronic/vibronic excited-state species are mainly used in ionization of analytes. The generated electronic/vibronic excited-state species can directly interact with the sample in proximity to the MS inlet (Figure 2.18). Various ionization mechanisms may be involved in the ionization processes in DART. The nature of the carrier gas, analyte concentration, and polarity of ions determine the ionization pathway. For example, nitrogen predominantly produces low energy excited states and only ionizes the analytes that have ionization potentials lower than the energy of the vibronic excited state. The analyte molecules undergo the so-called Penning ionization to form molecular ions. Fragment ions will only be produced if excess energy is available during ionization.

2.7.3.3 Low-temperature Plasma Probe

The so-called *low-temperature plasma* (LTP) probe [126–128] has been used as the ion source to desorb/ionize condensed-phase samples such as explosives [127], drugs [127],

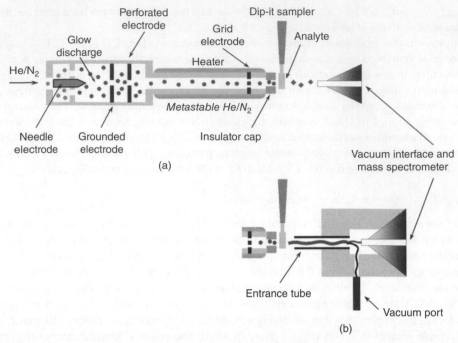

Figure 2.18   *(a) Scheme of DART-ion source; and (b) scheme of a gas-ion separator (Vapur interface) equipped with a vacuum pump [124]. Reproduced with permission from Cody, R.B., Laramée, J.A., Durst, H.D. (2005) Versatile New Ion Source for the Analysis of Materials in Open Air under Ambient Conditions. Anal. Chem. 77: 2297–2302. Copyright (2005) American Chemical Society. See colour plate section for colour figure*

and low-molecular-weight analytes with diverse polarities [128]. The LTP setup resembles that used in DESI; however, the LTP probe is used instead of the electrospray emitter. The LTP is generated in the course of electric discharge produced by an alternating current. The LTP probe is made of a glass tube. It incorporates an internal grounded electrode surrounded by an outer electrode. The discharge gas (He, Ar, N_2, or air) passes along the glass tube, between the two electrodes. The alternating voltage (2.5–5 kV) with a frequency of 2–5 kHz is applied to generate the dielectric barrier discharge. The temperature of the LTP is low ($\sim 30\,°C$), so the LTP does not cause much damage to the sample during ionization. For this reason, LTP can be utilized in monitoring chemical reactions.

2.7.3.4   *Flowing Atmospheric-pressure Afterglow*

Unlike LTP – generated with an alternating-current electric field – the *flowing atmospheric-pressure afterglow* (FAPA) ion source uses direct-current gas discharge [129, 130]. The typical operating voltage is in the order of several hundred volts, while the currents are several tens of milliamperes. The plasma region is designed to be far away from the sample, so that the damage caused by the plasma is reduced. A miniaturized version of FAPA has been used as an interface for coupling separation techniques such as capillary electrophoresis with MS [130]. Thus, it has the potential to facilitate hyphenation of separation techniques with MS.

2.8 Concluding Remarks

Numerous ionization techniques for MS have been introduced over the past few decades. One can easily find suitable ionization techniques for analytes with diverse physicochemical properties. For example, EI and CI are suitable for analysis of small organics with high volatility and low polarity, while ESI, FAB, and MALDI often are used to analyze polar compounds with a wide range of molecular weights. Ionization techniques operated at atmospheric pressure can be used to analyze small and large molecules with different polarities. However, several points such as the time spent in ionization and sampling should be considered when selecting ion sources for TRMS measurements. Ionization times of different ion sources vary from femtoseconds to milliseconds. The residence time of ions in the ion source also depends on the accelerating/extraction voltage and the dimensions of the ion source components. Furthermore, the sampling time imposed by the interface also greatly contributes to the temporal resolution. Various ion sources are interfaced with dynamic samples in different ways (see Chapter 4). This trend is especially clear when comparing off-line and on-line techniques, prominently represented by MALDI and ESI. In the case of off-line techniques, the time spent on introduction of the sample target to the ion source of the instrument may be the limiting factor for many studies which require real-time monitoring. On the other hand, the on-line techniques provide high flexibility when it comes to the delivery of aliquots of fast changing mixtures.

References

1. Bleakney, W. (1929) A New Method of Positive Ray Analysis and Its Application the Measurement of Ionization Potentials in Mercury Vapors. Phys. Rev. 34: 157–160.
2. Nier, A.O. (1947) Electron Impact Mass Spectrometry. Rev. Sci. Instrum. 18: 415–422.
3. Munson, M.S.B., Field, F.H. (1966) Chemical Ionization Mass Spectrometry. I. General Introduction. J. Am. Chem. Soc. 88: 2621–2630.
4. Yamashita, M., Fenn, J.B. (1984) Electrospray Ion Source. Another Variation on the Free-Jet Theorem. J. Phys. Chem. 88: 4451–4459.
5. Fenn, J.B., Mann, M., Meng, C.K., Wong, S.F., Whitehouse, C.M. (1989) Electrospray Ionization for Mass Spectrometry of Large Biomolecules. Science 246: 64–71.
6. Horning, E.C., Carroll, D.I., Dzidic, I., Haegele, K.D., Horning, M.G., Stillwell, R.N. (1974) Atmospheric Pressure Ionization (API) Mass Spectrometry. Solvent-mediated Ionization of Samples Introduced in Solution and in a Liquid Chromatograph Effluent Stream. J. Chromatogr. Sci. 12: 725–729.
7. Karas, M., Hillenkamp, F. (1988) Laser Desorption Ionization of Proteins with Molecular Masses Exceeding 10,000 Daltons. Anal. Chem. 60: 2299–2301.

8.   Tanaka, K., Waki, H., Ido, Y., Akita, S., Yoshida, Y., Yoshida, T., Matsuo, T. (1988) Protein and Polymer Analyses up to m/z 100 000 by Laser Ionization Time-of-Flight Mass Spectrometry. Rapid Commun. Mass Spectrom. 2: 151–153.

9.   Sunner, J., Dratz, E., Chen, Y.-C. (1995) Graphite Surface-assisted Laser Desorption/ionization Time-of-Flight Mass Spectrometry of Peptides and Proteins from Liquid Solutions. Anal. Chem. 67: 4335–4342.

10. Alberici, R.M., Simas, R.C., Sanvido, G.B., Romão, W., Lalli, P.M., Benassi, M., Cunha, I.B.S., Eberlin M.N. (2010) Ambient Mass Spectrometry: Bringing MS into the "Real World". Anal. Bioanal. Chem. 398: 265–294.

11. Svec, H.J., Junk, G.A. (1967) Electron-impact Studies of Substituted Alkanes. J. Am. Chem. Soc. 89: 790–796.

12. Gross, J. H. (2011) Mass Spectrometry: A Textbook, 2nd Edition. Springer, Berlin.

13. Watson, J.T. Introduction to Mass Spectrometry, 3rd Edition. Lippincott Williams & Wilkins, Philadelphia.

14. Pfeiffer, A.N., Cirelli, C., Smolarski, M., Dörner, R., Keller, U. (2011) Timing the Release in Sequential Double Ionization. Nature Phys. 7: 428–433.

15. Meier, K., Seibl, J. (1974) Measurement of Ion Residence Times in a Commercial Electron Impact Ion Source. Int. J. Mass Spectrom. Ion Phys. 14: 99–106.

16. Lias, S.G., Bartmess, J.E., Liebman J.F., Holmes, J.L., Levin, R.D., Mallard, W.G. (1988) Gas-phase Ion and Neutral Thermochemistry. J. Phys. Chem. Red. Data 17: Suppl. 1.

17. Ghaderi, S., Kulkarnl, P.S., Ledford Jr, E.B., Wilkins, C.L., Gross, M.L. (1981) Chemical Ionization in Fourier Transform Mass Spectrometry. Anal. Chem. 53: 428–437.

18. Westmore, J.B., Alauddin, M.M. (1986) Ammonia Chemical Ionization Mass Spectrometry. Mass Spectrometry Reviews, 5: 381–465.

19. Griffith, K.S., Gellene, G.I. (1993) A Simple Method for Estimating Effective Ion Source Residence Time. J. Am. Soc. Mass Spectrom. 4: 787–791.

20. Carroll, D.I., Dzidic, I., Stillwell, R.N., Horning, M.G., Horning, E.C. (1974) Sub-picogram Detection System for Gas Phase Analysis based upon Atmospheric Pressure Ionization (API) Mass Spectrometry. Anal. Chem. 46: 706–710.

21. Crowther, J.B., Covey, T.R., Silvestre, D., Henion, J.D. (1985) Direct Liquid Introduction LC/MS: Four Different Approaches. LC Mag. 3: 240–254.

22. Covey, T.R, Lee, E.D., Henion, J.D. (1986) High-speed LC/MS/MS for the Determination of Drugs in Biological Samples. Anal. Chem. 58: 2453–2460.

23. Thomson, B.A. (1989) Atmospheric Pressure Ionization and Liquid Chromatography/Mass Spectrometry – Together at Last. J. Am. Soc. Mass Spectrom. 9: 187–193.

24. Carroll, D.I., Dzidic, I., Stillwell, R.N., Haegele, K.D., Horning, E.C. (1975) Atmospheric Pressure Ionization Mass Spectrometry. Corona Discharge Ion Source for Use in A Liquid Chromatograph-Mass Spectrometer-Computer Analytical System. Anal. Chem. 47: 2369–2373.

25. Sabo, M., Matejcik, S. (2013) A Corona Discharge Atmospheric Pressure Chemical Ionization Source with Selective NO^+ Formation and its Application for Monoaromatic VOC Detection. Analyst 138: 6907–6912.

26. Kolakowski, B.M., Grossert J.S., Ramaley, L. (2004) Studies on the Positive-ion Mass Spectra from Atmospheric Pressure Chemical Ionization of Gases and Solvents Used

in Liquid Chromatography and Direct Liquid Injection. J. Am. Soc. Mass Spectrom. 15: 311–324.

27. Ketkar, S.N., Penn, S.M., Fite, W.L. (1991) Influence of Coexisting Analytes in Atmospheric Pressure Ionization Mass Spectrometry. Anal. Chem. 63: 924–925.

28. Sunner, J., Nicol, G., Kebarle, P. (1988) Factors Determining Relative Sensitivity of Analytes in Positive Mode Atmospheric Pressure Ionization Mass Spectrometry. Anal. Chem. 60:1300–1307.

29. Dass, C. (2007) Chapter 2, Modes of Ionization. In: Fundamentals of Contemporary Mass Spectrometry. John Wiley & Sons, Inc., Hoboken.

30. Zeleny, J. (1914) The Electrical Discharge from Liquid Points, and a Hydrostatic Method of Measuring the Electric Intensity at their Surfaces. Phys. Rev. 3: 69–91.

31. Taylor, G. (1964) Disintegration of Water Droplets in an Electric Field. Proc. R. Soc. A 280: 383–397.

32. Kebarle, P., Tang, L. (1993) From Ions in Solution to Ions in the Gas Phase. The Mechanism of Electrospray Mass Spectrometry. Anal. Chem. 64: 972A–986A.

33. Dole, M., Mack, L.L., Hinks, R.L. (1968) Molecular Beams of Macroions. J. Chem. Phys. 49: 2240–2249.

34. Wong, S.F., Meng, C.K., Fenn, J.B.(1988) Multiple Charging in Electrospray Ionization of Poly(ethylene glycols). J. Phys. Chem. 92: 546–550.

35. Covey, T.R., Bonner, R.F., Shushan, B.I., Henion, J.D. (1988) The Determination of Protein, Oligonucleotide and Peptide Molecular Weights by Ion-spray Mass Spectrometry. Rapid Commun. Mass Spectrom. 2: 249–256.

36. Gomez, A, Tang, K. (1994) Charge and Fission of Droplets in Electrostatic Sprays. Phys. Fluids 6: 404–414.

37. Wilm, M.S., Mann, M. (1994) Electrospray and Taylor-Cone theory, Dole's Beam of Macromolecules at Last?. Int. J. Mass Spectrom. Process. 136: 167–180.

38. Wilm, M.S., Mann, M. (1996) Analytical Properties of the Nanoelectrospray Ion Source. Anal. Chem. 68: 1–8.

39. Kebarle, P., Verkerk, U.H. (2009) Electrospray: from Ions in Solution to Ions in the Gas Phase, What We Know Now. Mass Spectrom. Rev. 28: 898 917.

40. Peschke, M.,Verkerk, U.H., Kebarle, P. (2004) Features of the ESI Mechanism that Affect the Observation of Multiply Charged Noncovalent Complexes and the Determination of the Association Constant by the Titration Method. J. Am. Soc. Mass Spectrom. 15: 1424–1434.

41. Smith, J.N., Flagan, R.C., Beauchamp, J.L. (2002) Droplet Evaporation and Discharge Dynamics in Electrospray Ionization. J. Phys. Chem. A 106: 9957–9967.

42. Konermann, L., Ahadi, E., Rodriguez, A.D., Vahidi, S. (2013) Unraveling the Mechanism of Electrospray Ionization. Anal. Chem. 85: 2–9.

43. Ahadi, E., Konermann, L. (2012) Modeling the Behavior of Coarse-Grained Polymer Chains in Charged Water Droplets: Implications for the Mechanism of Electrospray Ionization. J. Phys. Chem. B 116: 104–112.

44. Kebarle, P. (2000) A Brief Overview of the Present Status of the Mechanisms Involved in Electrospray Mass Spectrometry. J. Mass Spectrom. 35: 804–817.

45. Iribarne, J.V., Thompson, B.A. (1976) On the Evaporation of Small Ions from Charged Droplets. J. Chem. Phys. 64: 2287–2289.

46. Robb, D.B., Covey, T.R., Bruins, A.P. (2000) Atmospheric Pressure Photoionization: An Ionization Method for Liquid Chromatography–Mass Spectrometry. Anal. Chem. 72: 3653–3659.

47. Raffaelli, A., Saba, A. (2003) Atmospheric Pressure Photoionization Mass Spectrometry. Mass Spectrom. Rev. 22: 318–331.

48. Lias, S.G. (2003) Ionization Energy Evaluation. In: Linstrom, P.J., Mallard, W.G. (eds) NIST Chemistry WebBook, NIST Standard Reference Database Number 69. National Institute of Standards and Technology, Gaithersburg, http://webbook.nist .gov (accessed September 9, 2015).

49. Macfarlane, R.D., Torgerson, D.F. (1976) Californium-252 Plasma Desorption Mass Spectroscopy. Science 191: 920–925.

50. Magee, C.W., Harrington, W.L., Honig, R.E. (1978) Secondary Ion Quadrupole Mass Spectrometer for Depth Profiling – Design and Performance Evaluation. Rev. Sci. Instrum. 49, 477–485.

51. Bennmghoven, A., Schtermann, W.K. (1978) Detection, Identification, and Structural Investigation of Biologically Important Compounds by Secondary ion Mass Spectrometry. Anal. Chem. 50: 1180–1184.

52. Barber, M., Bordoli, R.S., Sedgewick, R.D., Tyler, A.N. (1981) Fast Atom Bombardment of Solids as an Ion Source in Mass Spectrometry. Nature 293: 270–275.

53. Barber, M., Bordoli, R.S., Elliott, G.J., Sedgwick, R.D., Tyler, A.N. (1982) Fast Atom Bombardment Mass Spectrometry. Anal. Chem. 54: 645A–657A.

54. Pauw, E.D., Agnello, A., Derw, F. (1991) Liquid Matrices for Liquid Secondary Ion Mass Spectrometry–Fast Atom Bombardment: an Update. Mass Spectrom. Rev. 10: 283–301.

55. Honig, R.E., Woolston, J.R. (1963) Laser-Induced Emission of Electrons, Ions, and Neutral Atoms from Solid Surfaces. Appl. Phys. Lett. 2: 138–139.

56. Fenner, N.C., Daly, N.R. (1966) Laser Used for Mass Analysis. Rev. Sci. Instrum. 37: 1068–1070.

57. Hillenkamp, F., Unsöld, E., Kaufmann, R., Nitsche, R. (1975) A High Sensitivity Laser Microprobe Mass Analyzer. Appl. Phys. 8: 341–348.

58. Denoyer, E., Grieken, R.V., Adams, F., Natusch, D.F.S. (1982) Laser Microprobe Mass Spectrometry. 1. Basic Principles and Performance Characteristics. Anal. Chem. 54: 26A–41A.

59. Karas, M., Bachmann, D., Hillenkamp, F. (1985) Influence of the Wavelength in High-Irradiance Ultraviolet Laser Desorption Mass Spectrometry of Organic Molecules. Anal. Chem. 57: 2935–2939.

60. Spencer, N.D., Moore, J.H. (2001) Encyclopedia of Chemical Physics and Physical Chemistry: Fundamentals. IOP Publishing, London.

61. Karas, M., Bachmann, D., Bahr, U., Hillenkamp, F. (1987) Matrix-assisted Ultraviolet Laser Desorption of Non-volatile Compounds. Int. J. Mass Spectrom. Ion Proc. 78: 53–68.

62. Strupat, K., Karas, M., Hillenkamp, F. (1991) 2,5-Dihidroxybenzoic Acid: A New Matrix for Laser Desorption–Ionization Mass Spectrometry. Int. J. Mass Spectrom. Ion Processes 72: 89–102.

63. Kandiah, M., Urban, P.L. (2013) Advances in Ultrasensitive Mass Spectrometry of Organic Molecules. Chem. Soc. Rev. 42: 5299–5322.

64. Cohen, S.L., Chait, B.T. (1996) Influence of Matrix Solution Conditions on the MALDI-MS Analysis of Peptides and Proteins. Anal. Chem. 68: 31–37.
65. Beavis, R.C., Chaudhary, T., Chait, B.T. (1992) α-Cyano-4-hydroxycinnamic Acid as a Matrix for Matrix Assisted Laser Desorption Mass Spectrometry. Org. Mass Spectrom. 27: 156–158.
66. Hillenkamp, F., Karas, M., Beavis, R.C., Chait, B.T. (1991) Matrix-assisted Laser Desorption/Ionization Mass Spectrometry of Biopolymers. Anal. Chem. 63: 1193A–1203A.
67. Fitzgerald, M.C., Parr, G.R., Smith, L.M. (1993) Basic Matrices for the Matrix-assisted Laser Desorption/Ionization Mass Spectrometry of Proteins and Oligonucleotides. Anal. Chem. 65: 3204–3211.
68. Urban, P.L., Amantonico, A., Zenobi, R. (2011) Lab-on-a-plate: Extending the Functionality of MALDI-MS and LDI-MS Targets. Mass Spectrom. Rev. 30: 435–478.
69. Lin, Y.-S., Chen, Y.-C. (2002) Laser Desorption/Ionization Time-of-Flight Mass Spectrometry on Sol–Gel-derived 2,5-Dihydroxybenzoic Acid Film. Anal. Chem. 74: 5793–5798.
70. Ho, K.-C., Lin, Y.-S., Chen, Y.-C. (2003) Laser Desorption/Ionization Mass Spectrometry on Sol–Gel-derived Dihydroxybenzoic Acid Isomeric Films. Rapid Commun. Mass Spectrom. 17: 2683–2687.
71. Tanaka, K. The Origin of Macromolecule Ionization by Laser Irradiation, http://www.nobelprize.org/nobel_prizes/chemistry/laureates/2002/tanaka-lecture.html (accessed September 11, 2015).
72. Arakawa, R., Kawaski, H. (2010) Functionalized Nanoparticles and Nanostructured Surfaces for Surface-assisted Laser Desorption/Ionization Mass Spectrometry. Anal. Sci. 12: 1229–1240.
73. Guinan, T., Kirkbride, P., Pigou, P.E., Ronci, M., Kobus, H., Voelcker, N.H. (2015) Surface-assisted Laser Desorption Ionization Mass Spectrometry Techniques for Application in Forensics. Mass Spectrom. Rev. 34: 627–640.
74. Dale, M.J., Knochenmuss, R., Zenob, R. (1996) Graphite/Liquid Mixed Matrices for Laser Desorption/Ionization Mass Spectrometry. Anal. Chem. 68: 3321–3329.
75. Chen, Y.-C., Sun, M.-C. (2001) Determination of Trace Quaternary Ammonium Surfactants in Water by Combining Solid Phase Extraction (SPE) with Surface-assisted Laser Desorption/Ionization (SALDI) Mass Spectrometry. Rapid Commun. Mass Spectrom. 15: 2521–2525.
76. Chen, Y.-C., Shiea, J., Sunner, J. (2000) Rapid Determination of Trace Nitrophenolic Organics in Water by Combining Solid Phase Extraction with Surface-assisted Laser Desorption Ionization (SALDI)/Time-of-Flight Mass Spectrometry. Rapid Commun. Mass Spectrom. 14: 86–90.
77. Chen, Y.-C., Tsai, M.-F. (2000) Using Surfactants to Enhance the Analyte Signals in Activated Carbon, Surface-assisted Laser Desorption Ionization (SALDI) Mass Spectrometry. J. Mass Spectrom. 35: 1278–1284.
78. Chen, Y.-C., Tsai, M.-F. (2000) Sensitivity Enhancement for Nitrophenols Using Cationic Surfactant-modified Activated Carbon for Solid Phase Extraction (SPE)/Surface-assisted Laser Desorption Ionization (SALDI) Mass Spectrometry. Rapid Commun. Mass Spectrom. 14: 2300–2304.

79. Chen, Y.-C. (1999) In Situ Determination of Organic Reaction Products by Combining Thin Layer Chromatography with Surface-assisted Laser Desorption Ionization (SALDI)/Time-of-Flight Mass Spectrometry. Rapid Commun. Mass Spectrom. 13: 821–825.

80. Wu, J.-Y., Chen, Y.-C. (2002) A Novel Approach by Combining Thin Layer Chromatography with Surface-assisted Laser Desorption/Ionization (SALDI) Time-of-Flight Mass Spectrometry. J. Mass Spectrom. 37: 85–90.

81. Wei, J., Buriak, J.M., Siuzdak, G. (1999) Desorption–Ionization Mass Spectrometry on Porous Silicon. Nature 399: 243–246.

82. Chen, C.-T., Chen, Y.-C. (2004) Molecularly Imprinted TiO_2-Matrix-assisted Laser Desorption/Ionization Mass Spectrometry for Selectively Detecting α-Cyclodextrin. Anal. Chem. 76: 1453–1457.

83. Chen, C.-T., Chen, Y.-C. (2004) Desorption/Ionization Mass Spectrometry on Nanocrystalline Titania Sol–Gel-deposited Films. Rapid Commun. Mass Spectrom. 18: 1956–1964.

84. Lo, C.-Y., Lin, J.-Y., Chen, W.-Y., Chen, C.-T., Chen Y.-C. (2008) Surface-assisted Laser Desorption/Ionization Mass Spectrometry on Titania Nanotube Arrays. J. Am. Soc. Mass Spectrom. 19: 1014–1020.

85. Chen, C.-T., Chen, Y.-C. (2005) Fe_3O_4/TiO_2 Core/Shell Nanoparticles as Affinity Probes for the Analysis of Phosphopeptides Using TiO_2 Surface-assisted Laser Desorption/Ionization Mass Spectrometry. Anal. Chem. 77: 5912–5919.

86. Chen, W.-Y., Chen, Y.-C. (2006) Affinity Based Mass Spectrometry by Using Iron Oxide Magnetic Particles as the Matrix and Concentrating Probes for SALDI MS Analysis of Peptides and Proteins. Anal. Bioanal. Chem. 386: 699–704.

87. Chiu, Y.-C., Chen, Y.-C. (2008) Carboxylate-functionalized Iron Oxide Nanoparticles in Surface-assisted Laser Desorption/Ionization Mass Spectrometry for the Analysis of Small Biomolecules. Anal. Lett. 41: 260–267.

88. McLean, J.A., Stumpo, K.A., Russell, D.H. (2005) Size-selected (2–10 nm) Gold Nanoparticles for Matrix Assisted Laser Desorption Ionization of Peptides. J. Am. Chem. Soc. 127: 5304–5305.

89. Chiu, T.-C., Chang, L.-C., Chiang, C.-K., Chang, H.-T. (2008) Determining Estrogens Using Surface-assisted Laser Desorption/Ionization Mass Spectrometry with Silver Nanoparticles as the Matrix. J. Am Soc. Mass Spectrom.19: 1343–1346.

90. Kawasaki, H., Yao, T., Suganuma, T., Okumura, K., Iwaki, Y., Yonezawa, T., Kikuchi, T., Arakawa, R. (2010) Platinum Nanoflowers on Scratched Silicon by Galvanic Displacement for An Effective SALDI Substrate. Chem. Eur. J. 16: 10832–10843.

91. Chiang, C.-K., Chen, W.-T., Chang, H.-T. (2001) Nanoparticle-based Mass Spectrometry for the Analysis of Biomolecules. Chem. Soc. Rev. 40: 1269–1281.

92. Nichols, K.P., Gardeniers, H.J.G.E. (2007) A Digital Microfluidic System for the Investigation of Pre-steady-state Enzyme Kinetics Using Rapid Quenching with MALDI-TOF Mass Spectrometry. Anal. Chem. 79: 8699–8704.

93. Nichols, K.P., Azoz, S., Gardeniers, H.J.G.E. (2008) Enzyme Kinetics by Directly Imaging a Porous Silicon Microfluidic Reactor Using Desorption/Ionization on Silicon Mass Spectrometry. (2008) Anal. Chem. 80: 8314–8319.

94. Takáts, Z., Wiseman, J. M., Gologan, B., Cooks, R.G. (2004) Mass Spectrometry Sampling under Ambient Conditions with Desorption Electrospray Ionization. Science 306: 471–473.
95. Cooks, R.G., Ouyang, Z., Takats, Z., Wiseman, J.M. (2006) Ambient Mass Spectrometry. Science 311: 1566–1570.
96. Gross, J.H. (2011) Chapter 13, Ambient Mass Spectrometry. In: Mass Spectrometry. A Textbook, 2nd Edition. Springer-Verlag, Berlin.
97. Huang, M.-Z., Yuan, C.-H., Cheng, S.-C., Cho, Y.-T., Shiea, J. (2010) Ambient Ionization Mass Spectrometry. Annu. Rev. Anal. Chem. 3: 43–65.
98. Monge, M.E., Harris, G.A., Dwivedi, P., Fernandez, F.M. (2013) Mass Spectrometry: Recent Advances in Direct Open Air Surface Sampling/ Ionization. Chem. Rev. 113: 2269–2308.
99. Ma, X., Zhang, S., Zhang, X. (2012) An Instrumentation Perspective on Reaction Monitoring by Ambient Mass Spectrometry. Trends Anal. Chem. 35: 50–66.
100. Glish, G.L., Vachet, R.W. (2003) The Basics of Mass Spectrometry in the Twenty first Century. Nat. Rev. Drug Discov. 2: 140–150.
101. Takáts, Z., Wiseman, J.M., Cooks, R.G. (2005) Ambient Mass Spectrometry Using Desorption Electrospray Ionization (DESI): Instrumentation, Mechanisms and Applications in Forensics, Chemistry, and Biology. J. Mass Spectrom. 40: 1261–1275.
102. Roach, P.J., Laskin, J., Laskin, A. (2010) Nanospray Desorption Electrospray Ionization: An Ambient Method for Liquid-Extraction Surface Sampling in Mass Spectrometry. Analyst 135: 2233–2236.
103. Takats, Z., Wiseman, J.M., Gologan, B., Cooks, R.G. (2004) Electrosonic Spray Ionization. A Gentle Technique for Generating Folded Proteins and Protein Complexes in the Gas Phase and for Studying Ion–Molecule Reactions at Atmospheric Pressure. Anal. Chem. 76: 4050–4058.
104. Santos, V.G., Regiani, T., Dias, F.F.G., Romao, W., Jara, J.L.P., Klitzke, C.F., Coelho, F., Eberlins, M.N. (2011) Venturi Easy Ambient Sonic-Spray Ionization. Anal. Chem. 83: 1375–1380.
105. Shiea, J., Chang, D.-Y., Lin, C.-H., Jiang, S.-J. (2001) Generating Multiply Charged Protein Ions by Ultrasonic Nebulization/Multiple Channel-Electrospray Ionization Mass Spectrometry. Anal. Chem. 73: 4983–4987.
106. Chen, H., Venter, A., Cooks, R.G. (2006) Extractive Electrospray Ionization for Direct Analysis of Undiluted Urine, Milk and Other Complex Mixtures Without Sample Preparation. Chem. Commun. 2042–2468.
107. Chen, T.-Y., Lin, J.-Y., Chen, J.-Y., Chen, Y.-C. (2010) Ultrasonication-assisted Spray Ionization Mass Spectrometry for the Analysis of Biomolecules in Solution. J. Am. Soc. Mass Spectrom. 21: 1547–1553.
108. Chen, T.-Y., Chao, C.-S., Mong, K.-K.T., Chen, Y.-C. (2010) Ultrasonication -assisted Spray Ionization Mass Spectrometry for On-line Monitoring of Organic Reactions. Chem. Commun. 46: 8347–8349.
109. Hsieh, C.-H., Chang, C.-H., Urban, P.L., Chen, Y.-C. (2011) Capillary Action-supported Contactless Atmospheric Pressure Ionization for the Combined Sampling and Mass Spectrometric Analysis of Biomolecules. Anal. Chem. 83: 2866–2869.

110. Hsieh, C.-H., Chao, C.-S., Mong, K.-K.T., Chen, Y.-C. (2012) Online Monitoring of Chemical Reactions by Contactless Atmospheric Pressure Ionization Mass spectrometry. J. Mass Spectrom. 47: 586–590.

111. Lo, T.-J., Chen, T.-Y., Chen, Y.-C. (2012) Study of Salt Effects in Ultrasonication-assisted Spray Ionization Mass Spectrometry. J. Mass Spectrom. 47: 480–483.

112. Hsieh, C.-H., Meher, A.K., Chen, Y.-C. (2013) Automatic Sampling and Analysis of Organics and Biomolecules by Capillary Action-supported Contactless Atmospheric Pressure Ionization Mass Spectrometry. PLOS One 8: e66292.

113. Wong, S.-Y., Chen, Y.-C. (2014) Droplet-based Electrospray Ionization Mass Spectrometry for Qualitative and Quantitative Analysis. J. Mass Spectrom. 49: 432–436.

114. Meher, A. K., Chen, Y.-C. (2015) Polarization Induced Electrospray Ionization Mass Spectrometry for the Analysis of Liquid, Viscous, and Solid Samples. J. Mass Spectrom. 50, 444–450.

115. Shiea, J., Huang, M.-Z., Hsu, H.J., Lee, C.-Y., Yuan, C.-H., Beech, I., Sunner, J. (2005) Electrospray-assisted Laser Desorption/Ionization Mass Spectrometry for Direct Ambient Analysis of Solids. Rapid Commun. Mass Spectrom. 19: 3701–3704.

116. Huang, M.Z., Hsu, H.J., Lee, J.Y., Jeng, J., Shiea, J. (2006) Direct Protein Detection from Biological Media through Electrospray-assisted Laser Desorption Ionization/Mass Spectrometry. J. Proteome Res. 5: 1107–1116.

117. Nemes, P., Vertes, A. (2007) Laser Ablation Electrospray Ionization for Atmospheric Pressure, *In Vivo*, and Imaging Mass Spectrometry. Anal. Chem. 79: 8098–8106.

118. Lin, J.-Y., Chen, T.-Y., Chen, J.-Y., Chen, Y.-C. (2010) Multilayer Gold Nanoparticle-assisted Thermal Desorption Ambient Mass Spectrometry for the Analysis of Small Organics. Analyst 135: 2668–2675.

119. Cotte-Rodriguez, I., Mulligan, C.C., Cooks, R.G. (2007) Non-proximate Detection of Small and Large Molecules by Desorption Electrospray Ionization and Desorption Atmospheric Pressure Chemical Ionization Mass Spectrometry: Instrumentation and Applications in Forensics, Chemistry, and Biology. Anal. Chem. 79: 7069–7077.

120. Jjunju, F.P.M., Badu-Tawiah, A.K., Li, A., Soparawalla, S., Roqan, I.S., Cooks, R.G. (2007) Hydrocarbon Analysis Using Desorption Atmospheric Pressure Chemical Ionization. Int. J. Mass Spectrom. 345–347: 80–88.

121. Takats, Z., Cotte-Rodriguez, I., Talaty, N., Chen, H.W., Cooks, R.G. (2005) Direct, Trace Level Detection of Explosives on Ambient Surfaces by Desorption Electrospray Ionization Mass Spectrometry. Chem. Commun. 1950–1952.

122. McEwen, C.N., McKay, R.G., Larsen, B.S. (2005) Analysis of Solids, Liquids, and Biological Tissues Using Solids Probe Introduction at Atmospheric Pressure on Commercial LC/MS Instruments. Anal. Chem. 77: 7826–7831.

123. McEwen, C., Gutteridge, S. (2007) Analysis of the Inhibition of the Ergosterol Pathway in Fungi Using the Atmospheric Solids Analysis Probe (ASAP) Method. J. Am. Soc. Mass Spectrom. 18: 1274–1278.

124. Cody, R.B., Laramée, J.A., Durst, H.D. (2005) Versatile New Ion Source for the Analysis of Materials in Open Air under Ambient Conditions. Anal. Chem. 77: 2297–2302.

125. Hajslova, J., Cajka, T., Vaclavik, L. (2011) Challenging Applications Offered by Direct Analysis in Real Time (DART) in Food-quality and Safety Analysis. Trends Anal. Chem. 30: 204–218.
126. Harper, J.D., Charipar, N.A., Mulligan, C.C., Zhang, X., Cooks, R.G., Zheng, O. (2008) Low-temperature Plasma Probe for Ambient Desorption Ionization. Anal. Chem. 80: 9097–9104.
127. Albert, A., Engelhard, C. (2012) Characteristics of Low-temperature Plasma Ionization for Ambient Mass Spectrometry Compared to Electrospray Ionization and Atmospheric Pressure Chemical Ionization. Anal. Chem. 84: 10657–10664.
128. Andrade, F.J., Shelley, J.T., Wetzel, W.C., Webb, M.R., Gamez, G., Ray, S.J., Hieftje, G.M. (2008) Atmospheric Pressure Chemical Ionization Source. 1. Ionization of Compounds in the Gas Phase. Anal. Chem. 80: 2646–2653.
129. Shelley, J.T., Wiley, J.S., Hieftje, G.M. (2011) Ultrasensitive Ambient Mass Spectrometric Analysis with a Pinto-capillary Flowing Atmospheric-pressure Afterglow Source. Anal. Chem. 83: 5741–5748.
130. Jecklin, M.C., Schmid, S., Urban, P.L., Amantonico, A., Zenobi, R. (2010) Miniature Flowing Atmospheric-pressure Afterglow Ion Source for Facile Interfacing of CE with MS. Electrophoresis 31: 3597–3605.

3

Mass Analyzers for Time-resolved Mass Spectrometry

3.1 Overview

The mass analyzer is a key component of all mass spectrometers. It is responsible for ion separation, fragmentation, and in some cases ion detection. Ions generated in the ion source are transferred to the mass analyzer and sorted according to their mass-to-charge ratio (m/z) by means of electric or magnetic interactions. Because ion separation and fragmentation require long durations, ions typically spend most time in a mass analyzer after leaving ion source regions (with an exception of long ion manipulation time within ion mobility spectrometers). Thus, it is important in time-resolved molecular monitoring to understand not only the theory behind various ion separation techniques, but also the temporal properties of mass analyzers that affect time-resolved mass spectrometry (TRMS) experiments. This chapter introduces the underlying principles behind frequently used mass analyzers and ion detectors. It also summarizes different hybrid instruments, which combine the advantages of different mass analysis techniques.

Before detailing the characteristics of mass analyzers, it is worth noting that electric fields as well as collisions between ions with gas molecules affect ion motion. To ensure the highest possible sensitivity, modern mass spectrometers operate under a vacuum ranging between 10^{-5} and 10^{-12} Pa, depending on the mass analyzer used. Electric potentials applied to metal electrodes are usually used to guide ions from the ion source all the way to the detector. The vacuum environment not only prevents interactions between ions and air molecules, but also prevents high voltage components from electric discharge. Although kinetic theory on ion–molecule interactions and electromagnetism are beyond the scope of this book, readers should keep in mind that both the electric gradient and hydrodynamic gas flow determine the trajectory of ions. Therefore, working conditions such as gas pressure, physical dimensions, and electric potentials, are optimized for every component of a mass spectrometer to ensure the best analytical performance in terms of spectrum acquisition speed, mass accuracy, mass resolving power, and sensitivity.

Time-Resolved Mass Spectrometry: From Concept to Applications, First Edition.
Pawel Lukasz Urban, Yu-Chie Chen and Yi-Sheng Wang.
© 2016 John Wiley & Sons, Ltd. Published 2016 by John Wiley & Sons, Ltd.

3.2   Individual Mass Analyzers

The term "individual mass analyzer" is introduced here to distinguish instruments incorporating a single mass analyzer from hybrid instruments comprising multiple analyzers. The concept of an individual mass analyzer is important because the spectrum acquisition time (SAT) of hybrid instruments results from the combined operation times of individual analyzers used in the hybrid. Based on the way they examine ions, mass analyzers can be categorized into beam-type and trapping-type mass analyzers. The two types of instrument have very different SATs. Beam-type mass analyzers separate ions in a single ion beam path, whereas trapping-type mass analyzers store ions in an ion trapping device before recording mass spectra. Understanding the temporal characteristics of individual mass analyzers is therefore vital for examination of the temporal characteristics of more complex (tandem) devices. Herein, four types of individual mass analyzer are introduced in the order of increasing SAT. These are: time-of-flight (TOF), quadrupole (Q), sector (S), and Fourier transform (FT) mass analyzers. Table 3.1 lists the typical SATs of these mass analyzers and ion sources which are often combined with them. The SAT is determined by multiple parameters, including speed of ions, scanning speed of electric potentials, speed of data acquisition systems, and operation mode of mass analyzers. The relationship between SAT and analytical performance of a mass spectrometer is further discussed in Chapter 5.

3.2.1   Time-of-flight Mass Analyzers

Among all mass analyzers, TOF typically has the shortest SAT (or highest spectrum acquisition speed) mainly because of its simple configuration and structure. The TOF mass analyzer is a beam-type instrument. It separates ions of identical kinetic energy according to their m/z within a flight tube [1, 2]. Although its underlying principle and configuration are simple, the TOF mass analyzer is highly regarded for its analytical performance. It is second only to FT mass analyzers in terms of mass resolving power and mass accuracy.

The concept of the TOF mass analyzer was first proposed by Stephens in 1946 [3], and the first instrument was demonstrated by Cameron and Eggers in 1948 [4]. Since then, several important advances in TOF technology have occurred, including time-lag focusing and two-stage acceleration techniques demonstrated by Wily and McLaren in 1955 [5], reflectron configuration demonstrated by Mamyrin *et al.* in 1973 [6], and curved-field reflectron demonstrated by Cornish and Cotter in 1993 [7].

Table 3.1 *Typical spectrum acquisition time of important types of individual mass analyzers and ion sources that are often combined with these analyzers*

Mass analyzer	Spectrum acquisition time (s)	Ion sources
Time-of-flight	$\sim 10^{-5}$	MALDI, LDI, SIMS
Quadrupole	$\sim 10^{-3}$	EI, CI, ESI
Sector	$\sim 10^{-1}$	EI, CI, FAB, SIMS
Fourier transform	$> 10^{-1}$	ESI, MALDI, LDI

The theory behind TOF-MS is based on a very basic physical principle – the law of conservation of energy [8]. Specifically, all ions gain the same potential energy (*PE*) at the instance of ionization in the ion source region. The *PE* is set using an electrically biased metal electrode installed at the front-end of the ionization region. The *PE* is then converted to translational energy (*TE*) before the ions enter a field-free region. Assuming that ions are produced at a metal sample electrode biased at potential V_s (in volts), the *PE* of the ions is:

$$PE = qV_s = zeV_s \tag{3.1}$$

where q is the charge of the ion, z is the charge number, and e is the elementary charge (1.6×10^{-19} C). Conversion of the *PE* to *TE* relies on creating an electric field with another metal electrode or series of electrodes being installed several millimeters from the sample electrode. These electrodes are called *extraction* or *acceleration electrodes*. The extraction electrodes are biased at lower potentials to extract positive ions, and vice versa for negative ion extraction. Figure 3.1a shows the configuration of a typical TOF mass spectrometer, in which the ion source consists of a sample electrode (sample) and multiple extraction electrodes (extractions I and II). Within this region, ions travel along the electric field toward the flight tube, which is typically kept at ground potential.

Since an ion with a mass m (u) and velocity v (m s^{-1}) inside the flight tube has a *TE* of $\frac{1}{2}mv^2$, the *PE* is converted to *TE* according to:

$$zeV_s = \frac{1}{2}mv^2 \tag{3.2}$$

The flight tube is also known as the field-free region because there is typically no further electric potential change within this region. One can readily obtain the velocity of ions by rearranging Equation 3.2:

$$v = \sqrt{\frac{2zeV_s}{m}} \tag{3.3}$$

An ion detector is installed at the end of the flight tube to record ions colliding with its surface. The flight time (t) an ion spends in a flight tube of length L is given by:

$$t = \frac{L}{v} = \sqrt{\frac{m}{2zeV_s}}L \tag{3.4}$$

Because t is proportional to $\sqrt{m}$, the velocity of a heavy ion is lower than that of a light ion. Therefore, heavy ions exhibit longer flight times than lighter ions. For example, ions with molecular weight of 100 Da achieve a velocity of $\sim 1.96 \times 10^5$ m s^{-1} if accelerated by an electric field of 20 kV. This velocity corresponds to a travel time of 6.6 μs across a 1.3-m long flight tube (the typical length of a commercial TOF tube). Under the same experimental conditions, the typical SAT of a singly charged molecule ranging from 0 to 100 kDa, analyzed using this instrument, is roughly 210 μs. This SAT enables recording to a maximum of roughly 5000 mass spectra per second. Such a spectrum scanning rate is only achievable if the speed of the electronic system used for data acquisition is sufficiently fast.

The mass resolving power of TOF-MS critically relies on the energy spread of ions with the same m/z. Energy spread is dictated by spatial and energy distributions of ions at the

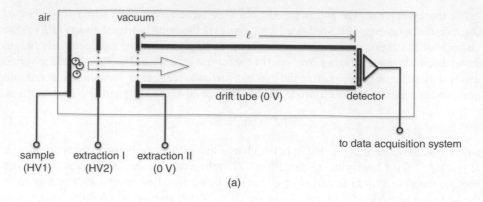

(a)

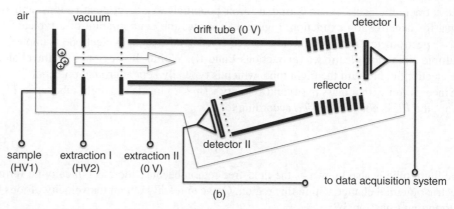

(b)

Figure 3.1 *Schematic of TOF mass spectrometers equipped with two-stage acceleration ion sources. (a) A linear TOF mass spectrometer; and (b) a reflectron TOF mass spectrometer*

ion source region. Spatial distribution of ions is affected mainly by the excitation method (e.g., dimensions of laser, electron or particle beams used for ionization). Spatial spread of ions in the ion source region can be reduced at the ion detection region by using multiple acceleration stages in the ion source region. The most widely used design is the two-stage acceleration field enabled by two extraction electrodes, as depicted schematically in Figure 3.1. By contrast, energy distribution is a grand canonical (statistical) ensemble. This characteristic is due to the random spatial and velocity distribution of molecules during ionization or at the very moment at which ion extraction starts. Energy spread makes ions with the same m/z travel through the flight tube with slightly different velocities. This phenomenon eventually results in the broadening of spectral features.

To compensate for the energy spread, the first stage of the two-stage acceleration field can be activated shortly after ion production or ejection from the sample. This event typically occurs in the nanosecond to low microsecond range. The method is known as *time-lag focusing* or *delayed extraction* technique. In TOF-MS equipped with matrix-assisted laser desorption/ionization (MALDI) ion source, delayed extraction may improve the mass resolving power by a factor of two- to fivefold [9, 10]. Mass resolving power defines the sharpness of spectral features, which is explained in Chapter 5.

Another effective way of compensating for energy spread is to increase the *flight distance* of ions. It can be achieved by introducing a *deceleration field* into the field-free region to reflect ions toward an ion detector positioned at the axis of reflection. Such a deceleration field is generated with a series of metal electrodes known as a *reflectron* or a *reflector* [6, 11], as shown in Figure 3.1b. Because the penetration depth of faster ions within the reflector is longer than that of slower ions, the differences in flight times of these ions toward the detector can be reduced. In addition to compensating for flight distance within the deceleration field, the reflector also increases overall flight distance; consequently, it improves mass resolving power. In comparison with linear TOF instruments, reflectron-TOF typically improves the mass resolving power by a few times up to one order of magnitude [6, 9].

In TOF-MS, ions are converted to electronic signals using a microchannel plate (MCP) detector [12]. Such detectors are made of semiconductor materials supplied with high electrical gradients during measurements. These materials can emit secondary electrons when subjected to high-energy collisions by ions. The detectors are fabricated in the form of thin disks, typically 20–40 mm in diameter and 0.5-mm thick, with 60% of the surface area occupied by microchannels of ($\sim$6–20 µm in diameter). The microchannels are tilted at roughly 10° to the surface normal to ensure ions collide with the surface of the detector instead of passing through microchannels directly. Such detector geometry allows for highly accurate flight distance determination, which is important for TOF-MS. When ions of sufficiently high velocity impinge on the surface of the microchannels, secondary electrons are ejected from the surface. When the secondary electrons are accelerated further into the microchannels due to the electric potential gradient, electron–surface collisions occur. Every such electron–surface collision produces more secondary electrons, and this reaction repeats. Such cascade processes continue and thereby increase the number of electrons produced during the ion–surface collision. While the degree of amplification depends on the strength of the electric field applied and the geometry of the microchannels, a typical MCP can convert one ion into roughly 10^4 secondary electrons (at the potential of +1000 V applied to the rear side of the MCP with respect to the front side). Additional amplification is possible by stacking two or three MCP disks with their microchannels aligned in a V- or Z-type configuration, respectively. However, the resultant gain is not proportional to the number of disks. The highest amplification is about 10^8 with three MCP disks. Notably, since each MCP has a response time of about 2–10 ns, increasing the number of disk results in a slower response time.

TOF-MS is advantageous in terms of spectrum acquisition speed and repetition rate. Because the flight time of ions in such instruments is typically in the range of a few microseconds, the spectrum-acquisition frequency can be in the kilohertz range (including the time required for data storage). As mentioned previously, TOF-MS instruments are the fastest among all mass spectrometers and are highly compatible with a variety of TRMS studies.

3.2.2 Quadrupole Mass Analyzers

Quadrupole mass spectrometry (Q-MS) has wide functionality making it the most popular form of MS. The family of quadrupole mass analyzers includes a variety of devices, such as quadrupole mass filter (QMF), two-dimensional (2D) ion trap (IT) or linear ion trap (LIT), and three-dimensional (3D) IT. The 3D IT device is also called the "Paul trap" in honor of Wolfgang Paul, the inventor of this device [13]. Wolfgang Paul was awarded the Nobel Prize in Physics in 1989 for developing these important techniques [14]. Quadrupole mass analyzers are more compact than other devices (e.g., TOF, sector, and FT mass analyzers), and most commercial products offer good tandem MS capability which is important for elucidating molecular structure (see Section 3.3.2). Quadrupole mass analyzers can be easily interfaced with continuous ion sources such as electron ionization (EI), chemical ionization (CI), and electrospray ionization (ESI), which makes them ideal analyzers for time-resolved molecular examination. Integration of pulsed ion sources, such as MALDI, into Q-MS is also possible but less efficient because of cycle incompatibility. Below, we introduce the principles of QMF, LIT, and 3D IT mass analyzers. In Section 3.3.1.1, a detailed description on the usage of three quadrupole mass analyzers in series (triple quadrupole, QqQ) is provided.

3.2.2.1   Quadrupole Mass Filter

The theory of QMF and IT was developed in the 1950s by Paul and Steinwedel [13, 15]. QMF is a beam-type instrument that utilizes electric fields combining direct-current (DC) and alternating-current (AC) potentials to manipulate ion motion [16]. The AC potential is in the radio-frequency (RF) range. Because the motion of different ions in electric fields depends on m/z, separation of ions can be achieved by changing the amplitude or the frequency of the electric field inside the quadrupole mass analyzer. Recording the mass spectrum is then achieved by correlating the measured signal with the components of an electric-field (DC, AC) scan across a certain range.

A QMF consists of four cylindrical or hyperbolic metal rods installed parallel to, and at a distance r_0 from the central axis (z direction). Figure 3.2 presents a schematic drawing of a QMF in the Cartesian coordinate system. The rods have identical dimensions and are equally spaced with respect to each other. The opposite rods are paired and connected to a source of the same electric potential. Thus, there are two pairs of electric potentials applied to the rods. During QMF operation, ions propagate along the z axis from one end of the system to the other, where they are detected by an ion detector. The fast-changing electric potentials of the rods affect ion trajectories, allowing only ions with a specific m/z to pass stably through the QMF. Other ions are either ejected radially away from the QMF or collide with the rods and are diminished.

In order to describe the electric potential of a QMF, we define Component I in the x-direction applied to the horizontal pair of rods, and Component II in the y-direction applied to the vertical pair. Each of the components consists of a DC and an AC term such that:

$$\text{Component I}: \quad \Phi_I = +(U - V \cos \omega t) \tag{3.5}$$

$$\text{Component II}: \quad \Phi_{II} = -(U - V \cos \omega t) \tag{3.6}$$

in which U is the DC bias potential, V is the amplitude of AC potential, ω is the AC frequency (typically $0.7-1.5$ MHz), and t is time. The two components are superimposed over one another to yield the resultant potential (Φ_0) in QMF:

$$\Phi_0 = \Phi_I - \Phi_{II} = +2(U - V \cos \omega t) \tag{3.7}$$

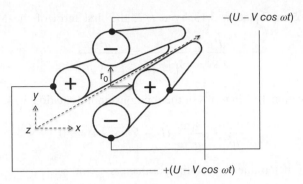

Figure 3.2 *Schematic diagram of a quadrupole mass filter. The opposite metal rods are paired and coupled to a source of the same potential. Opposite polarity potentials are supplied to the two pairs of rods*

Therefore, the electric potential Φ (x, y) anywhere in the mass filter can be expressed (using Φ_0) as:

$$\Phi(x, y) = \frac{\Phi_0}{2r_0^2}(x^2 - y^2)$$

(3.8)

Remember that Φ_0 is a function of time. To analyze ions in this time-varying electric potential, ions produced at the ion source with the same kinetic energy are introduced into the QMF and oscillate with this electric field. The force equations (F_x and F_y) of an ion at any position (x, y) within the QMF are:

$$F_x = ma_x = m\frac{d^2x}{dt^2} = -ze\frac{d\Phi}{dx}$$

(3.9)

$$F_y = ma_y = m\frac{d^2y}{dt^2} = -ze\frac{d\Phi}{dy}$$

(3.10)

Replacing the electric potential Φ in Equations 3.9 and 3.10 by Equations 3.7 and 3.8, we obtain:

$$\frac{d^2x}{dt^2} + \frac{2ze}{mr_0^2}(U - V\cos\omega t)x = 0$$

(3.11)

$$\frac{d^2y}{dt^2} - \frac{2ze}{mr_0^2}(U - V\cos\omega t)y = 0$$

(3.12)

By rearranging Equations 3.11 and 3.12, we obtain a Mathieu function [17]:

$$\frac{d^2u}{d\xi^2} + (a_u - 2q_u\cos 2\xi)u = 0$$

(3.13)

in which a_u and q_u represent stability parameters of DC and AC electric fields in quadrupole mass analyzers, respectively, and u represents either the x or the y coordinate.

After replacing ξ in Equation 3.13 by $\omega t/2$, the first term of the Mathieu function becomes:

$$\frac{d^2u}{d\xi^2} = \frac{d^2u}{d\left(\frac{\omega t}{2}\right)^2} = \frac{d^2u}{\left(\frac{\omega}{2}\right)^2 dt^2} \tag{3.14}$$

By rewriting Equation 3.11 in terms of the relationship in Equation 3.14, we obtain:

$$\frac{d^2x}{dt^2} + \frac{8ze}{mr_0^2\omega^2}(U - V\cos\omega t)x = 0 \tag{3.15}$$

Thus, we can now determine the stability parameters a_x and q_x:

$$a_x = \frac{8zeU}{mr_0^2\omega^2}, \tag{3.16}$$

$$q_x = -\frac{4zeV}{mr_0^2\omega^2}. \tag{3.17}$$

The same derivation applies to Equation 3.12 to yield a_y and q_y. Here, one obtains $a_x = -a_y$ and $q_x = -q_y$ because the x and y components have identical value but opposite polarities. When the amplitude of the ion oscillations in both the x and y directions is anywhere less than r_0, ions are able to propagate through the QMF region without colliding with the rods or being pushed away radially. Solving the Mathieu function for stable ion trajectories results in two independent stability criteria (i.e. a_u and q_u) in the x and y directions.

It is possible to illustrate stable regions along the x and y directions by combining them in a plot of a_u versus q_u, where $a_u = a_x = -a_y$ and $q_u = q_x = -q_y$, as shown in Figure 3.3a. In this figure, the gray areas are stable regions in the x direction and the hatched areas represent those in the y direction. Notably, there is more than one stable region in each direction. For example, the first, second, and third stable regions in the x direction are denoted by x_1, x_2, and x_3, respectively. The same analysis applies in the y direction. Thus, some areas in this plot satisfy stable conditions in both x and y directions. These are the overlapping stable regions.

In QMF, stable ion trajectory conditions are met when the values of a_u and q_u fall in the stable region in both x and y directions. Figure 3.3b shows the overlapping area of the x_1 and y_1 regions. The overlapping area includes two triangle-like regions symmetric about the q_u axis. Due to electronic circuit limitations and cost effectiveness, this is the most practical region to conduct QMF. The shaded area in Figure 3.3b illustrates the upper part of this overlapping stable region. Ions with a_u and q_u values outside the shaded area are thus filtered and excluded from the QMF. Such an exclusion happens when the x or y position of an ion is larger than r_0.

It is worth noting that since all parameters except z and m in a_u and q_u are variables, the stability diagram is m/z-dependent. It means that individual ions have their own stability diagram (the overlapping area between x_1 and y_1). If all parameters are kept constant except m, a larger molecule will have a smaller a_u and q_u. In order to isolate ions of particular m/z range, changing the values of U, V, or ω is necessary. Commercial QMFs normally scan mass spectra by changing U and V because altering ω involves implementation of a complicated electronic circuit.

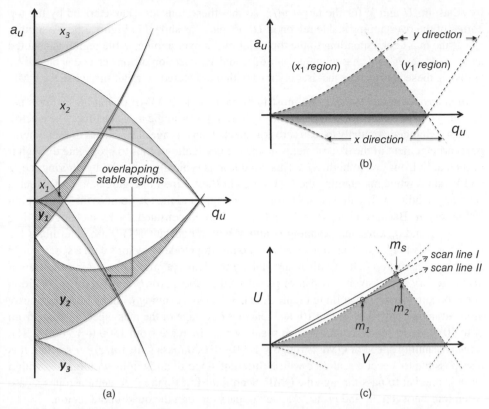

(a)   (b)   (c)

Figure 3.3 *Stable ion motion regions defined with respect to a_u and q_u. (a) The gray areas are the stable regions in the x direction while the hatched areas are those in the y direction. A QMF normally works in the first overlapping stable region. (b) The first overlapping stable region consists of two symmetrical triangle-like areas. (c) Two scan lines in U–V space showing different scanning modes, in which line I only allows m_s to pass through while line II allows m_1 to m_2 to pass. The dashed lines represent the boundaries of stable regions*

By varying U and V, we can achieve three basic QMF operation modes: ion guiding; selected ion monitoring; and continuous scanning.

Ion-guiding Mode. QMFs can serve as ion-guiding devices when the value of U is kept constantly at 0 V. In this case, a_u always equals to zero, and the ion with the smallest m/z within the stable region (Figure 3.3b) is determined only by V. Thus, all ions above this critical m/z can survive in the QMF. This type of electric field is suitable for ion transmission, especially when guiding ions through small orifices at the boundary of differentially pumped vacuum regions. Ion guiding with RF electric field (typically 1–2 MHz and 500 V amplitude) can also be accomplished using the same geometry but a different number of poles, i.e. with hexapoles, octopoles, or other $2n$ pole arrangements [18, 19].

Selected Ion Monitoring Mode. It is possible to select a pair of a_u and q_u values that allow transmission of ions of a specific m/z through the QMF. This condition can be achieved by adjusting U and V for the target m/z, so that these ions are characterized by the top corner of the overlapping stable region in U–V space, as shown in Figure 3.3c. Assuming $z = 1$, the mass corresponding to the top point of the overlapping stable region is denoted as m_s. Since every m/z has a unique pair of U and V, selection of ions can be achieved by adjusting these parameters; that is, only ions with the selected m/z can survive in the QMF.

Continuous-scanning Mode. The most important mode in QMF operation involves the scanning of a mass spectrum by transmitting ions of just a single m/z at a time. The electric fields are changed gradually to scan across a selected mass range. This means the spatiotemporal arrangement of the electric fields inside the QMF allows ions to propagate through it sequentially from low to high m/z. This condition is typically achieved by altering the U and V values while maintaining the U/V ratio constant. Line I in Figure 3.3c shows such a scanning condition. The line passes through the origin and the top point of the overlapping stable region. Because changes in U and V can be compensated for by a change in m/z to keep a_u and q_u constant, scanning U and V with a constant U/V ratio determines the resultant target m/z. Similar to selected ion monitoring mode, only ions with a specific m/z are able to pass through the QMF at any given time. This operation allows ions of different m/z to satisfy successively any pair of preselected a_u and q_u (or U and V). If the condition corresponding to the top point in Figure 3.3c is selected, scanning U and V simultaneously along scan line I to $10 \times U$ and $10 \times V$ enables coverage of the mass spectral range from m to $10 \times m$. In practice, during such scans, V may increase from 1500 to 15 000 V. The typical scanning speed of QMF is roughly 1000–10 000 mass units (m/z) per second. It is also possible to select a scan line with a different slope to allow ions with masses within an m/z interval to pass through the QMF. Scan line II in Figure 3.3c demonstrates a case when ions between m_1 and m_2 are allowed to pass through the quadrupole region.

QMFs normally detect ions using electron multipliers (EMs) because such detectors are easy to install and maintain. MCP detectors are not usually used here because ion flight time is not a critical factor determining mass accuracy and mass resolving power in Q-MS. The working principle of EMs is described in Section 3.2.3.

3.2.2.2   Ion-Trap Mass Analyzers

IT mass analyzers are trapping-type instruments. The theory behind IT mass analyzers is similar to that of the QMF, but ion traps are composed of electrodes with very different geometry to those of the QMF. The IT mass analyzer typically incorporates a ring electrode and a pair of identical parabola-shaped end-cap electrodes . The end-cap electrodes are installed symmetrically above and below the ring electrode to define the volume for ion manipulation. This type of device is also called 3D IT mass analyzer. Figure 3.4 presents a cross-sectional view of a 3D IT mass analyzer. Ions enter the analyzer from an opening at the apex of one of the end-cap electrodes. After ion selection or fragmentation, the ions are ejected away from the 3D IT mass analyzer through an opening at the apex of the opposite end-cap electrode, and pass through to an ion detector. These end-cap electrodes are also called entrance end-cap and exit end-cap. The insulating spacers are made of ceramic or plastic to isolate the electrodes electrically.

The electric field of 3D IT is described with a simplified Cartesian coordinate system (r, z). The ion manipulation occurs in a cylindrical region with a radius of r_0 and a height

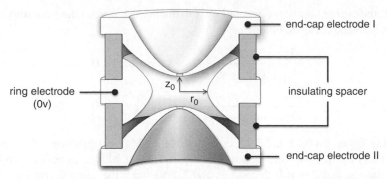

Figure 3.4   *A cross-sectional view of a 3D IT mass analyzer. The two end-cap electrodes are installed symmetrically on both sides of the ring electrode. The central openings at the apex of the end-cap electrodes are for receiving ions from the ion source and for ejecting ions toward an ion detector*

of $2z_0$. The two end-cap electrodes resemble one pair of rods in QMF, which are supplied with the same electric potential ($\Phi_{end\text{-}cap}$), whereas the ring electrode is supplied with an independent potential (Φ_{ring}) that serves a similar function as the potential applied to the other pair of rods in QMF. Therefore, quadrupolar potential inside the 3D IT (Φ_0) is produced by superimposing two components: $\Phi_0 = \Phi_{ring} - \Phi_{end\text{-}cap}$. According to this definition, Φ_0 corresponds to the quadrupolar potential in QMF.

A major difference between 3D IT and a QMF is that a 3D IT normally utilizes only one electric component to achieve quadrupolar potential for ion manipulation. This electric potential can be supplied either to the ring or the end-cap electrodes; rest electrodes are kept at ground potential during ion trapping. Commercial 3D IT mass spectrometers normally supply voltage to the ring electrode to reduce interference from the electric field on the flow of ions through the holes of end-cap electrodes. Thus, the resulting electric potential is given by:

$$\Phi_0 = \Phi_{end\text{-}cap} - \Phi_{ring} = 0 - \Phi_{ring} = U - V \cos \omega t \tag{3.18}$$

In this simplified Cartesian coordinate system, the electric potential anywhere inside the 3D IT can be expressed using Φ_0 as:

$$\Phi(r, z) = \frac{\Phi_0}{r_0^2 + 2z_0^2}(r^2 - 2z^2) + \frac{2\Phi_0 z_0^2}{r_0^2 + 2z_0^2} \tag{3.19}$$

When ions enter a 3D IT, they oscillate both axially and radially with changes in the electric field. The trajectory of ions can be interpreted by the z and the r components of the electric force to which they are subjected:

$$F_z = ma_z' = m\frac{d^2z}{dt^2} = -Ze\frac{d\Phi}{dz} \tag{3.20}$$

$$F_r = ma_r' = m\frac{d^2r}{dt^2} = -Ze\frac{d\Phi}{dr} \tag{3.21}$$

For 3D IT, the charge number is denoted by Z to distinguish it from the z coordinate. From Equations 3.18 and 3.19, we obtain:

$$\frac{d^2z}{dt^2} - \frac{4Ze}{m(r_0^2 + 2z_0^2)}(U - V\cos\omega t)z = 0 \tag{3.22}$$

$$\frac{d^2r}{dt^2} + \frac{2Ze}{m(r_0^2 + 2z_0^2)}(U - V\cos\omega t)r = 0 \tag{3.23}$$

Rearranging Equations 3.22 and 3.23, one can obtain a Mathieu function, which is similar to the formula for in QMF (cf. Equation 3.13). In this coordinate system, u represents z or r coordinates, and ξ is still defined as $\omega t/2$. Following the same derivation shown in Equation 3.14, Equation 3.22 can be rearranged to yield:

$$\frac{d^2z}{dt^2} - \frac{16Ze}{m(r_0^2 + 2z_0^2)\omega^2}(U - V\cos\omega t)x = 0 \tag{3.24}$$

Thus, one can express a_z and q_z as:

$$a_z = -\frac{16ZeU}{m(r_0^2 + 2z_0^2)\omega^2} \tag{3.25}$$

$$q_z = \frac{8ZeV}{m(r_0^2 + 2z_0^2)\omega^2} \tag{3.26}$$

Similar derivation can also be applied to the r component (Equation 3.23), resulting in the relationships $a_u = a_z = -2a_r$ and $q_u = q_z = -2q_r$. Figure 3.5 displays the first overlapped stability region (gray area) in the a_u–q_u plot. Unlike the two symmetrical triangular-shaped

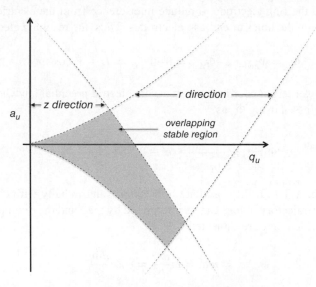

Figure 3.5   *Stable ion condition regions in 3D IT predicted with a_u and q_u. The gray area is the first overlapping stable region. The dashed lines represent the boundaries of stable regions*

stable regions of QMF (Figure 3.3a and b), the overlapping stable region of 3D IT mass analysis is not symmetrical with respect to the q_u axis.

When ions enter the 3D IT, their translational motions are confined to the trapping volume before they are ejected away from the 3D IT for detection. 3D IT is typically filled with helium (supplied at a pressure of ~ 1000 Pa) to lower (thermalize) the translational energy of ions, typically to < 1 eV. The movement of ions into the 3D IT is controlled by an electric potential external to the IT because there are no independent trapping electrodes in the 3D IT mass analyzer. Ion trapping is controlled by altering V to change q_z, whereas a_z is kept at 0 by setting U to 0 V (Equations 3.25 and 3.26). The critical trapping condition is when the ion position in the z direction reaches z_0, which results in a q_z of 0.908. Below this critical value, the ions will have stable ion trajectories inside the trapping volume. Thus, by keeping V and ω constant, it is possible to determine the critical ion mass m_c (assuming $Z = 1$). Any ion with higher mass than m_c can be trapped in the 3D IT for an arbitrary time period, which can range from sub-microseconds to sub-seconds.

Based on this principle, recording a mass spectrum can be achieved by ramping up V. Because any ion with a lower mass than m_c will be ejected from the trap, ramping up V shifts m_c upward. Therefore, ions with low to high m/z can be ejected sequentially from the trapping region towards the detector along the z axis (Figure 3.4). A mass spectrum can thus be obtained by correlating the detected signal with the corresponding V value. Several other scanning modes of 3D IT are similar to those for QMF except that only a_z or q_z is used in 3D IT.

Resonant excitation of ions in a 3D IT mass analyzer can be conducted by applying a low amplitude supplementary RF potential to one or both end-cap electrodes. The amplitude of the supplementary RF potential is typically a few volts, and its frequency is mass-dependent, typically $10-1000$ kHz. It correlates with the axial oscillation frequency of ions. The axial oscillation frequency is $\left(n + \frac{1}{2}\beta_z\right)\omega$, where n represents the order of the stable region and β_z is the secondary trapping parameter that represents a selected line in the over-lapping region (Figure 3.5). Various other operational modes of electric field for excitation of ions in IT mass analyzers are also available but they are beyond the scope of this book. Readers should explore the literature for further details [16].

The drawback of 3D IT mass analyzers is that the maximum number of ions that can be stored in it, i.e. ion capacity, is limited to roughly a few thousand. The ion capacity also determines the dynamic range of a mass analyzer. This limitation is greatly reduced with the second type of IT mass analyzer: 2D IT consisting of a QMF with additional electrodes or quadrupole rods installed at its entrance and exit ends [16, 20]. These additional electrodes or quadrupole rods are called *trapping electrodes* or *trapping quadrupoles*, respectively. The trapping quadrupoles are connected to RF/DC electric potentials with well-defined timing to control the flow of ions into and out of the quadrupole region. Figure 3.6 shows a 2D IT equipped with two sets of trapping quadrupoles. At the beginning of a scan, the trapping quadrupole of the entrance end (herein called the *entrance section*), is kept at a low electric potential to allow ions to flow freely into the quadrupole region. If the voltages of the main quadrupole (normally called the *central section*) are operated in the *ion-guiding mode* (see Section 3.2.2.1), all ions within the stable region (Figure 3.3b) will stably propagate along the electric field to the other end of the *central section* and approach the trapping quadrupole at the exit end (also called the *exit section*). If the exit section is at

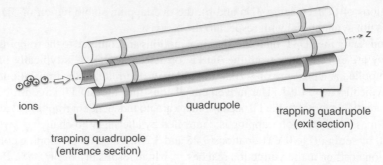

Figure 3.6   *Schematic drawing of a 2D IT mass analyzer that consists of three quadrupole sections. The first and the third sections are used to control the flow of ions into and out of the central quadrupole section*

a high electric potential, the ions will be reflected back to the central section and propagate towards the entrance section. These ions can exit the central section if the entrance section is still at a low electrical potential, but they can be reflected back to the central section again if the potential of the entrance section is lifted to block ion transmission. Thus, by keeping both the entrance and the exit sections at a high potential, the ions are trapped inside the central section of this quadrupole device. Changing the RF voltages of the 2D IT enables isolation, excitation, and fragmentation of the trapped ions for subsequent analysis.

In many cases, 2D IT provides greater analytical performance than regular QMF. For instance, a 2D IT mass analyzer offers higher sensitivity than QMF because it accumulates ions before recording the mass spectrum. The trapping is typically achieved by "opening" the entrance section, by lowering its DC potential for a long period of time, to receive ions continuously. Although, in principle, the number of ions in the central section can only increase by twofold owing to the leakage of ions at the entrance section after a round trip in the central section, further increase of the number of trapped ions is possible by lowering the translational energy of ions in the central section to a level below the potential of the entrance section. The cooling of ions can be achieved by thermalizing their translational energy using *buffer gases* in the central section. By using potential at the entrance section at a level slightly lower than the translation energy of incoming ions, ions can be trapped and accumulated. After accumulation, ions are excited radially based on the same principle as in QMF for recording spectrum. Because the radial motion of excited ions couple with their axial motion at the two ends of the central section, ions are ejected out of the 2D IT to an ion detector beyond the exit section when their radial motion is excited.

The main advantage of 2D and 3D ITs over QMF is improved functionality. For instance, the ions of interest can be isolated from others, and accumulated for subsequent experiments. After ejecting unwanted ions, the ions can be resonantly energized to large amplitude oscillations to promote ion–molecule collisions. Upon continuous collisions and excitation, the internal energy of ions increases until fragmentation occurs to produce characteristic products. If excitation is optimized for the precursor ion, the fragmentation products (of different m/z to the precursor ion) will not be further excited. The products can be analyzed directly or they can be isolated again to repeat the fragmentation process. Reactive collisions may also occur and the reactions between ions and gas molecules can be monitored. This operation can be achieved by the introduction of a reactive gas to initiate chemical reactions.

In IT, SAT ranges from tens of microseconds to hundreds of milliseconds (see Chapter 5). The exact time depends on ion abundance in the IT device, which subsequently determines signal intensity. Although signal intensity increases as ion abundance increases, spectral features broaden and significant positional shifts can occur if too many ions are present in the IT device. The peak broadening and shifts reduce the mass resolving power and accuracy of the mass spectral analysis. This phenomenon is referred to as the *space-charge effect* [21, 22]. Therefore, there is an optimum number of ions that can be confined in the trap. High flux ion beams from an ion source can accelerate ion accumulation and reduce SAT, which are favorable for tandem MS measurements.

3.2.3   Sector Mass Analyzers

Sector mass analyzers separate ions according to either static electric or magnetic interactions, or a combination of both in series in one device. Sector mass analyzers belong to beam-type instruments. The SAT of sector mass spectrometry (S-MS) ranges from hundreds of milliseconds to several seconds. Such speed is almost as low as that of the FT mass analyzer, but the mass resolving power and mass accuracy of S-MS is much lower. Sector instruments were popular in the 1980s because they offered the best mass resolving power and tandem MS capability at the time. However, in recent decades there has been a decline in the use of sector mass analyzers relative to other instruments (TOF-MS, Q-MS, and FT-MS). This trend is mainly due to the slow spectrum scanning rate of S-MS, and the integration of MALDI and ESI ion sources into sector mass analyzers is also more difficult than in other instruments. Nowadays only few manufacturers produce sector instruments, so only a brief introduction to sector mass spectrometers is given here.

Sector-type mass analyzers are named for the magnet (B) and electric (E) sectors, which bend the trajectory of ions through curvature paths induced by sector-shaped magnets or curved metal electrodes, respectively. S-MS utilizes the full expression of the Lorenz force equation, which describes magnetic (F_m) and electric (F_e) forces acting on an ion with charge q:

$$F = F_m + F_e = q\vec{v} \times \vec{B} + q\vec{E} \tag{3.27}$$

where $\vec{B}$ is the magnetic field strength, $\vec{E}$ is the electric field strength, and $\vec{v}$ is the ion velocity orthogonal to the vector of the magnetic field. Note that only velocity and magnetic field components orthogonal to each other induce magnetic force. B sectors only utilize the F_m component of Equation 3.27 while E sectors only utilize the F_e component. Commercial sector mass spectrometers normally combine two or more sectors in series. The configuration of sectors (i.e., EB, BE, BEB, BEEB or BEBE) depends on the tandem MS function, mass resolving power, and sensitivity needed.

In B sectors, an ion with velocity v experiences magnetic force F_m while entering the magnetic field region perpendicularly. Figure 3.7 shows a positive ion moving to the left while the magnetic field points directly into the paper. The principle is described by Fleming's right-hand rule [8]: a magnetic force acts on an electric charge moving perpendicularly to a uniform magnetic field in a direction that is perpendicular to both the directions of the magnetic field and the motion of the electric charge. In Figure 3.7, F_m pushes the ion

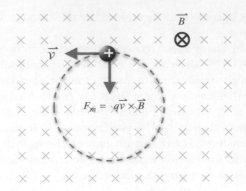

Figure 3.7   *The circular motion of an ion within a homogeneous magnetic field*

downward. If the magnetic field region is large enough, this force is continuously exerted on the moving ion, resulting in a circular ion path of radius r_B. In this case, F_m serves as a centrifugal force (F_c) inducing this ion into a circular motion. Thus, the radius r_B can be derived accordingly:

$$F_c = \frac{mv^2}{r_B} \tag{3.28}$$

$$F_c = F_m = \frac{mv^2}{r_B} = qvB \tag{3.29}$$

$$r_B = \frac{mv}{qB} \tag{3.30}$$

For an ion with an m/z of 100 and kinetic energy of 1 keV, ion velocity is 1500 m s^{-1}. If the magnetic field strength is 0.5 T, the resultant radius of circular motion is 60 cm. This radius indicates the minimum dimension of the vacuum system. Because available electromagnets produce magnetic field strengths of a few tenths of a tesla, the resultant radius of circular motion is too large to be entirely accommodated by a practical vacuum chamber. Thus, B sector instruments use a curved vacuum chamber of fixed radius. The trajectory of ions from the ion source are typically bent at 60° to their original direction. Figure 3.8 shows such a magnet sector mass analyzer. Two slits are installed at the entrance and the

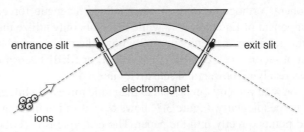

Figure 3.8   *A magnet sector mass analyzer bends the ion trajectory by 60°. The entrance and exit slits allow selected ions to pass through to the detector based on their momentum and charge*

exit of the magnet sector to determine r_B. According to Equation 3.30, fixing r_B ensures only ions of a particular momentum-to-charge ratio (i.e., mv/q) pass through the slits to reach the detector at one B value. As mentioned above, since ions of the same m/z move through the same path in a vacuum, it is possible to selectively detect ions of the same m/z by adjusting magnetic field strength.

In contrast to B sectors, E sectors utilize the F_e component of Equation 3.27. Such instruments select ions according to their kinetic energy rather than momentum (as in B sectors). The radius of the circular ion path formed in the static electric sector can be derived by combining centrifugal force (Equation 3.28) with electric force, similar to Equation 3.29 but by replacing F_m with F_e, to give:

$$F_c = F_e = \frac{mv^2}{r_e} = qE \tag{3.31}$$

$$r_e = \frac{mv^2}{qE} \tag{3.32}$$

where r_e represents the radius of ion trajectory under a curved electric field. Such electric fields are produced by a pair of curved metal plates, as shown in Figure 3.9.

The typical ion detector in sector instruments is an EM. Similar to an MCP, EMs are made of semiconductive materials that enable the ejection and amplification of secondary electrons after high-energy collisions by ions. In contrast to MCPs, which contain millions of microchannels, EMs are single channel ion detectors. EMs are typically horn-shaped detectors, roughly 30–50 mm in length. The entrance of the horn is ~10–20 mm in diameter, and the narrow end roughly 3–4 mm. The EM receives ions at the entrance and releases secondary electrons from the narrow end. The overall gain is roughly 10^5 when the detector is supplied with a voltage difference of 2500 V. Such detectors are more robust and last longer than MCPs, and their pulse width is between 20 and 30 ns. An EM detector is not suitable for TOF-MS because the horn-shaped entrance does not allow for precise determination of flight distance and ion flight times.

Mass resolving power of sector instruments can be improved by combining B and E sectors in series. This is because ions must attain the appropriate kinetic energy and momentum to pass through the two mass analyzers. Figure 3.10 shows an EB mass analyzer. By suitably designing the magnet and electric sectors, it is possible to use the second sector

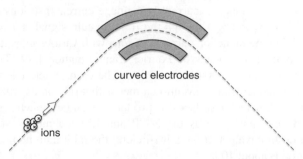

Figure 3.9 *An electric sector mass analyzer uses a pair of curved electrodes to bend the trajectory of ions with specific mass and kinetic energy*

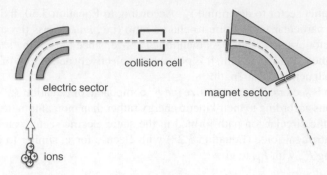

Figure 3.10 *The configuration of an EB mass analyzer containing an electric sector and a magnet sector installed in series. The collision cell between the two mass analyzers facilitates tandem mass analysis*

to correct the direction and energy spread caused by the first sector. Such instruments are known as *double focusing instruments*. If a collision cell is installed between the two sectors, the double focusing instruments are capable for tandem MS analysis. The theory of tandem mass analysis is described in Section 3.3.2. Aside from their slow scanning rates and limited ion source selection, sector instruments provide good mass resolving power, sensitivity, reproducibility, and dynamic range.

3.2.4   Fourier-transform Mass Analyzers

Fourier-transform (FT) mass analyzers belong to a special class of mass analyzer that detects ions non-destructively and periodically. Therefore, FT mass analyzers are trapping-type instruments. Because periodic and long detection times facilitate accurate ion recognition, FT mass analyzers offer the highest mass resolving power and accuracy of all instruments [23]. The concept of FT-MS was first described by Comisarow and Marshall in the 1970s [24]. The most important types of FT instruments available in the market include ion cyclotron resonance (ICR) [23] and orbital ion trap (orbitrap) [25] mass analyzers.

Unlike the unsynchronized ion motions in IT, ions inside the FT mass analyzers undergo synchronized periodic motions (oscillations or circulations). Since the frequency of periodic motion in FT mass analyzers is m/z-dependent, mass spectra can be derived from transients or time-domain signals induced by such periodic motions. This operation is achieved by recording induction currents (or image currents) of ions using electrodes installed near the ion trajectory. Although the time-domain signal is a complicated transient signal containing the induction currents of ions with various m/z, it can be resolved in the frequency-domain by means of Fourier transformation [26]. The characteristic frequencies obtained by Fourier transformation can be used to derive ion masses. Such a detection method is distinct from conventional means that rely on destructive high-energy collisions of ions with detector surfaces to produce secondary electrons for subsequent amplification and detection, such as the MCP and EM techniques mentioned earlier. Because the non-destructive detection is inefficient, the minimum number of charges for successful detection is about 100.

Under typical operating conditions, a time-domain signal of hundreds of milliseconds to tens of seconds is necessary to produce a mass spectrum. The long SAT of FT mass analyzers gives them the slowest duty cycle among mass analyzers. Another feature of FT-MS is that these mass analyzers have to be kept at ultrahigh vacuums ($< 10^{-10}$ mbar) to reduce ion–molecule collisions. Therefore, a long transmission path containing several differential pumping stages is necessary for such instruments. Thus, ions need to travel long distances from the ion source to the mass analyzer which imposes particularly long transmission times. In such cases, the duty cycle of FT-MS is typically <10%. In fact, the low duty cycle problem also occurs in Q-MS if a high-resolution scan is necessary. The working principles of ICR and orbital ion trap mass analyzers are described below.

3.2.4.1 Ion Cyclotron Resonance

The periodic motion of ions in an ICR mass analyzer is induced by Lorentz force which acts on the moving ions, as introduced previously in the magnet sector mass analyzers (Section 3.2.3). In ICR, the magnetic force (F_m) traps ions radially while electric potentials are used to confine ions axially. Therefore, ICR devices consist of bored magnets that accommodate the vacuum chamber of the mass analyzer. Because the mass resolving power of an FT-ICR increases as the magnetic field strength increases, high-field superconductive magnets are typically used (e.g., >3 T).

To explain ion motion in the ICR cell, we shall reconsider an ion moving with an initial velocity (v) inside a uniform magnetic field, as shown in Figure 3.7. The magnetic force results from the ion velocity and magnetic field strength: $F_m = q\vec{v} \times \vec{B}$. The only difference between ICR and the magnet sector for S-MS is that the magnetic field strength B in ICR is significantly stronger and unchangeable. By correlating the centrifugal force ($F_c = m\vec{v}^2/r$) and the magnetic force, we can obtain the *angular velocity* (ω_c) of the cyclotron motion from Equation 3.30:

$$\omega_c = \frac{v}{r_B} = \frac{qB}{m} \tag{3.33}$$

Dividing ω_c by 2π results in the cyclotron frequency (f_c) of this motion:

$$f_c = \frac{\omega_c}{2\pi} = \frac{qB}{2\pi m} \tag{3.34}$$

For a singly charged molecule with a molecular weight of 100 u, moving inside a 7 T magnet, f_c can be calculated using Equation 3.34 (q in coulombs; B in teslas; m in kilograms):

$$f_c = \frac{qB}{2\pi m} = \frac{1.6 \times 10^{-19} \times 7}{2\pi \times 0.1/(6.02 \times 10^{23})} \cong 1.075 \times 10^6 \tag{3.35}$$

According to this equation, f_c is directly proportional to the magnetic field strength and inversely proportional to the mass of the ion. Notably, f_c is independent of the velocity of an ion; hence, the kinetic energy spread of ions does not affect the f_c of a measured ion. That is, an ion with a higher speed than another ion of the same m/z exhibits a larger radius but their f_c values are the same. This important feature of ICR-MS stands behind its high mass resolving power.

The conventional ICR mass analyzer consists of an ICR cell. The cell is made of four identically curved metal electric plates installed symmetrically with respect to a central axis, and two metal hollow disks cover the two ends. Figure 3.11 illustrates the geometry

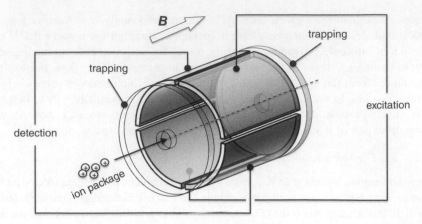

Figure 3.11   *Schematic representation of an ICR mass analyzer. The mass analyzer is installed in a homogeneous magnetic field along its central axis*

of such an ICR cell. The central axis is aligned precisely parallel to the magnetic field. Ions are ejected into this region from one side of the cell along this axis. When ions are inside the cell, one pair of opposing curved plates that are defined as excitation plates, is subjected to RF potentials to excite ion motion radially. The phase difference of the RF excitation voltage on the two excitation plates is 1π rad (180°). After excitation stops, the other pair of curved plates that are defined as detection plates, is connected to a differential current amplifier for induction current detection.

As mentioned above, the ions are confined radially by magnetic force, whereas their axial motion (flowing into or out of the cell) is controlled by electric potentials applied to the two hollow disks. The two disks serve as *trapping electrodes*, at either side of the cell. At the moment when ions enter the cell, electric potentials of a few tens of volts are supplied to the trapping electrodes to trap the ions. In order to improve ion trapping efficiency and to reduce the interference of trapping potential on ICR motion, the geometry and electric potential of trapping electrodes varies in different instruments [27].

Notably, ions with low kinetic energy in the ICR cell are unable to be detected. If the ion is maintained at room temperature, the ion velocity can be calculated by $v = \sqrt{3kT/m}$ to yield roughly $274\,\text{m s}^{-1}$. This ion velocity results in an ICR radius of only $\sim 40\,\mu\text{m}$. Because the diameter of a typical ICR cell is roughly 6 cm, this ion stays very close to the central axis of the cell. Since the intensity of the induction signal is inversely proportional to the distance between the ions and the detection plate, ions staying at the center of the cell cannot induce any significant current in the detection circuit. In addition, non-synchronized ion motion also seriously reduces induction current intensity. In order to detect induction signal, ions have to be excited simultaneously into a large ICR radius.

To achieve a large ICR radius, RF potentials are applied to the excitation plates (Figure 3.11) when ions are trapped inside the cell. In the case of the ions that are initially present at room temperature, the ion package is very small relative to the diameter of the cell. If the supplied RF potential is in resonance with the ICR frequency of the ions, the ions will be accelerated to a higher velocity. Because the f_c of ions with equal m/z is constant, increasing v of ions results in a proportional increase in r, as indicated by

Equation 3.33. Thus, when the RF frequency is in resonance with the ICR frequency of the ion, excitation will bring ions into orbits with large radii based on the relationship [23]:

$$r_B = \frac{VT_E}{dB} \tag{3.36}$$

where V is the RF amplitude, T_E is the excitation time, and d is the diameter of the ICR cell. For an ICR cell with $d = 6$ cm installed inside a 7 T magnet, an excitation amplitude of 1 V can bring the ion to $r_B = 3$ cm in 6.3 ms. After a desirable radius is achieved (typically half of the ICR cell radius), the excitation potentials are turned off to allow the ions to circulate freely in the cell. At the same time, when excitation stops, the detection circuit is activated to collect the induction current. The detection circuit cannot work while the excitation potentials are applied to the excitation plates because the RF potentials will obscure the ion signal. In addition, the detection circuit will also pick up intense RF excitation signal, which could damage the detection circuit. Thus, excitation and detection of ions is performed sequentially after ions enter the ICR cell.

The induction current is amplified and recorded by a data acquisition system to obtain the transient. The transient is converted to a frequency-domain signal by Fourier transformation and subsequently to m/z-domain spectra. If only ions of single m/z are present, the transient is simply a one-component sine wave. If ions of n-different m/z are present, the transient is a complicated waveform containing n-component sine waves having different frequencies. The concept can be demonstrated using simulated sine waves as induction signal of ions in the time domain. Figure 3.12 shows the relationships between the time- and frequency-domain data which could be recorded during analysis of one, two, and six ions with different m/z. In the one-ion case shown in the top panel of Figure 3.12, the time-domain spectrum contains transient of a simple sine wave with a frequency of f_c. After Fourier transformation, this transient results in a single feature in the frequency domain. When a second ion with a frequency of $1.5f_c$ is presented, as shown in the middle panel of the figure, the new transient combines the two sine waves. Such a transient results in two features in the frequency domain after Fourier transformation. In the six-ion case demonstrated in the bottom panel, the simulated frequencies include f_c, $1.2f_c$, $1.5f_c$, $1.7f_c$, $2.0f_c$, and $2.5f_c$. Obviously, transient is highly complicated when multiple ions are present in the ICR cell. However, the ions can be unambiguously distinguished after Fourier transformation.

An important property of the FT calculation is that the resolving power and the accuracy of frequency determination in the frequency domain relies critically on the length of the transient. Ideally, ion motion should be kept unchanged during the period of data acquisition. Thus, it is very important to keep the induction current as long as possible by maintaining the orbiting radii of ions. Because kinetic energy of ions rapidly reduces when colliding with residual gas molecules, ions orbiting with large ICR radii rapidly return to the center of the ICR cell when the pressure inside the ICR cell is high. In order to reduce the ion–molecule collision in the ICR cell, the pressure inside the cell has to be kept below 10^{-8} Pa. In the case of singly charged ions with mass of 100 u in a 7 T magnet, a transient of a few hundred milliseconds offers a mass resolving power in the range of 10^5. A resolving power above 10^6 is available if the ion signal can last for 1–2s.

Because excitation voltages can control ion motion, ion selection and fragmentation can be achieved inside the ICR cell. For ion selection, for example, the excitation voltage can

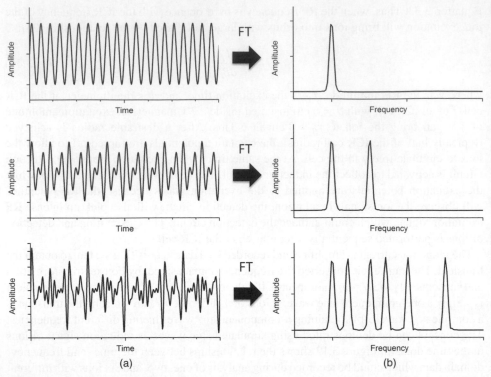

Figure 3.12 *Fourier transformation of transients in time-domain (a) to frequency-domain spectra (b) in three cases. Top: one-ion case (f_c). Middle: two-ion case (f_c, $1.5f_c$). Bottom: five-ion case (f_c, $1.2f_c$, $1.5f_c$, $1.7f_c$, $2.0f_c$, $2.5f_c$)*

be used to excite unwanted ions to a larger ICR radius than the radius of the cell. This excitation results in unwanted ions being rejected and target ions remaining unaffected. After the selection process, pulsed helium, argon, or nitrogen gas can be streamed into the cell while exciting the ions to a large radius to induce ion–molecule collisions. During the time of excitation, ions are excited constantly to induce multiple collisions, this builds up the internal energy until they fragment. Since fragmented ions have a different m/z from the precursor ions, they will not be excited by the RF potential at the same frequency as the f_c of the precursor. After fragmentation and evacuation of the collision gas from the ICR cell, excitation and detection of fragment ions are performed.

3.2.4.2   Orbital Ion Trap

An orbital ion trap or orbitrap, utilizes only a circular DC electric field to confine ions. The field is produced by a spindle-shaped metal central electrode biased with a high DC potential, and two identical metal outer electrodes at ground potential forming a barrel-shaped trapping region. Each of the outer electrodes covers half of the central electrode. The design was developed by Makarov in 2000 [25]. Figure 3.13 schematically illustrates an orbitrap mass analyzer. Ions enter this mass analyzer through a small hole in one of the outer electrodes, and then revolve around the central electrode. The two outer electrodes are

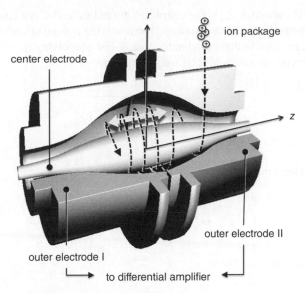

center electrode

ion package

z

outer electrode II

outer electrode I

to differential amplifier

Figure 3.13   *Schematic representation of an orbitrap mass analyzer. The central electrode is biased at a high voltage and the two outer electrodes are at ground potential. Ions enter this region at one side of the outer electrode and encircle the central electrode. The periodic motion of ions along the central electrode (indicated by the double-headed arrow) is recorded to obtain the time-domain signal*

connected to a differential amplifier to collect the induction current of periodically moving ions within this mass analyzer.

The spindle-shaped central electrode produces an electric field to trap ions both radially and axially. For example, the central electrode is biased at a high negative potential when analyzing positive ions. When ions enter the mass analyzer near one edge of the central electrode (Figure 3.13), they experience an electric field that pulls them towards the central electrode. When the centrifugal force is in balance with the electric force, ions revolve around the central electrode without colliding with any electrode. The electrode geometry inside orbitrap mass analyzers results in an ion trajectory that not only revolves around the central electrode but also directs the ions toward the center of the orbitrap. When the ions pass the center and reach the opposite end of the orbitrap, the concave shape of the central electrode produces an electric field that pulls the ions back towards the center of the mass analyzer. This electric field results in periodic ion motion back and forth between the two ends of the central electrode. Such a motion produces induction currents in the circuit of the two outer electrodes, which is subsequently amplified and processed to build a spectrum.

Because the mass analyzer is symmetrical along the central axis, the electric field U inside the trapping volume can be described in the r-z plane of a cylindrical coordinate (r, ϕ, z) system, assuming that the z axis is the center of symmetry [25]:

$$U(r, z) = \frac{k}{2}\left(z^2 - \frac{r^2}{2}\right) + \frac{k}{2}(R_m)^2 \ln\left(\frac{r}{R_m}\right) + C \tag{3.37}$$

where k is the field curvature, R_m is the characteristic radius, and C is a constant. The value of k depends on the geometry of the orbitrap. The revolving motion of ions can be described by three periodic motions in this coordinate system. For a revolving motion with radius r', the three periodic components and their frequencies are:

- rotation around the z axis:

$$\omega_\phi = \omega_z \sqrt{\frac{\left(\frac{R_m}{r'}\right)^2 - 1}{2}} \tag{3.38}$$

- radial motion along the r axis:

$$\omega_r = \omega_z \sqrt{\left(\frac{R_m}{r'}\right)^2 - 2} \tag{3.39}$$

- axial oscillation along the z axis:

$$\omega_z = \sqrt{\frac{q}{m} k} \tag{3.40}$$

As can be seen from the above expressions, both rotation frequency ω_ϕ and radial motion frequency ω_r depend on r', while the axial oscillation frequency ω_z only depends on m, q, and k. Because k is known and it is constant for an orbitrap mass analyzer, the m/z of ions can be deduced by Equation 3.40. As described previously, the pair of outer electrodes enables detection of the axial oscillation in the time domain, and the values of ω_z can be obtained by a fast Fourier transformation. The maximum mass resolving power provided by the orbitrap mass analyzer is $\sim 4 \times 10^5$ (see Chapter 5 for further discussion of this parameter).

Although orbitrap mass analyzers can trap ions, they are unable to eject unwanted ions or to conduct ion fragmentation. First, they are unable to selectively eject ions because the current design has no mechanism to selectively excite ions of certain m/z. Second, the fragmentation of ions within an orbitrap is also difficult because it is not possible to introduce a laser beam into the orbitrap for ion excitation. Ion activation via ion–molecule collision is also disadvantageous because the collision reduces the kinetic energy of ions, and it subsequently reduces the radius of circular motion around the z axis until ions collide with the central electrode. Even if the precursor ion decomposes before colliding with the central electrode, its product ions have lower kinetic energy than the precursor ion and cannot be trapped. Finally, there is no efficient pumping system to evacuate the orbitrap when introducing buffer gas for ion–molecule collision. As detailed for the ICR method, increasing pressure sharply reduces the length of the transient and subsequently reduces analytical sensitivity and mass resolving power.

The temporal property of an orbitrap is similar to that of an ICR mass analyzer. The mass resolving power relies on the length of the transient, so the SAT required to record a high-resolution spectrum is typically $1-10$ s. The SAT can be reduced to < 1 s if low or medium mass resolving power is acceptable.

3.3   Integrated Analytical Techniques

3.3.1   Hybrid Mass Spectrometers

Modern mass spectrometers normally consist of multiple analyzers used in tandem to increase overall functionality. There are two types of tandem designs. The first is called *tandem mass spectrometry in space*. In this setup, two or more mass analyzers are connected in series to conduct distinct functions. Depending on the property of each spectrometer, precursor ion selection, fragmentation, and product ion analysis can be performed in different mass analyzers. This design is very useful because beam-type mass analyzers are unable to conduct multiple analyses in one scan. The second type of tandem MS is called *tandem mass spectrometry in time*. In such instruments, tandem MS is done in a single trapping device (e.g., IT, ICR), which conducts ion selection, fragmentation, and analysis sequentially in the trapping region. Some hybrid mass spectrometers are described below.

3.3.1.1   Triple-quadrupole Mass Spectrometer

A triple-quadrupole (QqQ) mass spectrometer consists of three quadrupole devices joined together in series. As the acronym suggests, the three quadrupole devices are not entirely identical. The first (QMF I, "Q") and the third (QMF II, "Q") quadrupole devices are standard QMFs, whereas the second quadrupole (denoted with a lower-case "q") is smaller in size than the other two quadrupoles, and is typically operated in the ion-guiding mode (an RF ion guide). This RF ion guide is installed in a collision cell filled with gas, typically helium or other noble gases, to facilitate ion–molecule collisions that subsequently induce fragmentation of ions. Figure 3.14 shows the design of a typical QqQ mass spectrometer, in which the metal-made lenses installed between adjacent quadrupoles are typically supplied with electric potential that is used to reduce ion transmission loss. The RF ion guide transfers all ions from QMF I towards the QMF II, and it can be biased with a DC potential to adjust the ion velocity within this region. The above is a standard configuration of tandem MS in space, in which different components consecutively perform precursor ion selection, fragmentation, and product ion analysis.

Because tandem MS is in high demand, QqQ instruments have become popular. The three quadrupole devices aligned in series provide several scanning modes for various

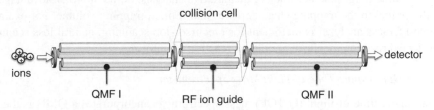

Figure 3.14   *Schematic representation of the QqQ mass analyzer. It consists of an RF ion guide installed between two QMFs. The ion guide is contained in a small chamber that serves as a collision cell for tandem MS. The electro-optical lenses in between every section are used to increase transmission efficiency*

applications [16]:

- Spectrum scanning and selected ion monitoring
  This mode is executed by scanning or fixing the electric potentials of QMF I in the absence of a collision gas in the collision cell, and operating the QMF II as an ion guide. Under these conditions, the transmission efficiency of ions in the collision cell and the QMF II are optimized by selecting proper electric potentials.
- Product-ion scanning
  Fragmentation products from an ion are analyzed in this mode. Analysis is conducted by selecting the precursor ions with QMF I and conducting ion–molecule collisions in the collision cell. The resultant fragmentation products are analyzed in QMF II. Adjusting the kinetic energy (e.g., tens to hundreds of electronvolts) of precursor ions in the collision cell is used to control the extent of fragmentation.
- Precursor-ion scanning
  To identify the origin of a specific fragmentation product, QMF I can be scanned while fixing QMF II so that it only transmits a fragmentation product of interest. In this mode, signal is available only when the precursor ions isolated by QMF I can produce the selected fragment product in the course of fragmentation in the collision cell.
- Neutral loss scanning
  This mode enables identification of a certain fragmentation pathway in the collision cell, such as dehydroxylation (-OH) or decarboxylation ($-CO_2$). Neutral loss scanning can be achieved by simultaneously scanning QMF I and QMF II while fixing the mass difference between the ions monitored by the two QMFs. Because QMF II is responsible for detecting fragments, it is normally set to transmit ions with lower m/z than that transmitted by QMF I. The difference in the two preset m/z corresponds to the mass of the neutral fragment.
- Selected reaction monitoring/multiple reaction monitoring
  A certain chemical reaction can be monitored by fixing the m/z of ions transmitted by QMF I and QMF II. The selected reaction product from a selected precursor ion can be monitored over a long period of time. It is also possible to monitor multiple reactions sequentially by switching QMF I and QMF II to move simultaneously from one reaction to other reactions. This method is called multiple reaction monitoring.

Although IT mass analyzers have the capability of conducting tandem MS in time with a single mass analyzer, QqQ-MS is still superior to IT-MS in some aspects, including stability, dynamic range, and accuracy of quantitative analysis. Some of the disadvantages of IT-MS are due to the trapping characteristics and limited trapping volume. For example, a single IT mass analyzer is inefficient for precursor-ion scanning, neutral loss scanning, and selected reaction monitoring/multiple reaction monitoring.

3.3.1.2   *Quadrupole Time-of-flight Mass Spectrometer*

A quadrupole time-of-flight (Q-TOF) mass spectrometer incorporates a QMF as the first mass analyzer and a TOF as the second mass analyzer. A collision cell is usually available to facilitate tandem MS. Figure 3.15 illustrates the design of a Q-TOF mass spectrometer that consists of a reflectron TOF mass analyzer in its last stage. Such a setup is similar to QqQ-MS except that QMF II is replaced by the reflectron TOF mass analyzer. When entering

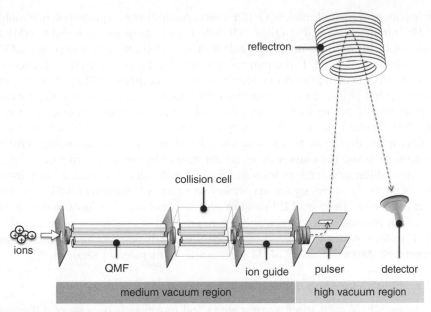

ions

QMF

collision cell

reflectron

ion guide   pulser   detector

medium vacuum region   high vacuum region

Figure 3.15 *Schematic representation of a Q-TOF mass analyzer. It typically consists of a QMF, a collision cell, an ion guide, and an orthogonal TOF mass analyzer*

the acceleration region of the TOF mass analyzer, the ions are accelerated orthogonally by a high-voltage (HV) pulse applied to the so-called *pulser electrode* or *pusher electrode*. Because a high vacuum environment is necessary for the TOF mass analyzer, an RF ion guide is used to transfer the ions across a differentially pumped region between the collision cell and the TOF region.

The advantage of Q-TOF over QqQ mass analyzers is that high-resolution mass spectra are available with high speed on account of the TOF mass analyzer. In fact, Q-TOF mass analyzers are now widely used in place of ITs in liquid chromatography (LC) detection systems, because of their superior analytical performance and highest spectrum acquisition speeds (see Chapters 5). For example, a QMF is a low-resolution mass analyzer that is unable to provide high-quality mass spectra. However, a QMF is an ideal device to serve as a precursor ion selector, collision cell, and ion-guiding device for fast and highly sensitive TOF mass analysis. Furthermore, since QMF is used as the first mass analyzer, Q-TOF mass spectrometers are typically equipped with continuous ion sources, including ESI, APCI, and APPI. This feature is important because individual TOF mass analyzers are incompatible with continuous and atmospheric pressure ion sources due to the discrete spectrum scanning nature of TOF mass analyzers and their high-vacuum requirement. In this sense, QMFs serve as the intermediate stages bridging the gap between atmospheric pressure ion sources and TOF mass analyzers. In addition, integrating a QMF into a TOF mass analyzer also greatly enhances the capability of the TOF mass analyzer for tandem mass analysis. Thus, a Q-TOF mass spectrometer can record a mass spectrum using the TOF mass analyzer while operating the QMF and the collision cell in the ion-guiding mode. It can also scan product ions by performing fragmentation in the collision cell on a precursor ion selected by the QMF.

An important variant of modern Q-TOF mass spectrometry is quadrupole-ion mobility-TOF MS, often abbreviated as Q-IM-TOF-MS. It consists of an ion mobility (IM) spectrometer replacing the quadrupole ion guide in the collision cell present in common Q-TOF mass spectrometers. The IM spectrometer consists of a stack of metal ring electrodes that produce an electric field to push ions slowly forward. Because the IM spectrometer region is filled with buffer gas, the ion–molecule collisions retard ion motion while the electric field continuously pushes ions to move forward. Since larger ions experience more collisions than smaller ions, the drift velocity of larger ions is slower. The IM spectrometer provides another dimension to separation, which differentiates ions according to their collision cross sections. The cross sections are determined by the steric structure of ions. The capability of differentiating ions according to their collision cross sections is an important feature in the study of tertiary and quaternary structures of proteins as well as isomers of smaller molecules. The use of IM spectrometry does not sacrifice functionality of MS in terms of ion activation or fragmentation. If fragmentation is necessary, the electric field of the IM spectrometer can be adjusted to ensure ions have adequate kinetic energy for fragmentation. More discussion of IM spectrometry is included in Section 6.3.

3.3.1.3   Fourier-transform Tandem Mass Spectrometers

All commercial FT mass spectrometers are hybrid instruments because of differences in performance and functionality of various MS techniques. For example, orbitrap mass analyzers are unable to perform ion isolation and fragmentation, so these ion manipulation processes have to be conducted externally by other mass analyzers. Furthermore, the spectrum scanning speeds of ICR and orbitrap mass analyzers are too low to provide adequate efficiency for hyphenated mass spectrometric analysis. For example, performing ion selection or fragmentation in ICR cells typically takes 10–100 times longer than in quadrupole devices (e.g., QMF, collision cell equipped with RF ion guide). Conducting tandem MS in ICR cells also raises gas pressure, which reduces mass resolving power. Finally, the long transmission distance from the ion source to the ultrahigh vacuum region in FT mass analyzers provides enough space to accommodate extra mass analyzers. Thus, in order to ensure the best FT-MS performance, ion selection and fragmentation is normally conducted in an external quadrupole device.

The most popular configuration of hybrid FT-MS is the integration of a QMF and a collision cell in front of the FT mass analyzer. Figure 3.16 shows a typical design of a hybrid FT-ICR-MS. In some commercial instruments, an LIT is used to trap ions and to fragment them. After passing through the collision cell/trap region, a long ion-guiding device transfers ions through two differentially pumped regions to the ICR cell (typically >1 m away). For safety reasons, the long vacuum system design is necessary in FT-ICR-MS to keep the electronic and vacuum pumps away from the fringe field of the magnet.

Accumulation of ions in the ion trapping device prior to ICR mass analysis is advantageous for FT-ICR-MS because more than 1000 charges are necessary to produce a reasonably high ICR signal, and ICR cells can only receive ion packages within short periods of time. Since the ions need to travel a long distance after they are pushed toward the ICR cell, light and heavy ions arrive in the ICR cell at different times [28, 29]. This phenomenon is known as the *time-of-flight effect*. Therefore, one needs to select the time point for applying the acceleration potential of the ion trapping device to extract ions and the time window for receiving ions by the trapping electrodes of the ICR cell.

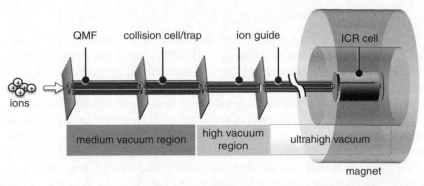

Figure 3.16   *Schematic representation of a hybrid FT-ICR mass spectrometer. It consists of a QMF, a collision cell/ion trap, and a long ion guiding device for transferring ions to the ICR cell*

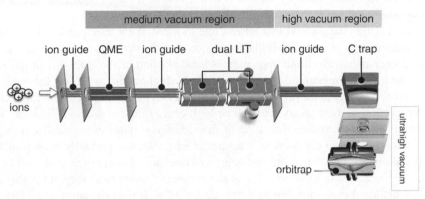

Figure 3.17   *Schematic representation of the orbitrap mass spectrometer. It consists of a QMF and a dual LIT. The C trap pushes ions into the orbitrap mass analyzer which is installed in the ultrahigh vacuum region*

The configuration of orbitrap mass spectrometers is similar to that of FT-ICR mass spectrometers. Figure 3.17 illustrates the design of an orbitrap mass spectrometer, which was commercialized in 2005. It consists of a QMF for precursor ion selection and a dual LIT that provides multiple functions, including thermalization of ion energy and inducing fragmentation. The second LIT can also serve as an independent mass analyzer because it contains two off-axis ion detectors capable of recording mass spectra. In this case, the SAT is fast because the orbitrap is disabled when recording a mass spectrum using LIT (as long as low spectral resolution is acceptable). If high resolution mass spectra are necessary, ions can be sent to the orbitrap for detection. Moving ions from the LIT to the orbitrap is performed by extracting ions from the dual LIT and through the ion guiding device towards the C trap. The C trap is responsible for manipulating ion kinetic energy before the ions enter the orbitrap. The C trap contains a curved compartment for accumulating ions and cooling their kinetic energy with buffer gas. The ions are pushed out of the C trap when an HV pulse is applied to the trap. The curvature of the C trap is optimized to focus the ion package towards the 1-mm entrance hole of the orbitrap mass analyzer.

3.3.2   Ion Activation Methods

In most cases, a mass spectrometer provides functionality beyond the acquisition of mass spectra. It normally serves multiple purposes. In TRMS, mass spectrometers are used to monitor the changes of reactants or products in chemical reactions. For example, a mass spectrometer can be used to selectively monitor the evolution of a precursor ion and its fragmentation products. Fragmentation products also assist the identification of precursor ions if many contaminants are present. In such cases, it is necessary to activate ions, either vibrationally or electronically, to induce fragmentation of precursor ions within the mass spectrometer to extract as much structural information as possible. An effective ion fragmentation technique should produce reproducible fragmentation patterns, so that it is easy to predict the molecular structure of ions. In some cases, it is also necessary to induce a chemical reaction of ions upon their collision with reagent gas molecules. In this section, we will focus on important ion activation techniques used in fragmentation. Ion activation for chemical reactions follows a similar principle but the reaction conditions are different (with respect to the pressure and the identity of the collision gas).

Ion activation is the method that allows one to increase the electronic, vibrational, or translational energy of ions. The initial excitation reaction depends on the amount and the rate of energy input. Electronic transition is the transition of the electrons in molecules from lower to higher electronic states. It requires the highest energy among the three excitation modes (roughly in the range of above 3 eV). It is typically induced by a rapid energy transfer reaction such as photoexcitation or high-energy ion–molecule collisions. Vibrational excitation concerns the excitation of the vibrational states of molecules. It requires only a few tenths of an electronvolt. Because macromolecules normally consist of many vibrational modes (i.e., the vibrational degree of freedom of linear molecules is $3N$-5 and non-linear molecules is $3N$-6, where N is the number of atoms) and they are highly associated, vibrational excitation can be achieved by infrared (IR) radiation (e.g., lasers) or low-energy ion–molecule collisions. The translational energy of ions associates with their velocity, so translational energy can be increased by accelerating ions under an electric field.

Increasing internal energy of ions, in most cases, may directly contribute to the cleavage of chemical bonds. It can increase the amplitude of vibration of chemical bonds until dissociation occurs. In contrast, increasing the translational energy of ions cannot directly induce fragmentation. The translational energy has to be converted to internal energy via ion–molecule collision before fragmentation. Additionally, increasing the translational energy of ions assists ions to surmount the activation energy of a chemical reaction upon collision of the ions with reactive species.

Ion activation reactions can take place in different compartments. The most common method used is to activate ions in collision cells, IT mass analyzers, or ICR cells. In these cases, ions flow into such regions before ion activation reaction starts. However, ion activation in MALDI-TOF-MS is different because there is no collision cell in single TOF mass analyzers. Therefore, ion activation and fragmentation in MALDI-TOF-MS are usually achieved using harsher desorption/ionization conditions, such as higher laser energy and higher acceleration field. The high acceleration field increases the internal energy of ions via ion–molecule collision during the desorption process. Fragmentation taking place in the MALDI ion source region is called in-source decay (ISD). On the other hand, fragmentation in MALDI-TOF-MS can also take place after the ion source region, such as in the field free region. In contrast to ISD, this type of fragmentation is called post-source decay (PSD).

It is convenient to categorize popular excitation methods according to their rate of energy input. Rapid or instantaneous ion activation occurs in roughly $10^{-15}-10^{-9}$ s, which is typically achieved using an ultraviolet/visible (UV/Vis) laser, charge–charge recombination, or high-energy ion–molecule collision. In these cases, sufficient energy is usually delivered to ions in a single step to cleave chemical bonds before energy redistributes over the entire ion. Slow or sequential ion activations are normally achieved with multiple IR laser photons or low-energy ion–molecule collisions to allow for the accumulation of internal energy in a stepwise manner [30–32]. Slow activation normally takes hundreds of microseconds or even seconds; this timescale is long enough to let the accumulated energy to redistribute over the entire ion before fragmentation.

3.3.2.1 Slow Activation

Slow activation of internal energy can be accomplished by increasing the velocity of ions to collide with a buffer gas (i.e., helium) [31, 33]. The energy of every collision in slow activation does not exceed 10 eV in the laboratory coordinate. During multiple collisions, the internal energy of the colliding ions gradually increases. Indeed, ion–molecule collisions can be described as a heating process that leads to the conversion of translational energy to the vibrational energies of collision partners. The center-of-mass energy (E_{COM}), or the translational energy that can be converted to internal energy, is the translational energy in the laboratory coordinate (E_{Lab}) multiplied by a proportionality factor [34–36]:

$$E_{COM} = \frac{M_c}{M_i + M_C} E_{Lab} \tag{3.41}$$

in which M_i and M_c are the masses of the ion and the buffer gas, respectively. Upon collision, the collision partners (ion and buffer gas) rapidly transfer their translational and vibrational energies until they are completely thermalized. When the internal energy is high enough to surmount the dissociation energy barrier, a fragmentation reaction occurs. This method is known as *collision-induced dissociation* (CID), or more precisely, the *low-energy CID*. Due to its simplicity and high efficiency, it is the most widely used method of fragmentation [32, 37]. However, if the ion is not continuously excited translationally by electric field during ion–molecule collisions, internal energy may be reduced in a subsequent collision because most of the buffer gas is at a lower temperature than the ion. Therefore, fragmentation is affected by competition between heating and cooling processes.

Activation of the vibrational energy of ions can also be induced by the absorption of IR radiations. A popular type of IR radiation source is far-IR laser. In fact, many molecules have a broad IR absorption band. The most widely used IR source is a continuous wave (c.w.) CO_2 laser, with the wavelength of 10.6 µm. This wavelength corresponds to an energy of 0.3 eV per laser photon. Because decomposition of a chemical bond requires >1 eV, laser excitation has to extended over hundreds of milliseconds to allow ions to absorb multiple IR photons. This method is known as *infrared multiphoton dissociation* (IRMPD). Another type of similar technique is *black-body infrared radiative dissociation*

(BIRD), which utilizes IR radiation from a heated vacuum chamber to excite trapped ions [38–40]. The BIRD method normally requires 1–20 s to dissociate ions.

Because IR heating and CID processes are slow enough to allow vibrational energy to be redistributed among all the vibrational degrees of freedom of ions, fragmentation would normally affect the weakest chemical bonds of the ions. In the case of proteins and peptides, the weakest bonds are amide bonds along the backbone chain [37]. Here, the fragmentation process leads to the production of the so-called b and y ions that correspond to fragments of N- and C-terminals, respectively.

3.3.2.2   Rapid Activation

It is possible to excite ions instantaneously and induce fragmentation before the energy can be redistributed, or when it is only distributed over a small portion of an ion. This option is in contrast to the slow activation processes described above. Such fast heating processes can be achieved by a single high-energy collision between an ion and its collision partner, irradiation of ions with a UV/Vis laser [41, 42], or by means of charge-neutralization reactions [43, 44].

A single high-energy collision can be performed by increasing the kinetic energy of ions by hundreds of or even a few thousand electronvolts using a strong electric field to accelerate ions before collision. This process is known as *high-energy CID* when the collision energy is above 1 keV. Another way of executing high-energy collisions is to induce dissociation of ions by colliding them against a surface. This technique is known as *surface-induced dissociation* (SID) [45–48]. According to Equation 3.41, E_{COM} in SID is increased dramatically because M_c approaches infinity, which makes the value of the proportionality factor 1. However, interactions of ions with a surface can cause loss of sensitivity because the precursor ions as well as the fragmentation products may not be able to desorb efficiently from the collision surface [49].

Notably, fragmentation induced by high-energy CID or SID still follows the same mechanism as low-energy CID. That is, high-energy CID and SID are basically the same as low-energy CID, except the collision energy in high-energy CID and SID is much higher [50, 51]. Consequently, fragmentation patterns obtained using these methods are – in most cases – similar to that of low-energy CID. However, in some cases, high-energy CID produces distinct fragmentation patterns from low-energy CID and may provide complementary structural information [52]. Most commercial mass spectrometers can only conduct one or other of the low- and high-energy CID methods due to limitations in instrument design. The only exception is the FT-ICR mass spectrometer because ions can be excited from a few to thousands of electronvolts in an ICR cell, but only low-energy CID is widely used.

In contrast to the IR lasers used in the slow activation process, UV/Vis lasers are capable of activating ions instantaneously. UV/Vis photons with the energy in the range 3–10 eV can excite the electronic energy of ions. The excess electronic energy can be converted to vibrational energy via internal conversion processes before fragmentation takes place. However, electronic excitation is highly selective to the wavelength of the incident photons. Therefore, it is important to select the proper photon source for individual precursor ions. It is also difficult to excite ions with two UV/Vis laser photons because the absorption cross sections of ions or molecules for the second photons are normally much smaller than those for the first photon.

Charge neutralization is a rapid and efficient ion activation method. It is exclusively applicable to multiply charged ions, especially those produced by ESI; charge neutralization to singly charged ions produces neutral species, which cannot be analyzed by a mass spectrometer. In order to perform charge neutralization reactions, charged particles of opposite polarity to the target ions are introduced into the region where the ions are held, normally an ion trap. Charge neutralization occurs when part of the charges of the target ions are neutralized with the charged particles. This neutralization reaction immediately releases energy, and induces the bond-breaking process before energy can be redistributed over the entire ion. The first method falling into this category is called *electron capture dissociation* (ECD), in which multiply charged positive ions react with free electrons inside an ICR cell [43, 53]. The ECD method is generally performed in an ICR cell because it can confine electrons at its center to overlap their trajectory with ions to facilitate neutralization reactions. Although ECD has been demonstrated to be feasible in IT using a permanent magnet near the trapping region to optimize electron energy and trajectory, such a design is not available commercially [54, 55]. Because ECD produces complementary fragmentation patterns to CID, it has become an important technique in proteomic analysis. A similar technique as ECD is now available in conventional ion trapping devices by replacing the electrons with negative ions. Such a derivative of the ECD method has been named *electron transfer dissociation* (ETD). In this case, electrons are transferred from carrier molecules to positive precursor ions within the IT [44].

References

1. Cotter, R. J. (1997) Time of Flight Mass Spectrometry: Instrumentation and Applications in Biological Research. American Chemical Society, Washington, DC.
2. Guilhaus, M. (1995) Principles and Instrumentation in Time-of-flight Mass-spectrometry – Physical and Instrumental Concepts. J. Mass Spectrom. 30: 1519–1532.
3. Stephens, W.E. (1946) A Pulsed Mass Spectrometer with Time Dispersion. Phys. Rev. 69: 691–691.
4. Cameron, A.E., Eggers, D.F. (1948) An Ion Velocitron. Rev. Sci. Instrum. 19: 605–607.
5. Wily, W.C., McLaren, I.H. (1955) Time-of-flight Mass Spectrometer with Improved Resolution. Rev. Sci. Instrum. 26: 1150–1157.
6. Mamyrin, B.A., Karataev, V.I., Shmikk, D.V., Zagulin, V.A. (1973) Mass-reflectron a New Nonmagnetic Time-of-flight High-resolution Mass-spectrometer. Zh. Eksp. Teor. Fiz. 64: 82–89.
7. Cornish, T.J., Cotter, R.J. (1993) A Curved-field Reflectron for Improved Energy Focusing of Product Ions in Time-of-flight Mass-spectrometry. Rapid Commun. Mass Spectrom. 7: 1037–1040.
8. Sears, F.W., Zemansky, M.W., Young, H.D. (1987) University Physics, 7th Edition. Addison-Wesley, Reading, MA.

9. Vestal, M.L., Juhasz, P., Martin, S.A. (1995) Delayed Extraction Matrix-assisted Laser-desorption TIme-of-flight Mass-spectrometry. Rapid Commun. Mass Spectrom. 9: 1044–1050.

10. Brown, R.S., Lennon, J.J. (1995) Mass Resolution Improvement by Incorporation of Pulsed Ion Extraction in a Matrix-assisted Laser-desorption Ionization Linear Time-of-flight Mass-spectrometer. Anal. Chem. 67: 1998–2003.

11. Gohl, W., Kutscher, R., Laue, H.J., Wollnik, H. (1983) Time-of-flight Mass-spectrometry for Ions of Large Energy Spread. Int. J. Mass Spectrom. 48: 411–414.

12. Wiza, J.L. (1979) Microchannel Plate Detectors. Nucl. Instrum. Meth. 162: 587–601.

13. Paul, W., Steinwedel, H. (1953) Ein neues Massenspektrometer ohne Magnetfeld. Z Naturforsch A 8: 448–450.

14. Paul, W. (1989) Electromagnetic Traps for Charged and Neutral Particles, http://www .nobelprize.org/nobel_prizes/physics/laureates/1989/paul-lecture.html (accessed March 29, 2015).

15. Paul, W., Steinwedel, H. (1956) Verfahren zur Trennung bzw zum getrennten nachweis von Ionen verschiedener spezifischer Ladung. German Patent DE 944900 C.

16. March, R.E., Todd, J.F.J. (2005) Quadrupole Ion Trap Mass Spectrometry, 2nd Edition. John Wiley & Sons, Inc., Hoboken.

17. Mathieu, P.M.E. (1868) Memoire sur le mouvement vibratoire d'une membrane de forme elliptique. J. Math. Pures Appl. 13: 137–203.

18. Gerlich, D. (1995) Ion-neutral Collisions in a 22-pole Trap at Very-low Energies. Phys. Scripta T59: 256–263.

19. Gerlich, D. (1992) Inhomogeneous RF Fields: A Versatile Tool for the Study of Processes with Slow Ions. In: Ng, C.Y., Baer, M. (eds) Advances in Chemical Physics, Volume 82, State-selected and State-to-state Ion-molecule Reaction Dynamics, Part I, Experiment. John Wiley & Sons, Inc., New York, pp. 31–52.

20. Dolnikowski, G.G., Kristo, M.J., Enke, C.G., Watson, J.T. (1988) Ion-trapping technique for ion molecule reaction studies in the center quadrupole of a triple Quadrupole Mass-spectrometer. Int. J. Mass Spectrom. Ion Proc. 82: 1–15.

21. Todd, J.F.J., Waldren, R.M., Mather, R.E. (1980) The Quadrupole Ion Store (Quistor). 9. Space-charge and Ion Stability – A Theoretical Background and Experimental Results. Int. J. Mass Spectrom. Ion Proc. 34: 325–349.

22. Todd, J.F.J., Waldren, R.M., Freer, D.A., Turner, R.B. (1980) The Quadrupole Ion Store (Quistor). 10. Space-charge and Ion Stability. B. On the Theoretical Distribution and Density of Stored Charge in RF Quadrupole Fields. Int. J. Mass Spectrom. Ion Proc. 35: 107–150.

23. Marshall, A., Hendrickson, C., Jackson, G. (1998) Fourier Transform Ion Cyclotron Resonance Mass Spectrometry: A Primer. Mass Spectrom. Rev. 17: 1–35.

24. Comisarow, M.B., Marshall, A.G. (1974) Fourier-transform Ion-cyclotron Resonance Spectroscopy. Chem. Phys. Lett. 25: 282–283.

25. Makarov, A. (2000) Electrostatic Axially Harmonic Orbital Trapping: A High-performance Technique of Mass Analysis. Anal. Chem. 72: 1156–1162.

26. Marshall, A.G., Verdun, F.R. (1990) Fourier Transforms in NMR, Optical, and Mass Spectrometry: A User's Handbook. Elsevier, Amsterdam.

27. Guan, S.H., Marshall, A.G. (1995) Ion Traps for Fourier-transform Ion-cyclotron Resonance Mass-spectrometry – Principles and Design of Geometric and Electric Configurations. Int. J. Mass Spectrom. 146: 261–296.
28. Dey, M., Castoro, J.A., Wilkins, C.L. (1995) Determination of Molecular-weight Distributions of Polymers by MALDI-ftms. Anal. Chem. 67: 1575–1579.
29. OConnor, P.B., Duursma, M.C., vanRooij, G.J., Heeren, R.M.A., Boon, J.J. (1997) Correction of Time-of-flight Shifted Polymeric Molecular Weight Distributions in Matrix Assisted Laser Desorption/Ionization Fourier Transform Mass Spectrometry. Anal. Chem. 69: 2751–2755.
30. Dunbar, R.C. (1994) Kinetics of thermal Unimolecular Dissociation by Ambient Infrared Radiation. J. Phys. Chem. 98: 8705–8712.
31. Shukla, A.K., Futrell, J.H. (2000) Tandem Mass Spectrometry: Dissociation of Ions by Collisional Activation. J. Mass Spectrom. 35: 1069–1090.
32. McLuckey, S.A., Goeringer, D.E. (1997) Slow Heating Methods in Tandem Mass Spectrometry. J. Mass Spectrom. 32: 461–474.
33. Laskin, J., Futrell, J.H. (2003) Collisional Activation of Peptide Ions in FT-ICR Mass Spectrometry. Mass Spectrom. Rev. 22: 158–181.
34. Gross, J.H. (2004) Mass Spectrometry: A Textbook, 1st Edition. Springer, Berlin.
35. Hoffmann, E.D., Charette, J.J., Stroobant, V. (1996) Mass Spectrometry : Principles and Applications. John Wiley & Sons, Ltd, Chichester.
36. Chapman, J.R. (1993) Practical Organic Mass Spectrometry: A Guide for Chemical and Biochemical Analysis, 2nd Edition. John Wiley & Sons, Ltd, Chichester.
37. Aebersold, R., Goodlett, D.R. (2001) Mass Spectrometry in Proteomics. Chem. Rev. 101: 269–295.
38. Schnier, P.D., Price, W.D., Jockusch, R.A., Williams, E.R. (1996) Blackbody Infrared Radiative Dissociation of Bradykinin and its Analogues: Energetics, Dynamics, and Evidence for Salt-bridge Structures in the Gas Phase. J. Am. Chem. Soc. 118: 7178–7189.
39. Price, W.D., Schnier, P.D., Williams, E.R. (1996) Tandem Mass Spectrometry of Large Biomolecule Ions by Blackbody Infrared Radiative Dissociation. Anal. Chem. 68: 859–866.
40. Dunbar, R.C., McMahon, T.B. (1998) Activation of Unimolecular Reactions by Ambient Blackbody Radiation. Science 279: 194–197.
41. Dunbar, R.C. (1987) Time-resolved Photodissociation of Chlorobenzene Ion in the ICR Spectrometer. J. Phys. Chem. 91: 2801–2804.
42. Dunbar, R.C. (2000) Photodissociation of Trapped Ions. Int. J. Mass Spectrom. 200: 571–589.
43. Zubarev, R.A., Kelleher, N.L., McLafferty, F.W. (1998) Electron Capture Dissociation of Multiply Charged Protein Cations. A Nonergodic Process. J. Am. Chem. Soc. 120: 3265–3266.
44. Syka, J.E.P., Coon, J.J., Schroeder, M.J., Shabanowitz, J., Hunt, D.F. (2004) Peptide and Protein Sequence Analysis by Electron Transfer Dissociation Mass Spectrometry. Proc. Natl. Acad. Sci. USA 101: 9528–9533.
45. Wysocki, V.H., Joyce, K.E., Jones, C.M., Beardsley, R.L. (2008) Surface-induced Dissociation of Small Molecules, Peptides, and Non-covalent Protein Complexes. J. Am. Soc. Mass Spectrom. 19: 190–208.

46. Dongre, A.R., Somogyi, A., Wysocki, V.H. (1996) Surface-induced Dissociation: An Effective Tool to Probe Structure, Energetics and Fragmentation Mechanisms of Protonated Peptides. J. Mass Spectrom. 31: 339–350.
47. Reimann, C.T., Quist, A.P., Kopniczky, J., Sundqvist, B.U.R., Erlandsson, R., Tengvall, P. (1994) Impacts of Polyatomic Ions on Surfaces – Conformation and Degree of Fragmentation of Molecular-ions Determined by Lateral Dimensions of Impact Features. Nucl. Instrum. Meth. B 88: 29–34.
48. Cooks, R.G., Ast, T., Mabud, A. (1990) Collisions of Polyatomic Ions with Surfaces. Int. J. Mass Spectrom. Ion Proc. 100: 209–265.
49. Grill, V., Shen, J., Evans, C., Cooks, R.G. (2001) Collisions of Ions with Surfaces at Chemically Relevant Energies: Instrumentation and Phenomena. Rev. Sci. Instrum. 72: 3149–3179.
50. Laskin, J., Denisov, E., Futrell, J. (2001) Comparative Study of Collision-induced and Surface-induced Dissociation. 2. Fragmentation of Small Alanine-containing Peptides in FT-ICR MS. J. Phys. Chem. B 105: 1895–1900.
51. Laskin, J., Denisov, E., Futrell, J. (2000) A Comparative Study of Collision-induced and Surface-induced Dissociation. 1. Fragmentation of Protonated Dialanine. J. Am. Chem. Soc. 122: 9703–9714.
52. Laskin, J., Futrell, J.H. (2003) Surface-induced Dissociation of Peptide Ions: Kinetics and Dynamics. J. Am. Soc. Mass Spectrom. 14: 1340–1347.
53. Zubarev, R.A., Haselmann, K.F., Budnik, B., Kjeldsen, F., Jensen, F. (2002) Towards an Understanding of the Mechanism of Electron-capture Dissociation: A Historical Perspective and Modern Ideas. Eur. J. Mass Spectrom. 8: 337–349.
54. Satake, H., Hasegawa, H., Hirabayashi, A., Hashimoto, Y., Baba, T. (2007) Fast Multiple Electron Capture Dissociation in a Linear Radio Frequency Quadrupole Ion Trap. Anal. Chem. 79: 8755–8761.
55. Baba, T., Hashimoto, Y., Hasegawa, H., Hirabayashi, A., Waki, I. (2004) Electron Capture Dissociation in a Radio Frequency Ion Trap. Anal. Chem. 76: 4263–4266.

4

Interfaces for Time-resolved Mass Spectrometry

4.1 Molecules in Motion

The development of photography techniques in the 19th century opened the door to brand-new possibilities. Using high-speed cameras, capturing motion by taking snapshots of real events at a high speed became feasible (Figure 4.1). The mass spectrometer is a chemist's "camera" that enables chemical "objects" (i.e., ions) with recognizable features (i.e., mass-to-charge ratios) to be captured. It also allows chemists to take "snapshots" of these "objects" with a high speed. Cameras are equipped with lenses and light-sensitive films (or electronic chips) while mass spectrometers are equipped with ion optics, mass analyzers, and detectors. Cameras record fluxes of photons with the aid of transparent optic elements; on the other hand, mass spectrometers combine ion sources with analyzers – utilizing electric and/or magnetic fields – to record fluxes of ions. Both devices are capable of recording information while preserving temporal resolution.

Analyzing labile solutions by mass spectrometry (MS) is not always a trivial task. Thus, in this chapter we will discuss development of interfaces used to introduce dynamic samples to mass spectrometers for analysis with temporal resolutions that are adequate for the studied processes. Interfacing is vital for the operation of time-resolved mass spectrometry (TRMS), and it influences the quality of analysis output. It also assures sufficient detection sensitivity. The choice of interface is dictated by the type of dynamic sample studied and the scientific question put forward by the analysts. Various technological advances have been made in this area to enable kinetic profiling of dynamic systems (for an overview, see also [1]). The interface is often integrated with the ion source, and usually it is hard to make a clear distinction between the two components. For example, desorption of an open stream of reaction mixture allowed one to probe minute quantities of reactants with high speed [2, 3]. In this case, the reaction took place in close proximity to the ionization zone. However, in other studies, the reactors were separated from the ionization stage with flow lines and manifolds. That option gives more flexibility for the spatial arrangement of

Time-Resolved Mass Spectrometry: From Concept to Applications, First Edition.
Pawel Lukasz Urban, Yu-Chie Chen and Yi-Sheng Wang.
© 2016 John Wiley & Sons, Ltd. Published 2016 by John Wiley & Sons, Ltd.

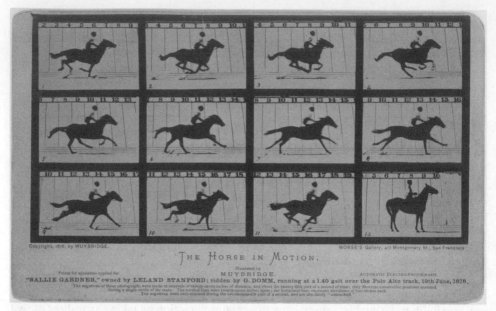

Figure 4.1 The Horse in Motion. "Sallie Gardner", owned by Leland Stanford; running at a 1:40 gait over the Palo Alto track, June 19, 1878/Muybridge. Reproduced from the website of the Library of Congress (http://www.loc.gov/pictures/item/97502309/). (No known restrictions on publication)

the reaction and monitoring systems; however, it also makes the instrumental setup larger, and – importantly – increases the likelihood of sample transfer-related artifacts and the inherent loss of temporal resolution. It is helpful to list the attributes of an ideal interface for TRMS. The interface should:

- assure fast and efficient transfer of sample aliquots to the ion source;
- ascertain minimum delay between sampling and analysis;
- minimize carry-over artifacts.

The small delay time requirement is especially important in the case of rapidly changing dynamic samples. It may be waived in the case of some single time point measurements (i.e., while recording quasi-steady samples). In fact, some studies, which target qualitative detection of short-lived species rather than temporal profiles of reactants of intermediates, impose less rigorous requirements for the experimental systems. The following sections provide examples of various TRMS systems emphasizing the role of interfaces.

Notably, numerous atmospheric pressure ionization/sampling techniques have been introduced over the past few years [4]. Some of these techniques are regarded as "ambient" to emphasize little or no sample preparation, and indicate the possibility of carrying out analysis at atmospheric pressure (cf. [5–8]). Although they have some advantages for research and development, many of them are not as robust and well-characterized (e.g., regarding quantitative capabilities) as the classical approaches used in the process and routine analyses (cf. [9]).

Examples of reports demonstrating development or applications of various TRMS approaches are listed in Table 4.1. The most common interfaces used in TRMS incorporate the following ion sources: electrospray ionization (ESI), desorption electrospray ionization

Table 4.1 *Selected examples of studies using different ion sources/interfaces that fall under the broad definition of time-resolved mass spectrometry. Please refer to the text in this and the following chapters for discussion of representative interfaces and their applications. A large portion of the examples refer to the early-stage demonstrations of the newly developed techniques; however, several examples of mechanistic and routine analyses have also been included*

Ion source/ interface	Studied process/purpose	Ref.
ESI	Real-time reaction monitoring	Lee *et al.* [81]
ESI	Hydrogen/deuterium exchange, study of conformational changes of proteins	Katta and Chait [213]
ESI	Study of protein conformation, hydrogen/deuterium exchange	Winger *et al.* [214]
ESI	Hydrogen/deuterium exchange, study of conformational changes of proteins	Katta and Chait [215]
ESI	Monitoring protein folding intermediates, hydrogen/deuterium exchange	Miranker *et al.* [216]
ESI	Study of protein conformation, hydrogen/deuterium exchange	Suckau *et al.* [217]
ESI	Study of phosphine-mediated reactions	Wilson *et al.* [49]
ESI	Detection of intermediates in the Suzuki reaction	Aliprantis and Canary [50]
ESI	Study of a protein complex	Sam *et al.* [218]
ESI	Study of a photochemical reaction	Arakawa *et al.* [84]
ESI	Monitoring photosubstitution of ruthenium(II) diimine complexes	Arakawa *et al.* [219]
ESI	Monitoring electrochemical processes	Bond *et al.* [167]
ESI	Study of protein folding	Hooke *et al.* [220]
ESI (LC-MS)	Kinetic study of enzymatic reactions	Hsieh *et al.* [58]
ESI	Monitoring the reaction of Fe(III)-bleomycin with iodosylbenzene	Sam *et al.* [51
ESI	Monitoring electrochemical processes	Zhou and Van Berkel [166]
ESI	Monitoring photolysis of (diamine)bis(2,2′-bipyridine)ruthenium(II) complexes	Arakawa *et al.* [221]
ESI	Study of protein folding (lysozyme)	Gross *et al.* [222]
ESI	Mechanistic study of intramolecular arylation of enamidines	Ripa and Hallberg [223]
ESI	Monitoring photocycloaddition of cyclic enones to C_{60}	Schuster *et al.* [224]
ESI	Monitoring photoallylation reactions of dicyanobenzenes by allylic silanes via photoinduced electron transfer	Arakawa *et al.* [225]
ESI	Monitoring photo-oxidation, isotopic labeling	Arakawa *et al.* [226]
ESI	Study of C-H activation by cationic iridium(III) complexes	Hinderling *et al.* [227]

(*continued overleaf*)

Table 4.1   (continued)

Ion source/ interface	Studied process/purpose	Ref.
ESI	Observation of photochemical switching	Kimura et al. [228]
ESI	Study of acid-induced denaturation of proteins	Konermann et al. [229]
ESI	Study of folding kinetics of proteins	Konermann et al. [230]
ESI	Characterization of enzymatic reaction intermediates	Paiva et al. [87]
ESI (LC-MS)	Study of protein folding (cytochrome c), hydrogen/deuterium exchange	Yang and Smith [231]
ESI, APCI	Monitoring photolysis of idoxifene	Brum and Dell'Orco [232]
ESI	Mechanistic study of Wittig reaction, detection of unstable intermediate	Wang et al. [233]
ESI	Pre-steady state kinetics of an enzymatic reaction	Zechel et al. [234]
ESI	Mechanistic study of oxidative self-coupling of areneboronic acids	Aramendía and Lafont [235]
ESI	Monitoring process-scale reactions	Dell'Orco et al. [236]
ESI	Study of gas-phase hydrogen/deuterium exchange (ubiquitin)	Freitas et al. [237]
ESI	Mechanistic study of palladium complex-catalyzed enantioselective Mannich-type reaction	Fujii et al. [238]
ESI	Screening of catalysts	Hinderling and Chen [239]
ESI	Ion–molecule reactions in an external ion reservoir	Hofstadler et al. [240]
ESI	Protein folding study, reconstitution of holomyoglobin	Lee et al. [241]
ESI	Study of protein folding (apomyoglobin), quench-flow, hydrogen/deuterium exchange	Tsui et al. [242]
ESI	Mechanistic study of Mn(salen)-catalyzed Jacobsen–Katsuki epoxidation	Adam et al. [243]
ESI	Study of thermally initiated reactions	Griep-Raming and Metzger [85]
ESI	Kinetic characterization of polymerization catalysts	Hinderling and Chen [244]
ESI	Detection of highly reactive ketenes from a flow pyrolyzer	Hong et al. [245]
ESI	Monitoring kinetics of chemical reaction, stopped-flow mixing	Kolakowski et al. [79]
ESI	Monitoring ozonation process	Kotiaho et al. [246]
ESI	Study of protein folding pathway	Nishimura et al. [247]
ESI	Study of myoglobin unfolding mechanism	Sogbein et al. [248]

Table 4.1 *(continued)*

Ion source/ interface	Studied process/purpose	Ref.
ESI	Study of gas-phase catalytic cycle for the oxidation of alcohols	Waters *et al.* [249]
ESI	Monitoring structural changes of proteins in ion trap	Badman *et al.* [250]
ESI	Monitoring reactions, hydrolysis of isatin	Brum *et al.* [82]
ESI	Identification of photogenerated intermediates	Ding *et al.* [251]
ESI	Study of oxomanganese-salen complexes	Feichtinger and Plattner [252]
ESI	Study of enzyme kinetics, ion trap	Ge *et al.* [253]
ESI	Continuous-flow analytical screening system	Hogenboom *et al.* [142]
ESI	Study of the kinetics of biochemical processes, stopped-flow	Kolakowski and Konermann [80]
ESI	Study of enzyme kinetics, inhibition, multiple reaction monitoring	Norris *et al.* [52]
ESI	Study of folding/unfolding (ubiquitin), electron capture dissociation	Breuker *et al.* [254]
ESI	Investigation of intermediates in radical chain reactions	Griep-Raming *et al.* [255]
ESI	Hydrogen/deuterium exchange, folding kinetics, continuous flow	Simmons and Konermann [256]
ESI (LC-MS)	Study of protein folding, hydrogen/deuterium exchange	Zhang *et al.* [257]
ESI	Mechanistic study of photochemical reactions, millisecond resolution	Ding *et al.* [258]
ESI	Mechanistic study of C-H activation by $[(N-N)Pt(CH_3)(L)]^+$	Gerdes and Chen [259]
ESI	Monitoring enzymatic reaction in the millisecond timescale	Li *et al.* [260]
ESI	Mechanistic study of Heck reaction	Masllorens *et al.* [261]
ESI	Study of reactive intermediates in radical cation chain reactions	Meyer *et al.* [262]
ESI	Monitoring folding of apomyoglobin, quench flow, hydrogen exchange	Nishimura *et al.* [263]
ESI	Protein folding study, myoglobin, hydrogen/deuterium exchange	Simmons *et al.* [264]
ESI	Continuous-flow apparatus for on-line kinetic studies	Wilson and Konermann [95]
ESI	Mechanistic study of allylic substitution	Chevrin *et al.* [265]
ESI	Mechanistic study of Michael additions	Comelles *et al.* [266]
ESI	Gas-phase reactions of organometallic ions	Dietiker and Chen [267]
ESI	Mechanistic study of nucleophilic substitution reactions	Domingos *et al.* [268]

(continued overleaf)

Table 4.1   (continued)

Ion source/ interface	Studied process/purpose	Ref.
ESI	Mechanistic study of the reaction of bis(2,4-dinitrophenyl) phosphate with hydrazine and hydrogen peroxide	Domingos et al. [269]
ESI	Hydrogen exchange MS to study protein dynamics and folding	Eyles and Kaltashov [270]
ESI	Monitoring electron transfer-initiated Diels–Alder reactions	Fürmeier and Metzger [89]
ESI, GC-MS	Monitoring catalytic reactions	Gerdes and Chen [271]
ESI	Monitoring catalytic intermediates	Markert and Pfaltz [272]
ESI	Mechanistic study of the oxidation of N,N-dimethyl-p-phenylenediamine	Modestov et al. [273]
ESI	Determination of enzyme/substrate specificity, kinetic evaluation	Pi and Leary [274]
ESI	Kinetic and mechanistic study of enzymatic reaction	Pi et al. [275]
ESI	Mechanistic study on coupling vinylic tellurides with alkynes (targeting Pd- and Te-containing cationic intermediates)	Raminelli et al. [276]
ESI	Mechanistic study of Heck reaction	Sabino et al. [277]
ESI	Mechanistic study of Baylis–Hillman reaction	Santos et al. [278]
ESI	Kinetics of enzyme reactions	Wilson and Konermann [279]
ESI	Monitoring enzymatic reaction, study of intermediates	Yu et al. [280]
ESI	Mechanistic study of synthesis of conformationally restricted analogues of γ-amino butyric acid	Ferraz et al. [281]
ESI	Mechanistic study of radical chain reactions	Fürmeier et al. [55]
ESI	Mechanistic and kinetic study of phosphomannose isomerase	Gao et al. [282]
ESI	Monitoring electrochemical processes	Kertesz and Van Berkel [159]
ESI	Kinetic and mechanistic study of enzymatic reaction	Pi et al. [283]
ESI	Mechanistic study of addition of organoboronic acids to allenes	Qian et al. [284]
ESI	Mechanistic study of coordination of iron(III) cations to β-keto esters	Trage et al. [285]
ESI	Monitoring enzymatic DNA hydrolysis	van den Heuvel et al. [286]
ESI	Monitoring protein disassembly and unfolding	Wilson et al. [287]
ESI	Detection of intermediates in Heck reaction	Enquist et al. [288]

Table 4.1   *(continued)*

Ion source/ interface	Studied process/purpose	Ref.
ESI	Detection of intermediates in aldol reaction	Marquez and Metzger [93]
ESI	Hydrogen/deuterium exchange, folding kinetics	Pan *et al.* [289]
ESI	Mechanistic study of homogeneously catalyzed Ziegler–Natta polymerization of ethene	Santos and Metzger [92]
ESI	Mechanistic study of Tröger's base formation	Abella *et al.* [290]
ESI	Detection of products/intermediates in microdroplets exposed to a reactive gas	Enami *et al.* [86]
ESI	Study of protein folding, pulsed thiol labeling (methyl methanethiosulfonate)	Jha and Udgaonkar [291]
ESI	Mechanistic study of organocatalytic α-halogenation of aldehydes	Marquez *et al.* [292]
ESI	Monitoring Mannich-type α-methylenation of ketoesters, mechanistic study	Milagre *et al.* [293]
ESI	Monitoring Michael–Michael–retro Michael addition	Wu *et al.* [294]
ESI	Study of the evolution of protein native structure in the gas phase	Breuker and McLafferty [295]
ESI	Study of protein folding, hydrogen/deuterium exchange, electron capture dissociation	Pan *et al.* [296]
ESI	Monitoring Brookhart polymerization	Santos and Metzger [297]
ESI	Mechanistic study of ruthenium olefin metathesis	Wang and Metzger [298]
ESI (LC-MS)	Photochemical oxidation of proteins with hydroxyl radical	Gau *et al.* [299]
ESI	Monitoring *Arthrobacter* 4-hydroxybenzoyl-coenzyme A thioesterase reaction in the millisecond time range	Li *et al.* [300]
ESI (LC-MS)	Trapping unstable metabolites	Motwani *et al.* [201]
ESI	Hydrogen/deuterium exchange, electron capture dissociation used to obtain more structural information on proteins	Pan *et al.* [301]
ESI	Hydrogen/deuterium exchange in the ion mobility instrument	Rand *et al.* [302]
ESI	Microreactor for millisecond timescale kinetics	Rob and Wilson [303]
ESI (LC-MS)	Detection of unstable reactive drug metabolites	Rousu *et al.* [202]
ESI	Liquid microjunction surface sampling probe, tagging	Van Berkel and Kertesz [122]
ESI	Study of catalytic activity in metathesis reactions	Wang *et al.* [90]
ESI	Monitoring motion of crown ethers along oligolysine peptide chains, gas-phase hydrogen/deuterium exchange	Weimann *et al.* [304]
ESI	Monitoring hydroformylation intermediates	Beierlein and Breit [305]

(continued overleaf)

Table 4.1   *(continued)*

Ion source/ interface	Studied process/purpose	Ref.
ESI	Study of enzyme kinetics and mechanism, quench-flow	Clarke *et al.* [306]
ESI (LC-MS)	Photochemical oxidation of proteins with sulfate radical anion	Gau *et al.* [307]
ESI	Monitoring molecular "self-sorting"	Jiang *et al.* [91]
ESI (LC-MS)	Detection of unstable reactive metabolites	LeBlanc *et al.* [203]
ESI	Microfluidic reactor for rapid (<4 s) digestion of proteins	Liuni *et al.* [308]
ESI	Hydrogen/deuterium exchange, top-down, 10 ms response	Pan *et al.* [309]
ESI	Recording thermal desorption profiles	Reynolds *et al.* [310]
ESI	Monitoring electrochemical oxidation of peptides	Roeser *et al.* [168]
ESI	Monitoring enzyme intermediate in the millisecond timescale	Roberts *et al.* [311]
ESI	Monitoring cross-coupling reactions	Schade *et al.* [312]
ESI	Monitoring air-/moisture-sensitive reactions, pressurized sample	Vikse *et al.* [83]
ESI	Monitoring of reactions taking place in microdroplets	Zhu and Fang [313]
ESI	Monitoring biochemical reactions	Bujara *et al.* [314]
ESI	Monitoring monomer exchange in dimeric protein	Chevreux *et al.* [315]
ESI	Charge-tagged acetate ligands for mechanistic studies, Suzuki and Heck phosphine-free reactions	Oliveira *et al.* [316]
ESI	Study of mechanism of *aza*-Morita–Baylis–Hillman reaction	Regiani *et al.* [317]
ESI (LC-MS)	Detection of unstable reactive drug metabolites	Rousu and Tolonen [205]
ESI	Mechanistic study of Sandmeyer's cyclization	Silva *et al.* [318]
ESI	Monitoring organic reactions, catalytic mechanism	Vikse *et al.* [77]
ESI	Monitoring self-assembly of hybrid polyoxometalates	Wilson *et al.* [319]
ESI	Monitoring cyclopropenium-activated chlorination reaction of alcohols	Zhao *et al.* [320]
ESI, (EC/LC/ESI-MS)	Conjugation and quantification of reactive metabolites	Jahn *et al.* [172]
ESI	Study of conformational dynamics of an enzyme, sub-second hydrogen/deuterium exchange	Liuni *et al.* [321]
ESI	Study of hydrogen/deuterium exchange in the sub-second regime	Rob *et al.* [322]
ESI	Study of free radicals by spin-trapping	Simões *et al.* [57]

Table 4.1   *(continued)*

Ion source/ interface	Studied process/purpose	Ref.
ESI	Transferring samples collected in mouse brain to the ESI emitter	Song *et al.* [323]
ESI (nanoLC-MS)	Time-resolved phosphoproteomic study	Verano-Braga *et al.* [324]
ESI	Monitoring hydrodehalogenation process	Ahmadi and McIndoe [325]
ESI	Reaction monitoring for drug synthesis	Roscioli *et al.* [326]
ESI, LC	Monitoring sub-millisecond protein folding, rapid mixing, oxidative labeling	Vahidi *et al.* [327]
ESI	In-spray solution mixing using theta capillaries	Fisher *et al.* [97]
ESI	Monitoring chemical reactions, gravitational sampling	Hsu *et al.* [328]
ESI	Fast screening of complex mixtures for ion-molecule reactions	Jarrell *et al.* [329]
ESI	Wire-in-a-capillary electrospray emitters prepared from theta-glass capillaries	Mortensen and Williams [98]
ESI	Monitoring Heck reaction, mechanistic study	dos Santos *et al.* [330]
ESI	Reaction monitoring and mechanistic studies (Negishi cross-coupling, hydrogenolysis, reductive amination)	Yan *et al.* [331]
ESI	Continuous-flow mixing apparatus for real-time MS	Zinck *et al.* [96]
ESI	Accelerated Hantzsch synthesis	Bain *et al.* [332]
ESI	Monitoring Soxhlet extraction	Chen and Urban [139]
ESI	Conformation studies, hydrogen/deuterium exchange, ion mobility	Donohoe *et al.* [333]
ESI	Detection of reaction intermediates in organometallic catalysis	Limberger *et al.* [334]
ESI	Investigation of protein folding using theta-glass capillaries	Mortensen and Williams [99]
ESI	Hydrogen/deuterium exchange, rapid monitoring of the exchange kinetics, gas phase	Rajabi [335]
ESI	Study of kinetic intermediates of holo- and apo-myoglobin, hydrogen/deuterium exchange	Schenk *et al.* [336]
ESI (LC-MS)	Kinetic study of enzymatic reaction, inhibition	Simithy *et al.* [337]
nanoESI	Monitoring assembly of a protein complex	Fändrich *et al.* [338]
nanoESI	Monitoring an organic reaction	Brivio *et al.* [339]
nanoESI	Study of the dynamics of protein complexes	Painter *et al.* [340]
nanoESI	Monitoring complexes of small heat shock proteins	Stengel *et al.* [341]
nanoESI (IM-MS)	Monitoring fibril assembly process	Smith *et al.* [342]

(continued overleaf)

Table 4.1   *(continued)*

Ion source/ interface	Studied process/purpose	Ref.
nanoESI	Organocatalysis, optimization of reactions	Fritzsche *et al.* [343]
nanoESI	Monitoring convection in liquid phase	Li *et al.* [344]
nanoESI	Monitoring cell culture	Olivero *et al.* [345]
Microspray	Monitoring chemical reactions with a "miniature" mass spectrometer	Browne *et al.* [346]
CSI	Characterization of unstable DNA species	Sakamoto and Yamaguchi [347]
DESI	"Reactive DESI"	Cotte-Rodríguez *et al.* [110]
DESI	"Reactive DESI"	Chen *et al.* [348]
DESI	"Reactive DESI", observation of intermediates of Schiff-base reaction	Zhang *et al.* [349]
DESI	Monitoring reactions with sub-millisecond time resolution	Miao *et al.* [2] Chen and Miao [3]
DESI	Probing reactions occurring in millisecond time intervals	Perry *et al.* [109]
DESI	Accelerated reaction in microdroplets	Girod *et al.* [116]
DESI	Detection of intermediates in Eschweiler–Clarke reaction	Xu *et al.* [350]
DESI	Detecting intermediates (catalytic C-H amination reaction cycle)	Perry *et al.* [351]
DESI	Mechanistic study of a heterogeneous catalytic reaction	Boeser *et al.* [352]
DESI	Identification of labile intermediates in an electrochemical reaction	Brown *et al.* [174]
nanoDESI	Monitoring electrochemical reactions	Liu *et al.* [175]
nanoDESI	Investigating molecular composition of mouse spinal cords	Hsu *et al.* [123]
Paper spray ionization	Monitoring cell culture	Liu *et al.* [353]
V-EASI	On-line reaction monitoring	Santos *et al.* [135]
V-EASI	Monitoring extraction in real time	Hu *et al.* [138]
V-EASI	Spatiotemporal characteristics of chemical wave	Ting and Urban [137]
V-EASI	Monitoring an enzymatic reaction (auxiliary method)	Hu *et al.* [136]
ESSI	Study of deprotonation reactions of peptide and protein ions	Touboul *et al.* [354]
ESSI	Atmospheric-pressure thermal activation for organic reactions	Chen *et al.* [124]
EESI	Detection of intermediate in electron-transfer-catalyzed dimerization of *trans*-anethole	Marquez *et al.* [100]
EESI	On-line reaction monitoring	Zhu *et al.* [126]
EESI	Monitoring conversion of fructose to 5-hydroxymethylfurfural	Law *et al.* [128]

Table 4.1   *(continued)*

Ion source/ interface	Studied process/purpose	Ref.
EESI	On-line reaction monitoring	McCullough *et al.* [127]
EESI	Kinetics of chemical reactions studied in the microsecond timescale	Lee *et al.* [101]
UASI	On-line monitoring of organic reactions	Chen *et al.* [132]
UASI	Monitoring accelerated chemical reactions	Lin *et al.* [133]
C-API	On-line monitoring of chemical reactions	Hsieh *et al.* [120]
Single-probe MS	Metabolomic analysis of individual living cells in real time	Pan *et al.* [355]
Inlet MS	Oxygen isotope discrimination of O_2-consuming reactions	Cheah *et al.* [356]
RP-MIMS	Evolution of methanol concentration	Creaser *et al.* [66]
MIMS, ESI	Monitoring biodegradation of 4-fluorobenzoic acid and 4-fluorocinnamic acid	Creaser *et al.* [357]
MIMS	Monitoring yeast metabolism	Roussel and Lloyd [358]
APCI, ESI	Mechanistic study of radical cation chain reactions, microreactor coupled on-line	Meyer *et al.* [262]
APCI, membrane	Monitoring of a concentrated pharmaceutical process reaction mixture	Clinton *et al.* [69]
APCI	Monitoring of organic reactions, flow injection analysis	Zhu *et al.* [150]
APCI	Monitoring environmental aerosols	Vogel *et al.* [182]
ESI, APCI, EI	Investigation of radical cation chain reactions	Meyer and Metzger [88]
DART	Monitoring reactions related to drug discovery	Petucci *et al.* [359]
DART, TLC	Monitoring reaction mixtures	Smith *et al.* [360]
DART	On-site monitoring of batch slurry reactions	Cho *et al.* [361]
DART	Monitoring solid phase organic synthesis	Sanchez *et al.* [362]
LTP	Real-time *in-situ* monitoring of chemical reactions	Ma *et al.* [151]
ICP	Electrothermal vaporizer for temporal separation of the isobars	Rowland *et al.* [146]
ICP	Sensitive immunoassay	Hu *et al.* [148]
Plasma jet	Study of the ionic species in the plume of atmospheric-pressure helium microplasma	Oh *et al.* [363]
Photoionization	Study of the kinetics of free radicals	Slagle *et al.* [364]
Photoionization	Study of the kinetics of free radicals	Slagle and Gutman [365]
Photoionization	Study of the kinetics of free radicals	Slagle *et al.* [366]
Photoionization	Study of dissociation dynamics of halotoluene ions	Olesik *et al.* [29]
Photoionization	Study of radical–radical reactions in the gas phase	Fockenberg *et al.* [367]

(continued overleaf)

Table 4.1   (continued)

Ion source/ interface	Studied process/purpose	Ref.
Photoionization	Determination of rate coefficients	Lee and Leone [19]
Photoionization	Study of low temperature rate coefficients	Lee et al. [18]
Photoionization	Determination of rate coefficients	Lee et al. [17]
Photoionization	Study of photodissociation of small peptide ions	Cui et al. [40]
Photoionization (tunable synchrotron vacuum ultraviolet photoionization)	Study of combustion, formation of polycyclic aromatic hydrocarbons	Qi et al. [368]
Photoionization	Monitoring reactions in the millisecond timescale, TOF instrument	Blitz et al. [23]
Photoionization	Study of the kinetics and isomeric product branching of gas-phase reactions	Osborn et al. [369]
Photoionization	Kinetic study of allyl radical self-reaction, synchrotron	Selby et al. [370]
Photoionization, synchrotron	Study of elementary reaction kinetics and the chemistry of low-pressure flames	Taatjes et al. [208]
Photoionization	Monitoring photolysis reactions	Baeza-Romero et al. [24]
Photoionization, synchrotron	Kinetic study of gas-phase radical reactions	Genbai et al. [371]
Photoionization	Study of a gas-phase reaction	Savee et al. [372]
Photoionization	Kinetic measurements of Criegee intermediate	Welz et al. [27]
Photoionization	Study of pressure-dependent laminar premixed flames	Zhou et al. [373]
Photoionization	Study of bimolecular radical reactions	Middaugh [374]
Photoionization, synchrotron	Monitoring combustion products	Lynch et al. [375]
SPI, CI	Monitoring organic compounds in disinfected water	Hua et al. [376]
Multiphoton ionization	Measurement of dissociation rates and branching ratios for naphthalene ion using ICR	Ho et al. [30]
Multiphoton ionization	Measurements of dissociation rates	Cui et al. [39]
Multiphoton ionization	Ionization and excitation of chloramines by femtosecond laser pulses	Rusteika et al. [377]
Non-resonant multiphoton ionization	Study of non-radiative dynamics in electronically excited hexafluorobenzene (femtosecond scale)	Studzinski et al. [378]

Table 4.1   *(continued)*

Ion source/ interface	Studied process/purpose	Ref.
Resonance-enhanced multiphoton ionization, single photon ionization, laser-induced electron-impact ionization	Monitoring complex gas mixtures	Mühlberger et al. [145]
Synchrotron (photoion-ization)	Study of flame chemistry	Cool et al. [379]
Synchrotron	Schottky MS of unstable nuclei	Franzke et al. [211]
Synchrotron	Isochronous MS of unstable nuclei	Hausmann et al. [212]
Synchrotron	Mass and lifetime measurements on newly produced exotic nuclei	Litvinov et al. [380]
ELISA	Time-resolved daughter ion MS	Støchkel et al. [41]
MALDI	Monitoring reaction at the resin surface	Fitzgerald et al. [381]
MALDI	Monitoring reactions on polymeric supports	Carrasco et al. [382]
MALDI	Monitoring enzymatic reaction, mix-quench	Gross et al. [46]
MALDI	Study of enzyme kinetics, quench-flow	Houston et al. [47]
MALDI	Pulsed alkylation, study of protein folding	Apuy et al. [383]
MALDI	Pulsed alkylation, study of protein complex stability	Apuy et al. [384]
MALDI	Ratiometric pulsed alkylation of a protein that interacts with toxic metals	Apuy et al. [385]
MALDI	Monitoring enzymatic reactions, immobilized reactants	Min et al. [386]
MALDI	Monitoring Schiff base formation	Brivio et al. [387]
MALDI	Off-line monitoring of reactions in segmented flow	Hatakeyama et al. [42]
MALDI	Monitoring enzymatic activity, screening enzyme inhibitors	Hu et al. [45]
MALDI	Pre-steady-state reaction kinetics, electrowetting-on-dielectric	Nichols and Gardeniers [388]
MALDI	Study of photodissociation of peptides	Yoon et al. [389]
MALDI	Mechanistic study of desorption dynamics	Minegishi et al. [390]

(continued overleaf)

Table 4.1   (continued)

Ion source/interface	Studied process/purpose	Ref.
MALDI, TLC	Monitoring chemical reactions	Chen *et al.* [391]
DIOS	Monitoring enzymatic reaction	Thomas *et al.* [44]
DIOS	Enzyme kinetics	Nichols *et al.* [392]
ELDI	Monitoring various chemical and biochemical reactions	Cheng *et al.* [143]
ELDI	"Reactive-ELDI"	Peng *et al.* [144]
PESI	Monitoring biological and chemical reactions in real time	Yu *et al.* [129]
PESI	Monitoring protease-catalyzed reactions in real time	Yu *et al.* [130]
Laser	Photodesorption of NH_3 adsorbed on a Cu(100)	Chuang and Hussla [393]
Laser	MS in the picosecond timescale	von der Linde and Danielzik [22]
Paper-assisted thermal ionization	Observation of reactive intermediates	Pei *et al.* [394]
FAB	Monitoring enzymatic reactions	Smith and Caprioli [395]
FAB	Determination of enzymatic reaction rates	Smith and Caprioli [396]
FAB	Monitoring conjugation reactions	Heidmann *et al.* [397]
Thermospray	Monitoring electrochemical processes	Hambitzer and Heitbaum [163]
Thermospray	Elucidation of electrochemical oxidation pathway	Volk *et al.* [164]
Thermospray	Elucidation of enzymatic and electrochemical oxidation pathways	Volk *et al.* [165]
EI	Observation of free radicals in decomposing gases	Lossing and Tickner [398]
EI	Observation of free radicals in decomposing gases	Lossing *et al.* [399]
EI	Flash photolysis	Meyer [13]
EI	Study of desorption of non-volatile organic salts following a laser pulse	Van Breemen *et al.* [21]
EI	*In-vivo* monitoring of blood, membrane probe	Brodbelt *et al.* [400]
EI	Monitoring dissociation of chlorobenzene by ICR in microseconds	Dunbar [317]
EI	Micropyrolysis; desorption, depolymerization, and thermal degradation observed in time	Chakravarty *et al.* [401]
EI	Monitoring nitrogen trichloride formation during wastewater treatment	Savickas *et al.* [402]
EI	Slow dissociation of *p*-iodotoluene ion observed with ICR	Dunbar and Lifshitz [35]

Table 4.1   (continued)

Ion source/ interface	Studied process/purpose	Ref.
EI	Slow dissociation of *tert*-butylbenzene ions observed with ICR	Faulk and Dunbar [32]
EI, MIMS	Monitoring chloramine reactions	Kotiaho et al. [403]
EI	Monitoring dissociation of ferrocene by ICR in microseconds	Faulk and Dunbar [33]
EI	Photodissociation rate measurements	Lin and Dunbar [36]
EI, MIMS	Monitoring biological reactions	Lauritsen and Gylling [404]
EI	Real-time sensing of reactants in rapid thermal chemical vapor deposition	Tedder et al. [405]
EI	Monitoring photocatalytic degradation of chlorinated volatile organic compounds	Alberici et al. [406]
EI, MIMS	Monitoring photocatalytic degradation of phenol and trichloroethylene	Nogueira et al. [65]
EI	Ion–molecule reaction in the millisecond timescale	Creaser et al. [206]
EI, MIMS	Monitoring chlorination of organic compounds	Rios et al. [407]
EI, MIMS	Monitoring simulated wastewater treatment plant	Creaser et al. [68]
EI	Kinetic studies on photolysis-induced reactions	Ludwig et al. [25]
EI	Real-time odor recorder	Somboon et al. [408]
EI	Monitoring ignition of nitrocellulose and RDX	Zhou et al. [409]
EI	Direct liquid introduction process MS, prototype thermal vaporizer	Owen et al. [196]
EI	Monitoring an esterification reaction by on-line direct liquid sampling MS	Owen et al. [9]
EI	Monitoring Soxhlet extraction	Chen and Urban [139]
EI, CI, MIMS	Monitoring chlorination products of organic amines	Kotiaho et al. [410]
CI, CI, MIMS	Monitoring reactions of epichlorohydrin	Johnson et al. [411]
MIMS	Mechanistic study of photosystem II	Hillier et al. [412]
MIMS	Mechanistic study of photosystem II	Messinger et al. [413]
MIMS	Mechanistic study of photosystem II	Hillier and Wydrzynski [414]
MIMS	Mechanistic study of photosystem II	Hendry and Wydrzynski [415]
MIMS	Mechanistic study of photosystem II	Nilsson et al. [416]

(*continued overleaf*)

Table 4.1   (continued)

Ion source/ interface	Studied process/purpose	Ref.
PTR	Monitoring emission of volatile organic compounds by plants	Harren and Cristescu [417]
PTR	Monitoring preparation of Espresso coffee	Sánchez-López et al. [187]
SIFT	Monitoring soil gas emissions	Milligan et al. [193]
SIFT	Monitoring air pollutants	Francis et al. [194]
SIFT	Monitoring peroxyacetyl nitrate	Hastie et al. [195]
AMS	Monitoring impact of wood combustion on ambient aerosols	Elsasser et al. [181]
PICT	Study of photodissociation of iodotoluene ions using ICR	Kim and Shin [37]
PICT	Study of photodissociation of iodotoluene ions using ICR	Shin et al. [38]
CTI	Monitoring synthesis of ammonia in microreactor	Xie et al. [418]
Charge exchange	Study of photodissociation kinetics of m-iodotoluene ions	Cho et al. [419]

AMS, aerosol mass spectrometry; APCI, atmospheric pressure chemical ionization; C-API, contactless atmospheric pressure ionization; CI, chemical ionization; CSI, cold spray ionization; CTI, charge-transfer ionization; DART, direct analysis in real time; DESI, desorption electrospray ionization; DIOS, desorption/ionization on silicon; EESI, extractive electrospray ionization; EI, electron ionization; ELDI, electrospray-assisted laser desorption ionization; ELISA, electrostatic ion storage ring, Aarhus (equipped with ESI source and accelerator); ESI, electrospray ionization; ESSI, electrosonic spray ionization; FAB, fast atom bombardment; LTP, low-temperature plasma; MALDI, matrix-assisted laser desorption/ionization; MIMS, membrane inlet mass spectrometry; nanoDESI, nanospray desorption electrospray ionization; nanoESI, nanospray electrospray ionization; PESI, probe electrospray ionization; PICT, phoionization charge transfer; PTR, proton-transfer reaction; SIFT, selected ion flow tube; SPI, single photon ionization; TLC, thin-layer chromatography; UASI, ultrasonication-assisted spray ionization; V-EASI, Venturi easy ambient sonic-spray ionization.

(DESI), Venturi easy ambient spray ionization (V-EASI), electron ionization (EI), and photoionization. These canonical ion sources were adapted for specific purposes in TRMS. Much emphasis was put on mixing reactants, exposing reaction mixtures to light (in the case of photochemical processes) or electric potentials (in the case of electrochemical processes) (cf. [10]). However, there also exist examples of more "exotic" systems constructed to enable measurements in the specific niche applications (see, e.g., [11]). Implementation of various interfaces in TRMS systems will be summarized in the following sections. Examples of their applications in different areas of chemistry and biochemistry will be discussed in Chapters 10–13.

4.2   Time-resolved Mass Spectrometry Systems

4.2.1   Photochemical Processes

A number of TRMS studies focus on photochemical reactions. While most of them can be regarded as fundamental studies, they have implications for atmospheric chemistry, astrochemistry, or the advanced understanding of combustion processes. The reaction-detection systems used in such type of research typically comprise reaction chambers, sources of light (lamp or laser used to initiate the process), ion sources (if gas-phase ions are not formed in the investigated reactions), and analyzers. Due to the transient nature of the studied species, it is desirable to position the reaction chamber as close to the mass analyzer as possible; however, in the case of microsecond-scale and faster processes the mass transport effects still need to be considered as the possible origin of measurement biases.

In the pioneering work conducted in the 1950s, Kistiakowsky and Kydd [12] demonstrated an innovative instrument which obtained spectra for gas samples effusing through a pinhole. Full spectra were registered every 50 μs. The system enabled TRMS studies of *flash photolysis*. The technique was soon picked up by other researchers, developed, and utilized to detect transient reactive free radicals. Meyer further verified the influence of various measurement parameters – including the effect of added oxygen, temperature, and flash intensity – on the progress of the studied reaction [13]. The interest in TRMS of flash photolysis endured over several decades [14]. Time-of-flight and quadrupole instruments were commonly utilized in such studies, while sector instruments were used to a lesser extent [14]. A typical setup consisted of an ultraviolet (UV) flash-lamp, a purge flow reactor (with pinhole or molecular beam sampling), an ion source (such as EI), a mass analyzer, a detector, pulse-counting electronics, computer data acquisition system, and vacuum pumps [14]. In one representative design, the gas sample was guided from the reactor, through the ion source, toward the diffusion pump [14]. Since the timescale of the studied process was very small, it was realized that the design of the interface is critical for minimizing possible sampling/transfer delays. Considering the imperfections of the instruments, the chemical kinetics may be convoluted with mass transport rates related to molecular velocity distribution, flow and molecular diffusion in the reactor, residence in the ion source, and residence in the vacuum chamber of the ion source [14]. Early on it also became apparent that the molecular velocity distribution of reactants can cause a non-negligible distribution of arrival times at the ion source, thus complicating the kinetic analysis [15]. In more recent reports, it was concluded that supersonic beam sampling possesses a greater time fidelity as compared with effusive beam sampling [16]. Supersonic sampling has been achieved with aid of the so-called *de Laval nozzle* (hourglass-shaped tube used to accelerate pressurized gas) installed at the kinetic reactor [17–19]. (Please refer to the review by Carr [14] for further discussion of the interfacing aspects in TRMS of flash photolysis.)

Various physical and chemical processes triggered by light can be studied by MS. Laser light is often used in conjunction with MS. In fact, several ion sources incorporate lasers (see also Section 4.2.8 and Chapter 2). The possibility to transmit photonic energy in short pulses of coherent light has also encouraged applications of laser MS in recording temporal information on the studied samples and processes. The ability to generate ultra-short laser pulses has sparked the idea to implement laser light in the initiation of ultra-fast chemical processes which could be studied using mass spectrometers as detectors of ions.

Over the past three decades the combination of pulse (e.g., femtosecond) lasers with MS has attracted considerable attention (for an overview, see [20]). TRMS has found applications in the studies of desorption processes induced by energetic light beams. As an illustration, Van Breemen *et al.* [21] investigated desorption of non-volatile organic salts following a laser pulse. The desorbed species were further analyzed by time-of-flight mass spectrometry (TOF-MS). von der Linde and Danielzik [22] demonstrated MS detection in the picosecond timescale. The experimental system incorporated two lasers. The pulse of one laser induced desorption of the sample surface material. The second pulse ionized the desorbed particles. This method facilitated analysis of heated inorganic surfaces.

In a dissimilar approach, Blitz *et al.* [23] constructed a time-resolved TOF mass spectrometer, and applied it in reaction monitoring at millisecond time scale. A laser was used to initiate reaction, while a discharge lamp or a pulsed laser were used for photoionization. Using this instrument, the authors were able to measure the kinetics of the reaction of SO· and ClSO· radicals with NO_2, as well as the kinetics of the formation of H_2CO from the reaction between CH_3CO· and O_2. In the study conducted by Baeza-Romero *et al.* [24], pulsed laser photolysis was implemented to initiate the reaction; the reactants were photoionized using pulsed laser vacuum ultraviolet (VUV) light, prior to the injection to a TOF-MS instrument. The reactor was made of a stainless steel tube incorporating a sampling orifice. The reactor was irradiated with the beam of a 248-nm excimer laser. A representative portion of reactant gas was sampled. The orifice coincided with the ion collection axis of the mass analyzer. The gaseous sample was ionized by a VUV laser beam supplied by a glass waveguide [24]. To ensure collection of high-quality data, sampling delays were corrected using a mathematical model. Reactions were studied with the rates approaching the sampling rate. The growth kinetics rates up to 7000 s^{-1} could be determined [24]. Ludwig *et al.* [25] applied TRMS and laser photolysis/flow reactor to determine the rate constant for the radical–radical reaction C_2H_5· + HO_2· → products. The authors used excimer laser photolysis to produce C_2H_5· and HO_2· radicals. The reaction was measured under almost pseudo-first-order conditions with an excess of HO_2·. Eventually, C_2H_5O· was found to be a direct reaction product. Laser light is also utilized in a biophysical method for studying fast protein folding events. For example, Hambly and Gross [26] presented a method that uses a 248 nm KrF excimer laser to cleave hydrogen peroxide at low concentrations to generate hydroxyl radicals. Due to the short lifetimes of the radicals, modification of the target protein is believed to occur in less than a microsecond. The products of this fast oxidation were detected by MS (see Section 12.4 for further discussion). Photoionization MS was also applied to study short-lived Criegee intermediates, produced as a consequence of laser photolysis of a precursor compound [27]. The lifetime of the target species is measured in milliseconds [27]. Such unstable species are often interrogated with other time-resolved techniques, especially vibrational spectroscopy [28]. The two analytical approaches have different selectivities and can be regarded as complementary.

Decomposition of gas-phase species on irradiation has extensively been studied using various model substrates. Olesik *et al.* [29] carried out photoelectron–photoion coincidence measurements to obtain the rates of energy-selective dissociations of halotoluene radical cations. Ho *et al.* [30] investigated dissociation rates and branching ratios for naphthalene ions by time-resolved photodissociation. The study led to calculation of bond energies. Photodissociation of various ions, including chlorobenzene [31], *tert*-butylbenzene [32], ferrocene [33], bromotoluene [34], and iodotoluene [35–38] was recorded using ion cyclotron resonance (ICR) analyzers. Microsecond-range photodissociation kinetics of small organic ions, including peptides, was also studied using a tandem instrument composed of a quadrupole ion trap and TOF analyzer [39, 40]. In the case of peptide ions, the experimental system incorporated three lasers with beams arranged in different ways to enable desorption, ionization, and dissociation, respectively [40].

Moreover, photoexcitation and subsequent dissociation of ions could be investigated using the electrostatic ion storage ring [41]. The implemented rapid switching approach enabled storage of daughter ions, and their detection. This technique enables characterization of dissociation pathways, and learning about the time-related contributions of different pathways to the overall dissociation reaction [41]. However, those measurements required implementation of a unique facility which is not available to most MS users.

4.2.2 Off-line Interfaces

During the past few decades, there has been a trend to design, construct, and implement on-line interfaces for MS. However, there also exist off-line interfaces, which can readily be utilized to conduct temporal monitoring of chemical systems. For example, Hatakeyama *et al.* [42] reported a method to perform the screening of synthetic reactions in segmented flow. The reactions were initiated inside a polymer "tee" (T-junction) by adding a reagent to the plugs containing complementary reagents. The output solution was transferred onto a MALDI plate. Deacetylation of ouabain hexaacetate was used as the model reaction while testing the system. Reactions involving various derivatives of ouabain were successfully monitored over time using different solvents (dimethylformamide, dioxane, acetonitrile, methanol, and water).

Similarly, enzymatic reactions can be performed directly on a MALDI plate, quenched, and the reaction products analyzed [43]. Various laser desorption/ionization (LDI)-based techniques facilitate such off-line measurements [44, 45]. When the operations of initiating and quenching the chemical reactions are carried out manually, the temporal resolution of the LDI-MS-based methods is typically in the order of a few minutes. However, by implementing flow mixing and quenching methodology, one can perform observations of sub-second phenomena with this kind of off-line MS detection (see, e.g., [46–48]).

While MALDI is a typical off-line ionization technique/interface, one can also implement classical interfaces to carry out off-line (or at-line) analyses of reaction mixtures. Aliquots can be obtained from a reaction mixture (a dynamic sample/matrix) at specific time points, and injected to the ion source of a mass spectrometer for analysis (e.g., [49–54]). In some cases, quenching is conducted [49, 50]. If the transient intermediates (e.g., radicals) are to be detected, it is important to assure that the reaction has not finished at the time of ionization [55]. Short-lived intermediates can be reacted with auxiliary compounds in order to enable subsequent measurements in the methodology referred to as *spin-trapping* (see, e.g., [56, 57]). The off-line analyses based on aliquoting of dynamic matrices, and subsequent separation, provide temporal resolutions in the order of a few minutes (see, e.g., [58]). Nonetheless, they are uncomplicated, and – in some cases – they may enable off-line sample treatment prior to detection by MS. For instance, inorganic ions, present in the reaction mixture, can readily be removed on an exchange resin to render the collected samples compatible with MS [59].

4.2.3 Membrane Interfaces

Various on-line methods exist that enable sampling followed by instantaneous introduction of molecules to the ion source of a mass spectrometer. In general, the common ways of introducing analytes to a mass spectrometer involve *direct inlets* and *membrane inlets* [60]. The so-called *membrane inlet mass spectrometry* (MIMS) technique is particularly suitable for the analysis of volatile organic compounds. It can reduce background interference, and increase detectability of the species of interest. The approach was introduced over 50 years ago [61], and has been widely used in on-line monitoring by MS, for example in environmental analysis, on-line process control as well as chemical and biological reaction monitoring [62, 63]. Analytes are transported by a three-stage process [62]:

1. transfer from the bulk sample to the surface of the membrane;
2. diffusion across the membrane;
3. evaporation of analytes at the other side of the membrane.

Since different compounds permeate the membranes with different rates, spectral interference can be reduced [60]. The membrane materials, used for that purpose, include polypropylene, poly(tetrafluoroethylene), cellulose, silicone rubber, dimethylvinyl silicone, polyethylene, and zeolite. The membrane can be interfaced to the mass spectrometer in different configurations – directly or with the aid of a *sweep gas*. Using MIMS, the sampling probe can be taken away from the mass spectrometer to which it is connected by a tube. Therefore, sampling of various environmental matrices can be conducted *in situ*. In fact, portable MIMS systems can be used as monitors, providing vital information which is not offered by other technologies (e.g., fluorescence, infrared spectroscopy) [64].

MIMS was used in real-time monitoring of photocatalytic degradation of organics [65]. A photoreactor comprising a UV lamp was coupled with the EI source of a quadrupole mass spectrometer via a silicone membrane, and peristaltic pump with a Tygon tube. The process was monitored during a period of several tens of minutes. Creaser *et al.* [66] demonstrated *reversed-phase MIMS* using a hollow-fiber Nafion membrane. They observed evolving methanol and ethanol concentrations in real time. The obtained profiles were comparable with off-line analyses conducted using gas chromatography. Interestingly, the Nafion membrane provided some selectivity as it preferentially transported methanol and ethanol. MIMS may be coupled with in-membrane pre-concentration and thermal desorption for sensitive analysis of semi-volatile organic compounds [67]. MIMS systems can readily be constructed by adapting gas chromatography–mass spectrometry (GC-MS) instruments after removing the chromatographic column [68].

In one study, Clinton *et al.* [69] established a single-stage membrane-based interface (Figure 4.2), and implemented it in the real-time MS monitoring of a concentrated pharmaceutical process reaction mixture. In this case, an *atmospheric pressure chemical ionization* (APCI) source was used in conjunction with a quadrupole instrument. The advantages of that approach included minimal analyst intervention and short sample preparation/analysis time [69].

4.2.4   Electrospray Ionization

In the 1980s, attempts were made to enable continuous introduction of liquid samples (especially aqueous buffer solutions) to ion sources of mass spectrometers. An early continuous flow interface was based on the *fast atom bombardment* (FAB) ion source [70]. However, it was the ESI interface that greatly facilitated temporal profiling of dynamic

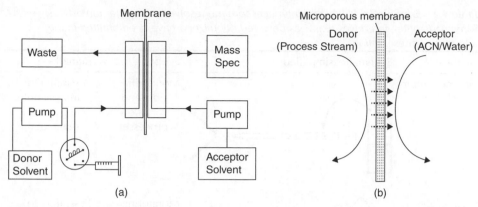

Figure 4.2 *Membrane inlet mass spectrometry: (a) schematic diagram of the membrane interface for APCI/MS; and (b) schematic diagram of the donor/acceptor flow across the microporous membrane [69]. Reprinted from Analytica Chimica Acta, 539, Clinton, R., Creaser, C.S., Bryant, D., Real-time Monitoring of a Pharmaceutical Process Reaction Using a Membrane Interface Combined with Atmospheric Pressure Chemical Ionisation Mass Spectrometry, 133–140. Copyright (2005), with permission from Elsevier*

samples in the liquid phase. Apart from the studies of chemical reactions (Chapter 11), the most prominent applications of time-resolved ESI-MS encompass studies of protein folding, enzyme–substrate interactions, and time-resolved hydrogen/deuterium exchange to gain insights on the dynamics of weakly structured protein regions [71] (see also Chapters 12 and 13). ESI-MS is also a powerful technique for rapid screening of catalysts [72, 73] and probing reactive intermediates (e.g., [74, 75]. When some of the species of interest are "invisible" to ESI-MS, derivatization with charged substrates can be used to render them amenable to detection by this technique [76, 77]. Temporal resolutions of the canonical approaches involving ESI-MS range from milliseconds to minutes [75].

There is a range of ways by which one can incorporate ESI in TRMS measurements (Table 4.2). For instance, one can conduct off-line analysis of discrete samples collected from a reaction chamber by ESI-MS. Samples obtained from a reactor can be purified, diluted, or separated. ESI also provides a convenient way to directly transfer liquid-phase dynamic samples into the gas phase while enabling ionization of the target analyte species (e.g., reactants). Some reaction-ESI-MS interfaces took advantage of stopped-flow incubation [78–80]. Here, the reaction is initiated by mixing two solutions (e.g., substrate and catalyst), and it is transferred to a reaction vessel. Subsequently, the resulting reaction mixture is infused to the ESI interface [80]. The time window covered by the early implementations of this approach extended from a few to a few tens of seconds [79]. Therefore, this method may be suitable for studying liquid-phase phenomena at moderate rates. (For an overview of the stopped-flow methods, see also [1].)

While the stopped-flow ESI technique represents a straightforward adaptation of the canonical spectroscopic methodology for MS measurements, many studies cited in this and other chapters of the present book report on continuous flow interface systems. Indeed, the possibility to feed the dynamic samples continuously to the ESI source is the forte of this ionization technique. In the case of real-time ESI-MS monitoring, it is important to

Table 4.2 *Examples of ESI-related interfaces for time-resolved studies (a non-exhaustive list). Some parts of the interfaces (e.g., gas-powered nebulizers) have been omitted in the schematic representations*

Interface	Schematic (simplified)	Scope	Remarks
Off-line sample collection		Monitoring slow and very slow processes	Possibility to treat samples (e.g., separation)
Valve system between batch reaction chamber and emitter/flow injection		Monitoring slow processes	Possibility to treat samples automatically (e.g., dilution)
Batch reaction chamber		Monitoring slow and medium-speed processes	Direct infusion, simple design; transfer of sample can be induced by hydraulic pressure of an auxiliary (e.g., nebulizing) gas
Fluidic reaction chamber in line with emitter		Monitoring fast and medium-speed processes	Various processes can be monitored (e.g., photo-chemical, electrochemi-cal, catalytic)
Fluidic micromixer, stopped flow		Monitoring slow and medium-speed processes	Suitable for reaction that require longer incubation

Table 4.2   *(continued)*

Interface	Schematic (simplified)	Scope	Remarks
Fluidic micromixer, continuous flow		Monitoring fast and medium-speed processes	Simplicity, mixing time can be controlled (with special mixers), possibility to incorporate multiple mixers
Microfluidic chip as mixer and microreactor		Monitoring fast and medium-speed processes	Miniaturization, low sample consumption, short response time, compact, possibility to incorporate multiple sample treatment stages
Plume impinging on pre-mixed reactants		Monitoring fast processes	Mixing time can be controlled
Dual lumen emitter		Monitoring ultra-fast processes	Simplicity, one emitter
Fused spray plumes		Monitoring ultra-fast processes	Simplicity, very short incubation time

assure a short delay time between sampling and spraying of the dynamic samples. This delay time depends on the volumetric flow rate of the sample in the sample conduit as well as the length of the conduit. In an early study, Lee *et al.* [81] demonstrated coupling of a thermostatted reaction vessel with an ESI source. Here, the delay time was believed to be shorter than 1 s. The sample reservoir was placed above a T-shaped union. Nebulizing gas (nitrogen) was used to sustain the spray in front of the MS orifice. Such a pneumatically assisted ESI interface is particularly suitable for reaction systems using a high percentage of aqueous buffer. It can also provide high stability and sensitivity [81]. Its modified versions could be found in various applications over the last three decades, some of which will be highlighted in the following sections and chapters. In one work, Brum *et al.* [82] used ESI-MS to study solution-phase reactions with the purpose of obtaining mechanistic and kinetic information: they monitored the hydrolysis of isatin (an indole derivative) in real time using the positive- and negative-ion modes. In another system, the reaction mixture was pushed by an overpressure of nitrogen gas, and diluted on-line before it is infused to the ESI port [83]. Thus, there was no need to collect aliquots from the reaction vessel by syringe or pipette – which would inevitably delay registration of mass spectra. ESI was also implemented in the monitoring of a photochemical reaction [84]. In one version of the protocol, a flow-through photoreaction cell was implemented while in another one, the ESI-generated charged droplets were directly photoirradiated. In this way, it was feasible to detect intermediates of a photochemical process. Interestingly, the reaction of interest can take place directly in the ESI plume [84]. Another adaptation of ESI involves implementation of heating elements to enable thermal activation of chemical reactions [85]. ESI droplets can also be exposed to a reactive gas (O_3) to initiate a reaction occurring in the low millisecond range [86]. (See Chapter 11 for examples of other applications of ESI in the monitoring of chemical reactions.)

Microscale mixers have been constructed to enable mixing reactants in proximity to an ESI source. Early mixers incorporated a fixed "tee" providing a single reaction time point [71]. They could be adopted for mixing two substrate solutions or substrate and catalyst solutions immediately before the ESI region (e.g., [55, 87–91]). Such mixers can be joined in series. One of the mixers can be used to dilute the reaction mixture right before the ionization [92, 93]. This TRMS approach was very successful in the monitoring of relatively fast processes (normally, seconds down to milliseconds). However, such flow mixers can incorporate an adjustable position mixing site allowing for recording reaction mixture composition at different time points [94]. In an intriguing design, one of the inlet capillaries (needle) is pulled back during analysis, and intensity–time profiles are recorded for specific reactive species (Figure 4.3, [95]). The change in the distance between the end of the inner capillary (needle) and the end of the ESI emitter influences the dead volume of the reaction zone, and determines the duration of the reaction. In this mixer, a spatial gradient of the reaction components is formed with different reaction points being present along the reaction tube (between the mixing region and the ion source). This early version of the apparatus could measure rate constants of up to ~ 100 s^{-1}. The approach has further been developed and implemented in various applications of TRMS (see Chapters 12 and 13). An improved continuous-flow mixer for on-line measurement by ESI-MS was reported recently [96]. It uses sheath flow of nitrogen gas to maintain stable spray and attain a high signal-to-noise ratio. It can readily be mounted onto a MS instrument replacing the conventional ESI interface.

In an ingenious approach recently reported, double-barrel theta capillaries were used for rapid mixing of two solutions in front of the ionization region [97–99]. When using such unconventional electrospray emitters (Figure 4.4), the complete mixing of reactants, infused through the two channels, occurred within a few microseconds [98]. It was possible to measure the apparent reaction time of the reduction of 2,6-dichloroindophenol by L-ascorbic acid to be 274 ± 60 µs. Introduction of this method can be regarded as a milestone in the monitoring of ultra-fast processes by ESI-MS.

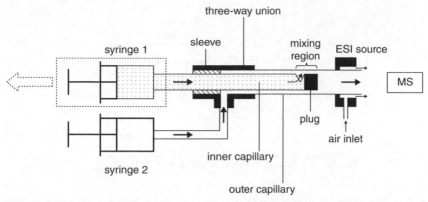

Figure 4.3 *Schematic cross-sectional diagram of the experimental apparatus used for time-resolved ESI-MS experiments. Syringes 1 and 2 deliver a continuous flow of reactants; mixing of the two solutions initiates the reaction of interest. The inner capillary can be automatically pulled back together with syringe 1 (as indicated by the dashed arrow), thus providing a means to control the average reaction time. Solid arrows indicate the directions of liquid flow. Small arrows in the ESI source region represent the directions of air flow [95]. Reprinted with permission from Wilson, D.J., Konermann, L. (2003) A Capillary Mixer with Adjustable Reaction Chamber Volume for Millisecond Time-Resolved Studies by Electrospray Mass Spectrometry. Anal. Chem. 75: 6408–6414. Copyright (2003) American Chemical Society*

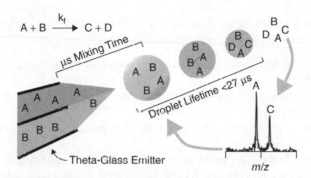

Figure 4.4 *Rapid mixing using theta capillaries [98]. Reprinted with permission from Mortensen, D.N., Williams, E.R. (2014) Theta-Glass Capillaries in Electrospray Ionization: Rapid Mixing and Short Droplet Lifetimes. Anal. Chem. 86: 9315–9321. Copyright (2014) American Chemical Society*

ESI droplets can be considered as microvessels for the study of fast reactions in solution in short time intervals. In one representative study, different arrangements of reactant sprayers were used to detect intermediate in electron-transfer-catalyzed dimerization of *trans*-anethole [100]. Most recently, Lee *et al.* [101] took advantage of the fusion of high-speed liquid droplets to record the kinetics of liquid-phase chemical reactions in the microsecond timescale (Figure 4.5). The reaction was quenched when the species entered a heated transfer capillary. This approach resembles, to some extent, the concept of fused

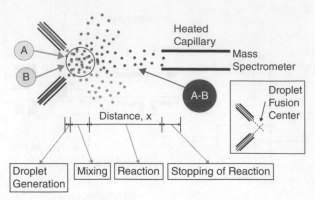

Figure 4.5   *Schematic (not to scale) of experimental setup for studying reaction kinetics in fused droplets. (Inset) The droplet fusion center is the intersection of the two droplet streams. Most fusion events take place in a circle (dotted) of about 500 μm surrounding the droplet fusion center [101]. Reproduced from Lee, J.K., Kim, S., Nam, H.G., Zare, R.N. (2015) Micro-droplet Fusion Mass Spectrometry for Fast Reaction Kinetics. Proc. Natl. Acad. Sci. USA 112: 3898–3903 with permission from PNAS*

droplet ESI [102] or extractive electrospray ionization (EESI) [103] – used for ionization of complex samples at atmospheric pressure (see Section 4.2.6 and Chapter 2). However, fusion of microdroplets is used here primarily to initiate a reaction rather than to promote ionization.

When describing TRMS methods for studies of fast processes, it is helpful to realize the differences between the key temporal characteristics:

- *Merger time* – the time point when the substrate (or an inducer, catalyst) solutions are brought together.
- *Duration (time) of mixing* – the time interval during which the two merged solutions become homogeneous.
- *Duration (time) of incubation* – the time interval elapsed from the completion of mixing to quenching.
- *Quenching time* – the time point when the reaction is terminated intentionally. The quenching step may be achieved by introduction of a quenching solution or by desolvation of microdroplets.
- *Duration (time) of reaction* – the time interval required for the reaction to occur (reach its completion).

The *duration of incubation* is most relevant for kinetic characterization of the studied reactions. Mixing devices merge substrate solutions, and they initiate the mixing process. However, in the laminar flow, mixing time may not be negligible (see Section 7.3). Measures need to be in place to accelerate mixing of reactant solutions, so that the *duration of incubation* is not overestimated. Quenching can be induced by addition of an appropriate solvent that stops the reaction, or even by desolvation/ionization, as in the above example. When monitoring transient intermediates of reactions, the *duration of incubation* can be much shorter than the time necessary for the reaction to reach its completion. Please note that in some cases *duration of reaction* may be used as the synonym of *duration of incubation*. Moreover, the term *time* is often used interchangeably with *duration*. Thorough study of the specific reaction systems and experimental conditions is necessary to discern these important descriptors.

The cloud of gas-phase ions moving toward the mass analyzer is a very dilute matrix while ions of the same charge are unlikely to react with each other. So, unless the generated ions undergo *in-source* or *post-source decay*, or are purposefully reacted with gas molecules/ions or irradiated, they can make their way to the detector. To estimate or confirm the *duration of incubation* in ultra-fast mixing approaches, one often utilizes indicator reactions with well-characterized kinetics, such as the reaction of 2,6-dichloro-phenolindophenol and ascorbic acid (e.g., [98, 101]).

We should point out that the temporal resolution of the above studies, which rely on T-shaped mixers, theta capillaries or spray droplet fusion, produce steady samples which in principle do not require real-time MS monitoring relying on the temporal resolution of the mass analyzer/detector. Therefore, slower instruments may be used to detect the products of fast reactions sustained by those interfaces. Here, the interface determines the *duration of incubation* of the reactants. In addition, in the spray droplet fusion approach, mixing does not only rely on diffusion. Most likely, it is accelerated by turbulence due to the physical impact and of droplet fusion process (cf. *inertial mixing* [101, 104]). The ultrafast (microsecond scale) chemical phenomena can be captured by fusing droplets because the *duration of incubation* of reactants (A and B, Figure 4.5) can be reduced to a minimum. Desolvation proceeds promptly after the merger of the microdroplets, thus the *duration of incubation* must be very short. Lack of co-reactants in the gas phase stops further transformations of reaction intermediates. The EESI methodology is further discussed in Section 4.2.6.

Another consideration, when using ESI-based approaches, is the possible influence of electrospray droplet shrinkage on the studied processes. Evaporation of solvent inevitably leads to an increase of reactant concentrations. Although no such influence was observed in a model equilibrium process [105], one cannot rule out the occurrence of confounding effects due to electrospray droplet shrinkage in the case of some fast, and concentration-sensitive, processes. Overall, the conditions of the reactions carried out in the ultra-fast interfaces described here may not be stable and defined well throughout the whole reaction/analysis process, which can be regarded as their drawback.

It should also be noted that miniaturized ESI interfaces are readily incorporated into microfluidic chips [106], providing opportunities for TRMS measurements (see Chapter 7).

4.2.5 Desorption Electrospray Ionization

More than a decade ago, the group of Graham Cooks introduced a modified version of ESI, which they called *desorption electrospray ionization* (DESI) [107] (see also Chapter 2). In this technique, the ESI plume is directed onto the sample surface. A very rapid chemical analysis of the sample surface is possible [5]. While the technique was originally presented as a way to analyze samples supported on solids, it has been modified to enable analysis of liquid samples [108]. This development facilitated implementations of DESI in TRMS. For example, in the work mentioned earlier in this chapter, Miao *et al.* [2] demonstrated the possibility to monitor reactions with sub-millisecond time resolution. Here, two reactant solutions mix rapidly to form a free liquid jet which is then ionized by

DESI at different positions corresponding to different *durations of incubation/reaction*. This approach resembles to some extent the micromixer method described above (Section 4.2.4). However, in this case, the reaction stream flows freely; it is not directly nebulized in front of the MS orifice. Instead, the reaction components are sampled by an independent ESI spray set in the proximity of the reactant stream (cf. Table 4.2). Since the pre-mixed reactant solution is already in the liquid phase, it is debatable whether this method falls under the original definition of DESI – which is most often used for analysis of molecules on solid supports. In other work, Perry *et al*. [109] also applied DESI-MS to probe reactions occurring in millisecond time intervals. The possibility to carry out a chemical reaction almost simultaneously with ionization was exploited by scientists applying the so-called *reactive DESI* [110–116]. In this variant, derivatization can be conducted on analyte molecules, augmenting analytical sensitivity and selectivity of detection. For example, Song and Cooks [113] applied reactive DESI to detect hydrolysis products and metabolites of chemical warfare agents. According to the authors, $H_2BO_3^-$ ions – generated in the spray – reacted with the condensed-phase analytes to form anionic adducts, which provided superior specificity. Derivatization can also be conducted in an electrochemical cell connected in line with the DESI source [117] (see also Section 4.2.10).

Nanospray desorption electrospray ionization (nanoDESI) [118] can be viewed as a "miniaturized version" of DESI. It incorporates two capillaries which serve different purposes. The first capillary delivers solvent onto the sample surface while the second capillary collects the solubilized analytes, and transfers them towards the orifice of the mass spectrometer. Please note that the latter part of the nanoDESI system closely resembles the nanospray electrospray ionization (nanoESI) developed in the 1990s [119], and its modified version called contactless atmospheric pressure ionization (C-API), developed recently [120] while the role of the liquid junction seems to be similar to the function of the liquid microjunction surface sampling probe [121, 122]. In one representative report, Hsu *et al*. [123] integrated an inverted light microscope with nanoDESI-MS to characterize the chemical dynamics of mouse spinal cords. Changes to actin-sequestering proteins β-thymosins during tissue development could be observed.

4.2.6   Other Interfaces Derived from Electrospray Ionization

There exist numerous other ion sources and interfaces, which can readily be implemented in the monitoring of chemical reactions. Some of them are extensions of conventional ESI (see Section 4.2.4). For example, Chen *et al*. [124] presented a method using *electrosonic spray ionization* (ESSI), coupled with *atmospheric-pressure thermal activation*, to carry out organic reactions. Using this method, they revealed a new pathway for the Fischer indole synthesis and a novel sodium ion-mediated pinacol rearrangement. They also speculated that the protonation/deprotonation capability of ESSI and the thermal activation methodology could find applications in the development of "green chemistry" pathways. Jecklin *et al*. [125] compared three different electrospray-based ionization techniques (conventional ESI, nanoESI, and ESSI) for the investigation of non-covalent complexes with MS. The most accurate dissociation constant (K_D) value (compared with liquid-phase experimental data) was found for ESSI.

Another technique, EESI-MS can also be used to initiate reactions in the electrospray plume [100, 101], which occur in short time intervals (down to microseconds; see Section

4.2.4). Alternatively, it can be used in a "slower" monitoring of reactions taking place in batch reactors. Zhu *et al.* [126] applied EESI-MS in real-time monitoring of acetylation reaction. It is suggested that EESI-MS can be a valuable technique for on-line characterization and for full control of chemical and pharmaceutical processes. In other work, McCullough *et al.* [127] validated an EESI-MS method using a series of mock reaction mixtures, and subsequently used it to monitor the base hydrolysis of ethyl salicylate to salicylic acid. EESI is compatible with viscous samples; therefore, it could readily be implemented in the monitoring of a reaction taking place in an ionic liquid solvent [128].

Yu *et al.* [129] utilized *probe electrospray ionization* (PESI) to monitor some biological and chemical reactions in real time, such as acid-induced protein denaturation, hydrogen/deuterium exchange of peptides, and Schiff base formation. In their report, the method was characterized as simple, consuming little sample, and requiring little sample preparation. The PESI technique was further developed and applied in the monitoring of protease-catalyzed reactions [130]. *Ultrasonication-assisted spray ionization* (UASI) takes advantage of ultrasound to facilitate dispersion of sample microdroplets in front of the MS orifice [131]. Since ultrasound is occasionally used to accelerate chemical reactions, this technique is highly compatible with real-time monitoring of organic synthesis [132, 133]. *Cold spray ionization* (CSI) is a variant of ESI operating at low temperature (from ~ -80 to ~ 10 °C) [134]. It enables detection of labile organic species; for example, unstable reagents and reaction intermediates [134].

A remarkable ESI-related interface has been developed by Santos *et al.* [135]. The so-called *Venturi easy ambient sonic-spray ionization* (V-EASI) uses a *Venturi pump* assembly to aspirate the sample and nebulize it in front of the MS inlet. This interface is particularly valuable for numerous applications involving dynamic samples because it does not require the use of a mechanic pump. Therefore, the sample does not need to be loaded to a syringe prior to analysis, preventing possible perturbation to the studied processes and sample ageing. It is also routinely used, as an auxiliary technique, to characterize the progress of enzymatic reactions [136]. In one study, we implemented V-EASI to monitor a bioautocatalytic reaction which led to the formation of a spatial gradient inside a high-aspect-ratio reaction cell [137]. The reaction was catalyzed by two enzymes acting co-operatively to create a wave of adenosine triphosphate. This application of real-time MS analysis helped to unravel both temporal and spatial characteristics of the investigated biochemical system.

MS allows mechanisms of conventional analytical processes such as extraction to be investigated. A better understanding of these fundamental processes can enable their optimization so that they can match the requirements of future challenging tasks. For example, in a recent study, we constructed an automated system for mechanical disruption and extraction of microscale specimens coupled directly to V-EASI-MS [138]. The obtained ion currents were unique for every extracted specimen. The appearance of different chemical species at different times highlighted the possibility of monitoring stepwise release of biomolecules as they are liberated from the biological material. This information is lost while conducting conventional analysis with off-line extraction and subsequent single-time-point detection. The Soxhlet extraction is a popular sample preparation technique used in chemical analysis. It enables release of molecules that are embedded in complex – for example, biological – sample matrices. In most protocols, samples are analyzed after the extraction is complete. However, in order to gain an insight on the Soxhlet extraction process, it would be desirable to monitor it in real time. We found that it is possible to use ESI-MS (fitted with a peristaltic pump) to follow the progress of the Soxhlet extraction in real time [139]. Various technical obstacles need to be overcome in order to assure seamless operation of such systems. Biological extracts are concentrated solutions of many chemical species. Therefore, pumping extract samples continuously towards a mass spectrometer invites heavy contamination of the ion source. This problem can be addressed, to some extent, by implementing electronic control of the experiment, and minimizing contamination of the ion source by periodic (discontinuous) infusion of the extract samples. Notably, the extent of ion suppression is not equal at the beginning and at the end of the extraction process. Therefore, it is necessary to implement methods for on-line sample treatment (dilution, purification, separation) that do not compromise temporal resolution. Currently, separation-based techniques are more adequate for quantitative analysis of the extraction process, especially when the collected data are used for kinetic evaluations [139]. (See Chapter 8 for further discussion of quantitative MS measurements.)

4.2.7 Interfaces for High-throughput Screening

The superior scanning speed of MS (up to few thousand spectra per second) is particularly useful in high-throughput screening. In this case, the temporal resolution of the instrument allows mass spectra of samples, which are infused to the ion source at a high speed, to be obtained. For example, MS can readily be coupled on-line with segmented-flow fluidic systems to enable fast screening of hundreds of samples in a few minutes. Along these lines, Sun and Kennedy [140] devised a system capable of reformatting 384-well plate samples into nanoliter droplets segmented by an immiscible oil. The operation was accomplished at a speed of 4.5 samples s^{-1}. Subsequent analysis of the segments was conducted at 2 samples s^{-1}. Jin *et al.* [141] constructed a "swan-shaped" probe to transfer small volumes of arrayed droplets to the orifice of the mass spectrometer. The U-shaped section of the probe has a micrometer-sized hole for sampling while its tapered tip enables ESI. In other work, Hogenboom *et al.* [142] presented a continuous-flow analytical screening system to measure the interaction of biologically active compounds with soluble affinity proteins. The screened compounds were driven in a continuous flow through a reaction zone where they reacted with the affinity protein. A reporter ligand was subsequently added to the flow. The concentration of the unbound ligand was then determined by ESI-MS. If active compounds were present in the sample, this caused an increase in the ligand concentration, which was not depleted from the flow.

4.2.8 Interfaces Using Laser Light

Laser-aided interfaces find applications in fundamental studies in physical chemistry (see Section 4.2.1). However, they can also be used in the monitoring of environmental matrices as well as chemical and biochemical reactions: (MA)LDI-MS is the most prominent example (Section 4.2.2). Moreover, laser beams can be introduced into canonical ion sources such as ESI to enable efficient transfer of microscale aliquots of solid or liquid matrices to the ionization zone. Along these lines, Cheng *et al.* [143] used *electrospray-assisted laser desorption/ionization* (ELDI) to monitor epoxidation of chalcone in ethanol, chelation of ethylenediaminetetraacetic acid with copper and nickel ions in aqueous solution, chelation of 1,10-phenanthroline with iron(II) in methanol, and tryptic digestion of cytochrome *c* in aqueous solution. The surface of the sample solution was irradiated with a pulsed UV laser. The desorbed gaseous analytes were then introduced into an electrospray plume to generate gas-phase ions. Peng *et al.* [144] further introduced *reactive ELDI*, which allowed reactions to be carried out simultaneously with the ELDI process. To demonstrate the capabilities of this technique, they monitored the reaction of disulfide bond reduction. Mühlberger *et al.* [145] disclosed the design of a mobile TOF mass spectrometer for on-line analysis and monitoring of complex gas mixtures. The device incorporated three ionization techniques (resonance-enhanced multiphoton ionization, single photon ionization, laser-induced electron-impact ionization) with different ionization selectivities. An example of a possible application of this technique is time-resolved on-line monitoring of automobile exhausts and waste incineration flue gas. Interestingly, this prototype instrument enabled quasi-parallel application of three ionization techniques [145].

4.2.9 Interfaces Using Plasma State

Mass analyzers operate in the gas phase while samples delivered to MS interfaces are in the gas, liquid, or solid phase. However, some interfaces use the plasma state in order to enable desorption and/or ionization of analyte molecules. Trace inorganic analysis is often conducted using high-temperature plasma ion sources which take advantage of microwave heating – they are referred to as *inductively coupled plasma* (ICP) sources. While organic matter is rapidly decomposed at a temperature of a few thousand kelvins, samples are often mineralized prior to off-line ICP-MS measurements in order to reduce interference and contamination of the equipment. Chromatographic separations are occasionally conducted on complex samples before their introduction to ICP compartments. In such cases, temporal properties of ICP-MS instruments are utilized to record chromatograms with peaks representing analyte zones separated in the chromatographic columns. Nonetheless, other sample treatment/introduction systems can also be coupled with ICP-MS instruments giving rise to temporal ion profiles. For example, electrothermal vaporizer can be used to achieve temporal separation of the isobaric ions of ^{87}Rb and ^{87}Sr present in the sample [146]. The one-step vaporization routine conducted with 100 ms time resolution warrants separation of ions of interest within a few seconds. This preparatory step is much faster than using conventional ion-exchange chromatography [146]. Electrochemical methods can also be used to preconcentrate metal analytes prior to ICP-MS [147]. Such analyses yield time-resolved ICP-MS responses to the analytes eluted from the anodic stripping voltammetry stage. The time-resolving capabilities of modern ICP-MS instruments facilitate sensitive bioassays. Nanoparticles can be used as tags of biomolecules [148]. ICP-MS records transient signals of particles entering the ionization region. The dwell time in one demonstration of this approach was as short as 10 ms. The frequency of temporal peaks scaled with the densities of the samples containing gold nanoparticles which were pumped to the interface [148]. A similar strategy has been used for conducting so-called *mass cytometry* where biological cells are labeled with antibodies conjugated with heavy isotopes [149].

While ICP can be considered to be a "harsh" ionization technique, it is not directly applicable to analysis of organic molecules. However, there exist "softer" plasma-based sources which can readily be used for the monitoring of organic reactions. Prominent examples include APCI [150] – which uses corona discharge – as well as the so-called *low-temperature plasma* (LTP) probe [151]. For instance, Zhu *et al.* [150] demonstrated a quantitative mass spectrometric method that enables MS monitoring up to molar concentrations. They achieved this capability by coupling a flow injection analysis system with APCI-MS. Eventually, quantitative real-time monitoring of a model reaction was performed at a concentration of ~ 1.6 M which is considered to be very high for conventional MS methods [150].

4.2.10   Electrochemical Mass Spectrometry

While electrospray is a nebulizing device, it also entails some aspects of electrochemical reactions [152, 153]. Electrochemical control of ESI is necessary to achieve optimum interface performance. Sometimes intricate features of the interface design (e.g., presence of a ground loop) can affect its operation [154]. Deficiencies of the electrical design, might – in some cases – lead to decomposition of infused samples, for example, due to oxidation [155]. Thus, for artifact-free real-time monitoring by ESI-MS it is vital that the analyte species are ionized as they arrive at the emitter, while any undesirable electrochemical effects have little or no influence. On the other hand, electrochemical oxidation can be used to pre-treat effluents of chromatographic columns prior to MS detection [156].

Ion sources can be used as electrochemical reactors for oxidation or reduction reactions [157]. They may be fitted with electrochemical cells in order to monitor such processes. Conventional electrochemical measurements do not provide direct information on intermediate/product species. Moreover, the reaction intermediates present between electrodes may be unstable. To enable thorough characterization of electrochemical processes, such devices can be coupled with MS (e.g., [158, 159]). Since the 1970s many of the popular ion sources have been used for that purpose [160]. For example, in one early work, a porous electrode was introduced to sample and detect volatile products (NO and N_2O) during electrochemical reduction of NO_2^- [158]. In the *differential electrochemical MS*, electrochemical cells are interfaced with MS using a porous Teflon membrane [161]. Such systems can facilitate characterization of reaction intermediates as well as electrode processes with respect to the applied potentials [161].

In an investigation of electrocatalytic reduction of nitrate on platinum, MS was utilized to reveal the release of volatile products of electrode reactions (N_2O and N_2) at the electrode [162]. Positively and negatively charged ions of basic and acidic products could be measured. Formation of non-volatile dimers and trimers during electrooxidation of *N,N*-dimethylaniline could be recorded [163]. The electrolyte was forced away from the working electrode into the heated capillary tube of the ion source. The potential range of -0.4 to 1.0 V was swept at a rate of 0.016 V s^{-1}. During this stage, the electrochemical reaction was characterized by the recorded mass intensity–potential profiles [163]. An electrochemical cell coupled on-line with thermospray-MS also enabled elucidation of the electrochemical oxidation pathway of uric acid [164]. The effect of potential on the relative intensity of an ammonolysis intermediate, produced during the oxidation of 9-methyluric acid, could be observed in the sub-second timescale [164]. On-line thermospray tandem MS was also used in a comparative study of enzymatic and electrochemical oxidation pathways of uric acid [165].

Different types of flow cells could be coupled on-line with ESI-MS, including thin-layer electrode cell, tubular electrode cell, and built-in cell (cf. [160, 166]). For example, a simple flow cell was assembled using two pieces of metal tubing, and used to perform on-line detection of electrolysis products [167]. The on-line monitoring approach enables straightforward characterization of electrochemical processes (including unstable products) at various potentials [168]. For example, intermediates and products generated in the course of linear sweep voltammetry could be recorded by ESI-MS with a single quadrupole despite complex reaction paths [159]. Temporal and spatial profiling of electrode processes can also be conducted by using the technique called *scanning capillary microscopy* coupled with ESI-MS [169].

Conversion of an intercalating antitumor agent was investigated on an electrochemical microchip coupled with ESI [170]. While matrix effects (ion suppression) can influence such monitoring, cross-validation can be done using off-line analysis by liquid chromatography–mass spectrometry (LC-MS) [168]. In fact, chromatographic separation can be conducted before and after electrochemical reaction: to investigate the electrochemical conversion of individual substances present in a complex mixture and to separate the oxidation products in order to know their polarities [171]. An on-line electrochemistry (EC)/LC/ESI-MS system was devised which enables oxidation and conjugation reactions. Reactive metabolites generated in the electrochemical cell were trapped with a conjugation reagent [172]. In other work, the electrochemical metabolism pathway of amodiaquine was studied with an electrochemical microchip coupled with LC-MS [173].

While coupling electrochemical cells to the DESI source, diverse electrode configurations can be implemented, including: thin-layer electrode cell; tubular electrode cell; and desorption from static droplets, in which electrochemical reactions take place (cf. [160]). Recently, Brown *et al.* [174] presented a "waterwheel" DESI-MS interface for monitoring intermediates of electrochemical reactions with millisecond resolution. It was coupled with an orbital ion trap mass analyzer in order to attain high mass resolving power and accuracy. It is known that temporal resolution is not a strong point of high-mass-resolution mass analyzers (cf. Chapter 5). However, the high temporal resolution of the ion monitoring in this approach is due to the fast sampling enabled by DESI. This characteristic is comparable with that of other interfaces described above, which lead to formation of quasi-steady samples (see Section 4.2.4). In another report, Liu *et al.* [175] demonstrated on-line coupling of nanoDESI-MS with a traditional electrochemical flow cell (Figure 4.6) as well as an interdigitated array electrode. They used oxidation of dopamine, reduction of benzodiazepines, and electrochemical derivatization of thiol groups as model processes to demonstrate the performance of this technique. Overall, it is expected that developments in instrumentation will render electrochemical-MS suitable to address various challenging problems in chemical analysis [160].

4.2.11   Aerosol Mass Spectrometry

Aerosol mass spectrometry (AMS) is used to monitor the chemical composition of particulate matter in the atmosphere. Commercial AMS instruments can provide size and chemical mass loading data on aerosol particles in real time [176]. Such instruments integrate sampling and MS analysis sub-systems. They can be installed permanently or used as components of mobile laboratories [176]. Both quadrupole and TOF AMS devices can provide quantitative data on the chemical composition of volatile/semi-volatile submicrometer aerosols [177]. Importantly, AMS can provide non-refractory aerosol mass

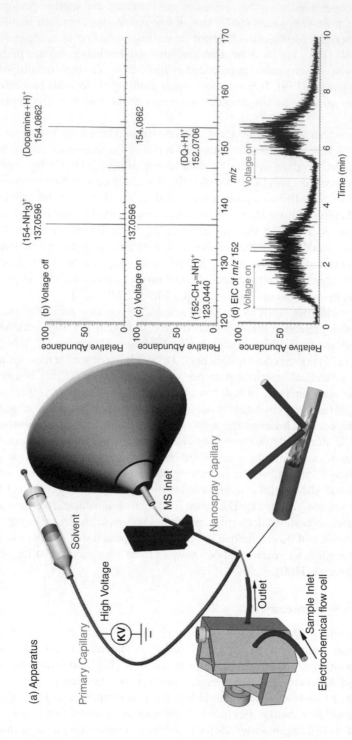

Figure 4.6 *Online coupling of an electrochemical flow cell with nanospray desorption electrospray ionization MS. (a) Scheme showing the configuration of the EC/nanospray desorption electrospray ionization MS with an electrochemical flow cell. Nanospray desorption electrospray ionization MS spectra acquired when the dopamine solution flowed through the thin-layer electrochemical cell with an applied potential of (b) 0.0 and (c) 1.5 V. (d) EIC of m/z 152 acquired when the dopamine solution flowed through the thin-layer electrochemical cell [175]. Reprinted with permission from Liu, P., Lanekoff, I.T., Laskin, J., Dewald, H.D., Chen, H. (2012) Study of Electrochemical Reactions Using Nanospray Desorption Electrospray Ionization Mass Spectrometry. Anal. Chem. 84: 5737–5743. Copyright (2012) American Chemical Society.*

concentrations, chemically speciated mass distributions and single particle information [178]. The automated data analysis routines implemented in AMS instruments can classify the peaks in mass spectra into groups of compounds relevant to air quality [177].

Chemical–biological mass spectrometers – which are designed for fast analysis of bioparticles – incorporate multiple stages of sample introduction [179]. Some of these stages are dissimilar to those found in conventional MS systems. At the inlet side, particles can be sorted according to their size using multiple pump systems and nozzles. Vaporization of particles can be conducted by heating the injected sample to a high temperature (>300 °C) [179]. In one design, sample particles are guided by an array of aerodynamic lenses to the vacuum chamber [178]. On their arrival in the vaporizer, they are transferred to the gas phase. The liberated species (chemical molecules) are instantly ionized by EI. The newly produced ions can be guided to the inlet of the quadrupole analyzer [180] or TOF tube [178]. In this particular case, the TOF instrument (Figure 4.7) operates with a kilohertz repetition rate.

In one study, AMS was used to investigate the impact of wood combustion on ambient aerosols [181]. Temporal variations of the wood combustion organic aerosol during a

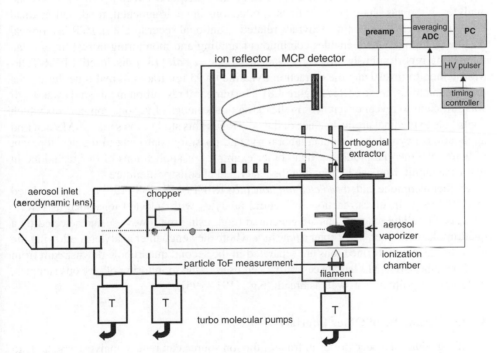

Figure 4.7 *Schematic of the time-of-flight aerosol mass spectrometer (TOF-AMS). Aerosol is introduced into the instrument through an aerodynamic lens focusing the particles through a skimmer and an orifice onto the vaporizer. Particle vapor is ionized and the ions are guided into the TOF-MS, which generates mass spectra at ~83.3 kHz repetition rate. For particle size measurement the particle beam is chopped with a mechanical chopper and the detection is synchronized with the chopper opening time [178]. Aerosol Science & Technology: A New Time-of-Flight Aerosol Mass Spectrometer (TOF-AMS) – Instrument Description and First Field Deployment. 39: 637–658. Copyright 2005. Reston, VA. Reprinted with permission*

day/night cycle have been observed. The data obtained by AMS showed good correlation with those recorded by GC-MS [181]. In another work, an aerosol concentrator and APCI-MS were used to monitor organic acid species in particles in the negative-ion mode [182]. It was possible to observe a very good correlation with the data obtained by a TOF-AMS instrument [182]. In another noteworthy approach, thermal desorption chemical ionization MS was implemented in the monitoring of the chemical composition of ultrafine (<10 nm) particles [183]. (For a review of on-line MS of aerosols, see [184].)

Mobile mass spectrometers are being developed with transient-concentration measurement capabilities. Temporal resolution of these devices is sufficient to resolve kinetic rates of adsorption of important air pollutants such as exhausts of diesel engines [185]. It is expected that the next few years will bring further miniaturization of mass spectrometers designed for chemical analysis of air in real time, which will enhance the existing environmental monitoring infrastructure.

4.2.12 Proton-transfer Reaction Mass Spectrometry

Proton-transfer reaction (PTR)-MS is based on the reaction between hydronium ions and analytes in the gas phase [186]. It finds applications in environmental, food, and medical analysis, as well as screening activities related to national security [186]. PTR is a typical on-line technique which enables continuous sampling and monitoring in real time. In an interesting report published recently, Sánchez-López *et al.* [187] disclosed a PTR-TOF-MS method to investigate the extraction dynamic of 95 ion traces in real time during the preparation of Espresso coffee (Figure 4.8). The temporal resolution in this study was 1 s. It was sufficient to reveal differences in the extraction kinetics of various components which contribute to the final aroma balance of the beverage. This study shows that TRMS can find applications beyond the chemical laboratory, and possibly contribute to the development of better commercial products. Many other examples of applications of this technique in real-time monitoring can be found in the analytical chemistry literature.

Another technique called *selected ion flow tube* (SIFT)-MS [188–190] is closely related to PTR-MS. It involves reactions of neutral analytes with selected ions such as H_3O^+, NO^+, or O_2^+. The reagent ions are generated (e.g., using microwaves), and selected by a quadrupole mass filter. They enter a *flow tube* where they encounter analytes. Following the reaction, the newly formed ions are separated in the second quadrupole downstream from the flow tube. SIFT-MS has successfully been used in the monitoring of gaseous samples, including air pollutants as well as breath (e.g., [191–195]).

4.2.13 Examples of Other Interfaces

Installing additional accessories in front of the ion source can render analytes amenable to ionization and subsequent mass spectrometric analysis. On-line sample treatment is especially important when analyzing liquid-phase, complex, and/or concentrated samples. For example a thermal vaporizer was used to enable analysis of liquid samples by a process mass spectrometer designed for gas analysis [196]. This system has been successfully implemented in the monitoring of an esterification reaction [197]. The obtained data were in a good agreement with those recorded by in-line mid-infrared spectrometry. The setup incorporated a magnetic sector analyzer with two detectors: an electron multiplier detector

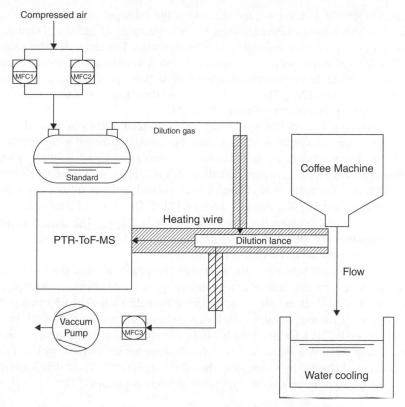

Figure 4.8 *Set up for sampling volatile organic compounds from the coffee flow. Volatiles were introduced into the dilution lance by a flow created with a vacuum pump and were then diluted 7.5 fold using dried compressed air containing a standard for mass calibration [187]. Reprinted with permission from Sánchez-López, J.A., Zimmermann, R., Yeretzian, C. (2014) Insight into the Time-resolved Extraction of Aroma Compounds during Espresso Coffee Preparation: Online Monitoring by PTR ToF-MS. Anal. Chem. 86: 11696–11704. Copyright (2014) American Chemical Society*

for low intensity ions and a Faraday cup (that measures electric current evoked by the impinging charged species) for high intensity and matrix ions [197].

In toxicological studies it is important to detect reactive electrophiles since they play critical roles in short-term toxicity and in pathogenesis (cf. [198]). However, they are unstable and present as small quantities. Therefore, trapping agents are developed which enable detection of such reactive intermediates using relatively slow analytical procedures [198]. The adducts can be analyzed by LC-MS. Glutathione is commonly used as the trapping agent while other agents are developed for specific reactions [199–205].

TRMS can provide information about possible artifacts related to the operation of standard analytical equipment (including ion sources and mass analyzers). Perfluorotri-n-butylamine (PFTBA) is often used as a calibration standard in EI-MS. Creaser *et al.* [206] observed products of the reaction of PFTBA ions with neutral species (residual water) in an ion trap. They monitored the buildup of the product of this reaction on a millisecond timescale. In this case, the reaction of the calibrant and water vapor occurred spontaneously in the mass analyzer leading to the appearance of additional signals.

Special interfaces are used for highly specialized tasks. For example, Sloane and Ratcliffe [207] devised an instrument for molecular beam sampling of transient combustion phenomena by TRMS. It was possible to detect various flame components with good spatial and temporal resolution [207]. Reaction kinetics and the chemistry of low-pressure flames was studied with aid of a photoionization method [208].

One less common but attention-grabbing application of TRMS is related to nuclear physics, in the area of synthesis of new unstable nuclei. In order to synthesize heavy atoms, Pb or Bi targets are irradiated with a stream of charged particles [209]. The newly produced heavy ions are directed through quadrupole lenses and velocity filters toward detectors. Their implantation energy is correlated with the subsequent radioactive decays in order to identify the generated nuclei [210]. Detection of new heavy elements is particularly difficult because they have very short half-lives. The data obtained from heavy ion detectors and silicon detectors are put together to match the characteristics of the new elements with the theoretical predictions.

In the technique called *Schottky mass spectrometry*, nuclei are injected into a storage ring [211]. They encircle the ring with characteristic frequencies which can be related to mass-to-charge (m/z) ratios. This technique can be applied to detect ions with lifetimes in the time range of seconds. Electron cooling is implemented to equalize velocities of all ions, which takes a few seconds [212]. On the other hand, in so-called *isochronous mass spectrometry*, electron cooling is avoided while the velocity difference between two particles of the same species is counterbalanced by a change in the orbit length [212]. Thus, characterization of nuclides which have lifetimes in the microsecond range is possible [212].

4.3   Concluding Remarks

Countless natural and artificial processes lead to the formation of temporal gradients of molecules. Collecting data on the chemical composition of dynamic samples/matrices is vital for fundamental or applied research. Several research groups and companies have introduced ingenious approaches for studying such fast-changing samples by MS. In the case of real-time monitoring, samples need to be transferred promptly to the ion source, analytes ionized, and the resulting ions separated in the mass analyzers. Therefore, some or all of the parts of the instrument may affect the temporal resolution of the analysis process. Many of the examples presented above concern the development and adaptations of various ion sources to accommodate the analysis of dynamic systems. All common ion sources find applications in different areas of TRMS. ESI-MS and DESI-MS systems are very popular, and they may certainly dominate future applications focused on the monitoring of liquid-phase processes. In some cases, the reaction chambers are highly integrated with the ion sources. Mixers and micromixers lead to the formation of spatial concentration gradients of reaction intermediates and products. After the quenching step (which can be the ionization itself), they give rise to quasi-steady samples. In this way, they can warrant high temporal resolution (millisecond or microsecond range) even when using slower mass analyzers. In some cases, it is the sample collection or treatment – rather than MS technology – that preserves temporal information about the chemical composition of labile samples.

References

1. Kaltashov, I.A., Eyles, S.J. (2012) Chapter 6, Kinetic Studies by Mass Spectrometry. In: Mass Spectrometry in Structural Biology and Biophysics: Architecture, Dynamics, and Interaction of Biomolecules, 2nd Edition. John Wiley & Sons, Inc., Hoboken.

2. Miao, Z., Chen, H., Liu, P., Liu, Y. (2011) Development of Submillisecond Time-resolved Mass Spectrometry Using Desorption Electrospray Ionization. Anal. Chem. 83: 3994–3997.

3. Chen, H., Miao, Z. (2014) Microsecond Time-resolved Mass Spectrometry. US Patent Application Publication No. US 2014/0147921 A1.

4. Van Berkel, G.J., Pasilis, S.P., Ovchinnikova, O. (2008) Established and Emerging Atmospheric Pressure Surface Sampling/Ionization Techniques for Mass Spectrometry. J. Mass Spectrom. 43: 1161–1180.

5. Cooks, R.G., Ouyang, Z., Takats, Z., Wiseman, J.M. (2006) Ambient Mass Spectrometry. Science 311: 1566–1570.

6. Weston, D.J. (2010) Ambient Ionization Mass Spectrometry: Current Understanding of Mechanistic Theory; Analytical Performance and Application Areas. Analyst 135: 661–668.

7. Harris, G.A., Galhena, A.S., Fernández, F.M. (2011) Ambient Sampling/Ionization Mass Spectrometry: Applications and Current Trends. Anal. Chem. 83: 4508–4538.

8. Monge, M.E., Harris, G.A., Dwivedi, P., Fernández, F.M. (2013) Mass Spectrometry: Recent Advances in Direct Open Air Surface Sampling/Ionization. Chem. Rev. 113: 2269–2308.

9. Owen, A.W., McAulay, E.A., Nordon, A., Littlejohn, D., Lynch, T.P., Lancaster, J.S., Wright, R.G. (2014) Monitoring of an Esterification Reaction by On-line Direct Liquid Sampling Mass Spectrometry and In-line Mid Infrared Spectrometry with an Attenuated Total Reflectance Probe. Anal. Chim. Acta 849: 12–18.

10. Santos, L.S., Knaack, L., Metzger, J.O. (2005) Investigation of Chemical Reactions in Solution Using API-MS. Int. J. Mass Spectrom. 246: 84–104.

11. Ma, X., Zhang, S., Zhang, X. (2012) An Instrumentation Perspective on Reaction Monitoring by Ambient Mass Spectrometry. Trends Anal. Chem. 35: 50–66.

12. Kistiakowsky, G.B., Kydd, P.H. (1957) A Mass Spectrometric Study of Flash Photochemical Reactions. I. J. Am. Chem. Soc. 79: 4825–4830.

13. Meyer, R.T. (1967) Flash Photolysis and Time-resolved Mass Spectrometry. I. Detection of the Hydroxyl Radical. J. Chem. Phys. 46: 967–972.

14. Carr, R.W. (1999) Flash Photolysis with Time-resolved Mass Spectrometry. In: Neckers, D.C., Volman, D.H., Von Bunau, G. (eds) Advances in Photochemistry, Volume 25. John Wiley & Sons, Inc., New York, pp. 1–57.

15. Moore, S.B., Carr, R.W. (1977) Molecular Velocity Distribution Effects in Kinetic Studies by Time-resolved Mass Spectrometry. Int. J. Mass Spectrom. Ion Phys. 24: 161–171.

16. Taatjes, C.A. (2007) How Does the Molecular Velocity Distribution Affect Kinetics Measurements by Time-resolved Mass Spectrometry? Int. J. Chem. Kinet. 39: 565–570.

17. Lee, S., Samuels, D.A., Hoobler, R.J., Leone, S.R. (2000) Direct Measurements of Rate Coefficients for the Reaction of Ethynyl Radical (C_2H) with C_2H_2 at 90 and 120 K Using a Pulsed Laval Nozzle Apparatus. J. Geophys. Res. 105: 15085–15090.

18. Lee, S., Hoobler, R.J., Leone, S.R. (2000) A Pulsed Laval Nozzle Apparatus with Laser Ionization Mass Spectroscopy for Direct Measurements of Rate Coefficients at Low Temperatures with Condensable Gases. Rev. Sci. Instrum. 71: 1816–1823.
19. Lee, S., Leone, S.R. (2000) Rate Coefficients for the Reaction of C_2H with O_2 at 90 K and 120 K Using a Pulsed Laval Nozzle Apparatus. Chem. Phys. Lett. 329: 443–449.
20. de Nalda, R., Bañares, L. (eds) (2013) Ultrafast Phenomena in Molecular Sciences: Femtosecond Physics and Chemistry (Springer Series in Chemical Physics). Springer, Berlin.
21. Van Breemen, R.B., Snow, M., Cotter, R.J. (1983) Time-resolved Laser Desorption Mass Spectrometry. I. Desorption of Preformed Ions. Int. J. Mass Spectrom. Ion Phys. 49: 35–50.
22. von der Linde, D., Danielzik, B. (1989) Picosecond Time-resolved Laser Mass Spectroscopy. IEEE J. Quant. Electron. 25: 2540–2549.
23. Blitz, M.A., Goddard, A., Ingham, T., Pilling, M.J. (2007) Time-of-flight Mass Spectrometry for Time-resolved Measurements. Rev. Sci. Instrum. 78: 034103.
24. Baeza-Romero, M.T., Blitz, M.A., Goddard, A., Seakins, P.W. (2012) Time-of-flight Mass Spectrometry for Time-resolved Measurements: Some Developments and Applications. Int. J. Chem. Kinet. 44: 532–545.
25. Ludwig, W., Brandt, B., Friedrichs, G., Temps, F. (2006) Kinetics of the Reaction $C_2H_5 + HO_2$ by Time-resolved Mass Spectrometry. J. Phys. Chem. A 110: 3330–3337.
26. Hambly, D.M., Gross, M.L. (2005) Laser Flash Photolysis of Hydrogen Peroxide to Oxidize Protein Solvent-accessible Residues on the Microsecond Timescale. J. Am. Soc. Mass Spectrom. 16: 2057–2063.
27. Welz, O., Savee, J.D., Osborn, D.L., Vasu, S.S., Percival, C.J., Shallcross, D.E., Taatjes, C.A. (2012) Direct Kinetic Measurements of Criegee Intermediate (CH_2OO) Formed by Reaction of CH_2I with O_2. Science 335: 204–207.
28. Su, Y.-T., Huang, Y.-H., Witek, H.A., Lee, Y.-P. (2013) Infrared Absorption Spectrum of the Simplest Criegee Intermediate CH_2OO. Science 340: 174–176.
29. Olesik, S., Baer, T., Morrow, J.C., Ridal, J.J., Buschek, J., Holmes, J.L. (1989) Dissociation Dynamics of Halotoluene Ions, Production of Tolyl, Benzyl and Tropylium ($[C_7H_7]^+$) ions. Org. Mass Spectrom. 24: 1008–1016.
30. Ho, Y.-P., Dunbar, R.C., Lifshitz, C. (1995) C-H Bond Strength of Naphthalene Ion. A Reevaluation Using New Time-resolved Photodissociation Results. J. Am. Chem. Soc. 117: 6504–6508.
31. Dunbar, R.C. (1987) Time-resolved Photodissociation of Chlorobenzene Ion in the ICR Spectrometer. J. Phys. Chem. 91: 2801–2804.
32. Faulk, J.D., Dunbar, R.C. (1991) Time-resolved Photodissociation of Three *tert*-Butylbenzene Ions. J. Phys. Chem. 95: 6932–6936.
33. Faulk, J.D., Dunbar, R.C. (1992) Time-resolved Photodissociation of Gas-phase Ferrocene Cation: Energetics of Fragmentation and Radiative Relaxation Rate at Nearthermal Energies. J. Am. Chem. Soc. 114: 8596–8600.
34. Kim, B., Shin, S.K. (1997) Time- and Product-resolved Photodissociations of Bromotoluene Radical Cations. J. Chem. Phys. 106: 1411–1417.
35. Dunbar, R.C., Lifshitz, C. (1991) Slow Time-resolved Photodissociation of *p*-Iodotoluene Ion. J. Chem. Phys. 94: 3542–3547.

36. Lin, C.Y., Dunbar, R.C. (1994) Time-resolved Photodissociation Rates and Kinetic Modeling for Unimolecular Dissociations of Iodotoluene Ions. J. Phys. Chem. 98: 1369–1375.
37. Kim, B., Shin, S.K. (2002) Time-resolved Photodissociations of Iodotoluene Radical Cations. J. Phys. Chem. A 106: 9918–9924.
38. Shin, S.K., Kim, B., Jarek, R.L., Han, S.-J. (2002) Product-resolved Photodissociations of Iodotoluene Radical Cations. Bull. Korean Chem. Soc. 23: 267–270.
39. Cui, W., Hadas, B., Cao, B., Lifshitz, C. (2000) Time-resolved Photodissociation (TRPD) of the Naphthalene and Azulene Cations in an Ion Trap/Reflectron. J. Phys. Chem. A 104: 6339–6344.
40. Cui, W., Hu, Y., Lifshitz, C. (2002) Time Resolved Photodissociation of Small Peptide Ions. Combining Laser Desorption with Ion Trap/Reflectron TOF Mass Spectrometry. Eur. Phys. J. D 20: 565–571.
41. Støchkel, K., Kadhane, U., Andersen, J.U., Holm, A.I., Hvelplund, P., Kirketerp, M.B., Larsen, M.K., Lykkegaard, M.K., Nielsen, S.B., Panja, S., Zettergren, H. (2008) A New Technique for Time-resolved Daughter Ion Mass Spectrometry on the Microsecond to Millisecond Time Scale Using an Electrostatic Ion Storage Ring. Rev. Sci. Instrum. 79: 023107.
42. Hatakeyama, T., Chen, D.L., Ismagilov, R.F. (2006) Microgram-scale Testing of Reaction Conditions in Solution Using Nanoliter Plugs in Microfluidics with Detection by MALDI-MS. J. Am. Chem. Soc. 128: 2518–2519.
43. Urban, P.L., Amantonico, A., Fagerer, S.R., Gehrig, P., Zenobi, R. (2010) Mass Spectrometric Method Incorporating Enzymatic Amplification for Attomole-level Analysis of Target Metabolites in Biological Samples. Chem. Commun. 46: 2212–2214.
44. Thomas, J.J., Shen, Z., Crowell, J.E., Finn, M.G., Siuzdak, G. (2001) Desorption/Ionization on Silicon (DIOS): A Diverse Mass Spectrometry Platform for Protein Characterization. Proc. Natl. Acad. Sci. USA 98: 4932–4937.
45. Hu, L., Jiang, G., Xu, S., Pan, C., Zou, H. (2006) Monitoring Enzyme Reaction and Screening Enzyme Inhibitor Based on MALDI-TOF-MS Platform with a Matrix of Oxidized Carbon Nanotubes. J. Am. Soc. Mass Spectrom. 17: 1616–1619.
46. Gross, J.W., Hegeman, A.D., Vestling, M.M., Frey, P.A. (2000) Characterization of Enzymatic Processes by Rapid Mix-quench Mass Spectrometry: The Case of dTDP-Glucose 4,6-Dehydratase. Biochemistry 39: 13633–13640.
47. Houston, C.T., Taylor, W.P., Widlanski, T.S., Reilly, J.P. (2000) Investigation of Enzyme Kinetics Using Quench-flow Techniques with MALDI-TOF Mass Spectrometry. Anal. Chem. 72: 3311–3319.
48. Gross, J.W., Frey, P.A. (2002) Rapid Mix-quench MALDI-TOF Mass Spectrometry for Analysis of Enzymatic Systems. Methods Enzymol. 354: 27–49.
49. Wilson, S.R., Perez, J., Pasternak, A. (1993) ESI-MS Detection of Ionic Intermediates in Phosphine-mediated Reactions. J. Am. Chem. Soc. 115: 1994–1997.
50. Aliprantis, A.O., Canary, J.W. (1994) Observation of Catalytic Intermediates in the Suzuki Reaction by Electrospray Mass Spectrometry. J. Am. Chem. Soc. 116: 6985–6986.
51. Sam, J.W., Tang, X.-J., Magliozzo, R.S., Peisach, J. (1995) Electrospray Mass Spectrometry of Iron Bleomycin II: Investigation of the Reaction of Fe(III)-Bleomycin with Iodosylbenzene. J. Am. Chem. Soc. 117: 1012–1018.

52. Norris, A.J., Whitelegge, J.P., Faull, K.F., Toyokuni, T. (2001) Kinetic Characterization of Enzyme Inhibitors Using Electrospray-ionization Mass Spectrometry Coupled with Multiple Reaction Monitoring. Anal. Chem. 73: 6024–6029.
53. Gilbert, B.C., Smith, J.R.L., Mairata i Payeras, A., Oakes, J., Pons i Prats, R. (2004) A Mechanistic Study of the Epoxidation of Cinnamic Acid by Hydrogen Peroxide Catalysed by Manganese 1,4,7-Trimethyl-1,4,7-triazacyclononane Complexes. J. Mol. Catal. A: Chem. 219: 265–272.
54. Greb, M., Hartung, J., Köhler, F., Špehar, K., Kluge, R., Csuk, R. (2004) The (Schiff base)vanadium(V) Complex Catalyzed Oxidation of Bromide – A New Method for the In Situ Generation of Bromine and Its Application in the Synthesis of Functionalized Cyclic Ethers. Eur. J. Org. Chem. 2004: 3799–3812.
55. Fürmeier, S., Griep-Raming, J., Hayen, A., Metzger, J.O. (2005) Chelation-controlled Radical Chain Reactions Studied by Electrospray Ionization Mass Spectrometry. Chem. Eur. J. 11: 5545–5554.
56. Makino, K., Hagiwara, T., Murakami, A. (1991) A Mini Review: Fundamental Aspects of Spin Trapping with DMPO. Int. J. Rad. Appl. Instrum. C 37: 657–665.
57. Simões, C., Domingues, P., Domingues, M.R.M. (2012) Identification of Free Radicals in Oxidized and Glycoxidized Phosphatidylethanolamines by Spin Trapping Combined with Tandem Mass Spectrometry. Rapid Commun. Mass Spectrom. 26: 931–939.
58. Hsieh, F.Y., Tong, X., Wachs, T., Ganem, B., Henion, J. (1995) Kinetic Monitoring of Enzymatic Reactions in Real Time by Quantitative High-performance Liquid Chromatography-Mass Spectrometry. Anal. Biochem. 229: 20–25.
59. Hess, T.F., Renn, T.S., Watts, R.J., Paszczynski, A.J. (2003) Studies on Nitroaromatic Compound Degradation in Modified Fenton Reactions by Electrospray Ionization Tandem Mass Spectrometry (ESI-MS-MS). Analyst 128: 156–160.
60. Cook, K.D., Bennett, K.H., Haddix, M.L. (1999) On-line Mass Spectrometry: A Faster Route to Process Monitoring and Control. Ind. Eng. Chem. Res. 38: 1192–1204.
61. Hoch, G., Kok, B. (1963) A Mass Spectrometer Inlet System for Sampling Gases Dissolved in Liquid Phases. Arch. Biochem. Biophys. 101: 160–170.
62. Creaser, C.S., Stygall, J.W., Weston, D.J. (1998) Developments in Membrane Inlet Mass Spectrometry. Anal. Commun. 35: 9H–11H.
63. Beckmann, K., Messinger, J., Badger, M.R., Wydrzynski, T., Hillier, W. (2009) On-line Mass Spectrometry: Membrane Inlet Sampling. Photosynth. Res. 102: 511–522.
64. Maher, S., Jjunju, F.P.M., Young, I.S., Brkic, B., Taylor, S. (2014) Membrane Inlet Mass Spectrometry for In Situ Environmental Monitoring. Spectroscopy Europe 26: 6–8.
65. Nogueira, R.F.P., Alberici, R.M., Mendes, M.A., Jardim, W.F., Eberlin, M.N. (1999) Photocatalytic Degradation of Phenol and Trichloroethylene: On-line and Real-time Monitoring via Membrane Introduction Mass Spectrometry. Ind. Eng. Chem. Res. 38: 1754–1758.
66. Creaser, C.S., Lamarca, D.G., Brum, J., Werner, C., New, A.P., dos Santos, L.M.F. (2002) Reversed-phase Membrane Inlet Mass Spectrometry Applied to the Real-time Monitoring of Low Molecular Weight Alcohols in Chloroform. Anal. Chem. 74: 300–304.

67. Creaser, C.S., Gómez Lamarca, D., Freitas dos Santos, L.M., New, A.P., James, P.A. (2003) A Universal Temperature Controlled Membrane Interface for the Analysis of Volatile and Semi-volatile Organic Compounds. Analyst 128: 1150–1156.

68. Creaser, C.S., Gómez Lamarca, D., Freitas dos Santos, L.M., LoBiundo, G., New, A.P. (2003) On-line Biodegradation Monitoring of Nitrogen-containing Compounds by Membrane Inlet Mass Spectrometry. J. Chem. Technol. Biotechnol. 78: 1193–1200.

69. Clinton, R., Creaser, C.S., Bryant, D. (2005) Real-time Monitoring of a Pharmaceutical Process Reaction Using a Membrane Interface Combined with Atmospheric Pressure Chemical Ionisation Mass Spectrometry. Anal. Chim. Acta 539: 133–140.

70. Caprioli, R.M., Fan, T., Cottrell, J.S. (1986) A Continuous-flow Sample Probe for Fast Atom Bombardment Mass Spectrometry. Anal. Chem. 58: 2949–2954.

71. Lento, C., Audette, G.F., Wilson, D.J. (2015) Time-resolved Electrospray Mass Spectrometry – a Brief History. Can. J. Chem. 93: 7–12.

72. Plattner, D.A. (2001) Electrospray Mass Spectrometry beyond Analytical Chemistry: Studies of Organometallic Catalysis in the Gas Phase. Int. J. Mass Spectrom. 207: 125–144.

73. Chen, P. (2003) Electrospray Ionization Tandem Mass Spectrometry in High-throughput Screening of Homogeneous Catalysts. Angew. Chem. Int. Ed. 42: 2832–2847.

74. Santos, L.S. (2008) Online Mechanistic Investigations of Catalyzed Reactions by Electrospray Ionization Mass Spectrometry: A Tool to Intercept Transient Species in Solution. Eur. J. Org. Chem. 2008: 235–253.

75. Zhu, W., Yuan, Y., Zhou, P., Zeng, L., Wang, H., Tang, L., Guo, B., Chen, B. (2012) The Expanding Role of Electrospray Ionization Mass Spectrometry for Probing Reactive Intermediates in Solution. Molecules 17: 11507–11537.

76. Adlhart, C., Chen, P. (2000) Fishing for Catalysts: Mechanism-based Probes for Active Species in Solution. Helv. Chim. Acta 83: 2192–2196.

77. Vikse, K.L., Ahmadi, Z., Manning, C.C., Harrington, D.A., McIndoe, J.S. (2011) Powerful Insight into Catalytic Mechanisms through Simultaneous Monitoring of Reactants, Products, and Intermediates. Angew. Chem. Int. Ed. 50: 8304–8306.

78. Northrop, D.B., Simpson, F.B. (1997) New Concepts in Bioorganic Chemistry. Beyond Enzyme Kinetics: Direct Determination of Mechanisms by Stopped-flow Mass Spectrometry. Bioorg. Med. Chem. 5: 641–644.

79. Kolakowski, B.M., Simmons, D.A., Konermann, L. (2000) Stopped-flow Electrospray Ionization Mass Spectrometry: A New Method for Studying Chemical Reaction Kinetics in Solution. Rapid Commun. Mass Spectrom. 14: 772–776.

80. Kolakowski, B.M., Konermann, L. (2001) From Small-molecule Reactions to Protein Folding: Studying Biochemical Kinetics by Stopped-flow Electrospray Mass Spectrometry. Anal. Biochem. 292: 107–114.

81. Lee, E.D., Mück, W., Henion, J.D., Covey, T.R. (1989) Real-time Reaction Monitoring by Continuous-introduction Ion-spray Tandem Mass Spectrometry. J. Am. Chem. Soc. 111: 4600–4604.

82. Brum, J., Dell'Orco, P., Lapka, S., Muske, K., Sisko, J. (2001) Monitoring Organic Reactions with On-line Atmospheric Pressure Ionization Mass Spectrometry: the Hydrolysis of Isatin. Rapid Commun. Mass Spectrom. 15: 1548–1553.

83. Vikse, K.L., Woods, M.P., McIndoe, J.S. (2010) Pressurized Sample Infusion for the Continuous Analysis of Air- and Moisture-sensitive Reactions Using Electrospray Ionization Mass Spectrometry. Organometallics 29: 6615–6618.

84. Arakawa, R., Tachiyashiki, S., Matsuo, T. (1995) Detection of Reaction Intermediates: Photosubstitution of (Polypyridine)ruthenium(II) Complexes Using On-line Electrospray Mass Spectrometry. Anal. Chem. 67: 4133–4138.

85. Griep-Raming, J., Metzger, J.O. (2000) An Electrospray Ionization Source for the Investigation of Thermally Initiated Reactions. Anal. Chem. 72: 5665–5668.

86. Enami, S., Vecitis, C.D., Cheng, J., Hoffmann, M.R., Colussi, A.J. (2007) Electrospray Mass Spectrometric Detection of Products and Short-lived Intermediates in Aqueous Aerosol Microdroplets Exposed to a Reactive Gas. J. Phys. Chem. A 111: 13032–13037.

87. Paiva, A.A., Tilton, R.F., Crooks, G.P., Huang, L.Q., Anderson, K.S. (1997) Detection and Identification of Transient Enzyme Intermediates Using Rapid Mixing, Pulsed-flow Electrospray Mass Spectrometry. Biochemistry 36: 15472–15476.

88. Meyer, S., Metzger, J.O. (2003) Use of Electrospray Ionization Mass Spectrometry for the Investigation of Radical Cation Chain Reactions in Solution: Detection of Transient Radical Cations. Anal. Bioanal. Chem. 377: 1108–1114.

89. Fürmeier, S., Metzger, J.O. (2004) Detection of Transient Radical Cations in Electron Transfer-initiated Diels-Alder Reactions by Electrospray Ionization Mass Spectrometry. J. Am. Chem. Soc. 126: 14485–14492.

90. Wang, H.-Y., Yim, W.-L., Klüner, T., Metzger, J.O. (2009) ESIMS Studies and Calculations on Alkali-metal Adduct Ions of Ruthenium Olefin Metathesis Catalysts and Their Catalytic Activity in Metathesis Reactions. Chem. Eur. J. 15: 10948–10959.

91. Jiang, W., Schäfer, A., Mohr, P.C., Schalley, C.A. (2010) Monitoring Self-sorting by Electrospray Ionization Mass Spectrometry: Formation Intermediates and Error-correction during the Self-assembly of Multiply Threaded Pseudorotaxanes. J. Am. Chem. Soc. 132: 2309–2320.

92. Santos, L.S., Metzger, J.O. (2006) Study of Homogeneously Catalyzed Ziegler-Natta Polymerization of Ethene by ESI-MS. Angew. Chem. Int. Ed. 45: 977–981.

93. Marquez, C., Metzger, J.O. (2006) ESI-MS Study on the Aldol Reaction Catalyzed by L-Proline. Chem. Commun. 1539–1541.

94. Rob, T., Wilson, D.J. (2012) Time-resolved Mass Spectrometry for Monitoring Millisecond Time-scale Solution-phase Processes. Eur. J. Mass Spectrom. 18: 205–214.

95. Wilson, D.J., Konermann, L. (2003) A Capillary Mixer with Adjustable Reaction Chamber Volume for Millisecond Time-resolved Studies by Electrospray Mass Spectrometry. Anal. Chem. 75: 6408–6414.

96. Zinck, N., Stark, A.-K., Wilson, D.J., Sharon, M. (2014) An Improved Rapid Mixing Device for Time-resolved Electrospray Mass Spectrometry Measurements. ChemistryOpen 3: 109–114.

97. Fisher, C.M., Kharlamova, A., McLuckey, S.A. (2014) Affecting Protein Charge State Distributions in Nano-electrospray Ionization via In-spray Solution Mixing Using Theta Capillaries. Anal. Chem. 86: 4581–4588.

98. Mortensen, D.N., Williams, E.R. (2014) Theta-glass Capillaries in Electrospray Ionization: Rapid Mixing and Short Droplet Lifetimes. Anal. Chem. 86: 9315–9321.

99. Mortensen, D.N., Williams, E.R. (2015) Investigating Protein Folding and Unfolding in Electrospray Nanodrops upon Rapid Mixing Using Theta-glass Emitters. Anal. Chem. 87: 1281–1287.

100. Marquez, C.A., Wang, H., Fabbretti, F., Metzger, J.O. (2008) Electron-transfer-catalyzed Dimerization of *trans*-Anethole: Detection of the Distonic Tetramethylene Radical Cation Intermediate by Extractive Electrospray Ionization Mass Spectrometry. J. Am. Chem. Soc. 130: 17208–17209.

101. Lee, J.K., Kim, S., Nam, H.G., Zare, R.N. (2015) Microdroplet Fusion Mass Spectrometry for Fast Reaction Kinetics. Proc. Natl. Acad. Sci. USA 112: 3898–3903.

102. Shiea, J., Chang, D.-Y., Lin, C.-H., Jiang, S.-J. (2001) Generating Multiply Charged Protein Ions by Ultrasonic Nebulization/Multiple Channel-electrospray Ionization Mass Spectrometry. Anal. Chem. 73: 4983–4987.

103. Chen, H., Venter, A., Cooks, R. G. (2006) Extractive Electrospray Ionization for Direct Analysis of Undiluted Urine, Milk and Other Complex Mixtures without Sample Preparation. Chem. Commun. 2042–2468.

104. Carroll, B., Hidrovo, C. (2013) Experimental Investigation of Inertial Mixing in Colliding Droplets. Heat Transf. Eng. 34: 120–130.

105. Wortmann, A., Kistler-Momotova, A., Zenobi, R., Heine, M.C., Wilhelm, O., Pratsinis, S.E. (2007) Shrinking Droplets in Electrospray Ionization and Their Influence on Chemical Equilibria. J. Am. Soc. Mass Spectrom. 18: 385–393.

106. Oleschuk, R.D., Harrison, D.J. (2000) Analytical Microdevices for Mass Spectrometry. Trends Anal. Chem. 19: 379–388.

107. Takáts, Z., Wiseman, J.M., Gologan, B., Cooks, R.G. (2004) Mass Spectrometry Sampling Under Ambient Conditions with Desorption Electrospray Ionization. Science 306: 471–473.

108. Miao, Z., Chen, H. (2009) Direct Analysis of Liquid Samples by Desorption Electrospray Ionization-Mass Spectrometry (DESI-MS). J. Am. Soc. Mass Spectrom. 20: 10–19.

109. Perry, R.H., Splendore, M., Chien, A., Davis, N.K., Zare, R.N. (2011) Detecting Reaction Intermediates in Liquids on the Millisecond Time Scale Using Desorption Electrospray Ionization. Angew. Chem. Int. Ed. 50: 250–254.

110. Cotte-Rodríguez, I., Takáts, Z., Talaty, N., Chen, H., Cooks, R.G. (2005) Desorption Electrospray Ionization of Explosives on Surfaces: Sensitivity and Selectivity Enhancement by Reactive Desorption Electrospray Ionization. Anal. Chem. 77: 6755–6764.

111. Huang, G., Chen, H., Zhang, X., Cooks, R.G., Ouyang, Z. (2007) Rapid Screening of Anabolic Steroids in Urine by Reactive Desorption Electrospray Ionization. Anal. Chem. 79: 8327–8332.

112. Nyadong, L., Green, M.D., De Jesus, V.R., Newton, P.N., Fernández, F.M. (2007) Reactive Desorption Electrospray Ionization Linear Ion Trap Mass Spectrometry of Latest-generation Counterfeit Antimalarials via Noncovalent Complex Formation. Anal. Chem. 79: 2150–2157.

113. Song, Y., Cooks, R.G. (2007) Reactive Desorption Electrospray Ionization for Selective Detection of the Hydrolysis Products of Phosphonate Esters. J. Mass Spectrom. 42: 1086–1092.

114. Wu, C., Ifa, D.R., Manicke, N.E., Cooks, R.G. (2009) Rapid, Direct Analysis of Cholesterol by Charge Labeling in Reactive Desorption Electrospray Ionization. Anal. Chem. 81: 7618–7624.

115. Zhang, Y., Chen, H. (2010) Detection of Saccharides by Reactive Desorption Electrospray Ionization (DESI) Using Modified Phenylboronic Acids. Int. J. Mass Spectrom. 289: 98–107.

116. Girod, M., Moyano, E., Campbell, D.I., Cooks, R.G. (2011) Accelerated Bimolecular Reactions in Microdroplets Studied by Desorption Electrospray Ionization Mass Spectrometry. Chem. Sci. 2: 501–510.

117. Li, J., Dewald, H.D., Chen, H. (2009) Online Coupling of Electrochemical Reactions with Liquid Sample Desorption Electrospray Ionization-Mass Spectrometry. Anal. Chem. 81: 9716–9722.

118. Roach, P.J., Laskin, J., Laskin, A. (2010) Nanospray Desorption Electrospray Ionization: an Ambient Method for Liquid-extraction Surface Sampling in Mass Spectrometry. Analyst 135: 2233–2236.

119. Wilm, M., Mann, M. (1996) Analytical Properties of the Nanoelectrospray Ion Source. Anal. Chem. 68: 1–8.

120. Hsieh, C.-H., Chang, C.-H., Urban, P.L., Chen, Y.-C. (2011) Capillary Action-supported Contactless Atmospheric Pressure Ionization for the Combined Sampling and Mass Spectrometric Analysis of Biomolecules. Anal. Chem. 83: 2866–2869.

121. Wachs, T., Henion, J. (2001) Electrospray Device for Coupling Microscale Separations and Other Miniaturized Devices with Electrospray Mass Spectrometry. Anal. Chem. 73: 632–638.

122. Van Berkel, G.J., Kertesz, V. (2009) Electrochemically Initiated Tagging of Thiols Using an Electrospray Ionization Based Liquid Microjunction Surface Sampling Probe Two-electrode Cell. Rapid Commun. Mass Spectrom. 23: 1380–1386.

123. Hsu, C.-C., White, N.M., Hayashi, M., Lin, E.C., Poon, T., Banerjee, I., Chen, J., Pfaff, S.L., Macagno, E.R., Dorrestein, P.C. (2013) Microscopy Ambient Ionization Top-down Mass Spectrometry Reveals Developmental Patterning. Proc. Natl. Acad. Sci. USA 110: 14855–14860.

124. Chen, H., Eberlin, L.S., Nefliu, M., Augusti, R., Cooks, R.G. (2008) Organic Reactions of Ionic Intermediates Promoted by Atmospheric-pressure Thermal Activation. Angew. Chem. Int. Ed. 47: 3422–3425.

125. Jecklin, M.C., Touboul, D., Bovet, C., Wortmann, A., Zenobi, R. (2008) Which Electrospray-based Ionization Method Best Reflects Protein-Ligand Interactions Found in Solution? A Comparison of ESI, nanoESI, and ESSI for the Determination of Dissociation Constants with Mass Spectrometry. J. Am. Soc. Mass Spectrom. 19: 332–343.

126. Zhu, L., Gamez, G., Chen, H.W., Huang, H.X., Chingin, K., Zenobi, R. (2008) Real-Time, On-line Monitoring of Organic Chemical Reactions Using Extractive Electrospray Ionization Tandem Mass Spectrometry. Rapid Commun. Mass Spectrom. 22: 2993–2998.

127. McCullough, B.J., Bristow, T., O'Connor, G., Hopley, C. (2011) On-line Reaction Monitoring by Extractive Electrospray Ionisation. Rapid Commun. Mass Spectrom. 25, 1445–1451.

128. Law, W.S., Chen, H., Ding, J., Yang, S., Zhu, L., Gamez, G., Chingin, K., Ren, Y., Zenobi, R. (2009) Rapid Characterization of Complex Viscous Liquids at the Molecular Level. Angew. Chem. Int. Ed. 48: 8277–8280.

129. Yu, Z., Chen, L.C., Erra-Balsells, R., Nonami, H., Hiraoka, K. (2010) Real-time Reaction Monitoring by Probe Electrospray Ionization Mass Spectrometry. Rapid Commun. Mass Spectrom. 24: 1507–1513.

130. Yu, Z., Chen, L.C., Mandal, M.K., Nonami, H., Erra-Balsells, R., Hiraoka, K. (2012) Online Electrospray Ionization Mass Spectrometric Monitoring of Protease-catalyzed Reactions in Real Time. J. Am. Soc. Mass Spectrom. 23: 728–735.

131. Chen, T.-Y., Lin, J.-Y., Chen, J.-Y., Chen, Y.-C. (2010) Ultrasonication-assisted Spray Ionization Mass Spectrometry for the Analysis of Biomolecules in Solution. J. Am. Soc. Mass Spectrom. 21: 1547–1553.

132. Chen, T.-Y., Chao, C.-S., Mong, K.-K. T., Chen, Y.-C. (2010) Ultrasonication-assisted Spray Ionization Mass Spectrometry for On-line Monitoring of Organic Reactions. Chem. Commun. 46: 8347–8349.

133. Lin, S.-H., Lo, T.-J., Kuo, F.-Y., Chen, Y.-C. (2014) Real Time Monitoring of Accelerated Chemical Reactions by Ultrasonication-assisted Spray Ionization Mass Spectrometry. J. Mass Spectrom. 49: 50–56.

134. Yamaguchi, K. (2003) Cold-spray Ionization Mass Spectrometry: Principle and Applications. J. Mass Spectrom. 38: 473–490.

135. Santos, V.G., Regiani, T., Dias, F.F.G., Romão, W., Paz Jara, J.L., Klitzke, C.F., Coelho, F., Eberlin, M.N. (2011) Venturi Easy Ambient Sonic-spray Ionization. Anal. Chem. 83: 1375–1380.

136. Hu, J.-B., Chen, T.-R., Chen, Y.-C., Urban, P.L. (2015) Microcontroller-assisted Compensation of Adenosine Triphosphate Levels: Instrument and Method Development. Sci. Rep. 5: 8135.

137. Ting, H., Urban, P.L. (2014) Spatiotemporal Effects of a Bioautocatalytic Chemical Wave Revealed by Time-resolved Mass Spectrometry. RSC Adv. 4: 2103–2108.

138. Hu, J.-B., Chen, S.-Y., Wu, J.-T., Chen, Y.-C., Urban, P.L. (2014) Automated System for Extraction and Instantaneous Analysis of Millimeter-sized Samples. RSC Adv. 4: 10693–10701.

139. Chen, S.-Y., Urban, P.L. (2015) On-line Monitoring of Soxhlet Extraction by Chromatography and Mass Spectrometry to Reveal Temporal Extract Profiles. Anal. Chim. Acta 881: 74–81.

140. Sun, S., Kennedy, R.T. (2014) Droplet Electrospray Ionization Mass Spectrometry for High Throughput Screening for Enzyme Inhibitors. Anal. Chem. 86: 9309–9314.

141. Jin, D.-Q., Zhu, Y., Fang, Q. (2014) Swan Probe: A Nanoliter-scale and High-throughput Sampling Interface for Coupling Electrospray Ionization Mass Spectrometry with Microfluidic Droplet Array and Multiwell Plate. Anal. Chem. 86: 10796–10803.

142. Hogenboom, A.C., de Boer, A.R., Derks, R.J.E., Irth, H. (2001) Continuous-flow, On-line Monitoring of Biospecific Interactions Using Electrospray Mass Spectrometry. Anal. Chem. 73: 3816–3823.

143. Cheng, C.-Y., Yuan, C.-H., Cheng, S.-C., Huang, M.-Z., Chang, H.-C., Cheng, T.-L., Yeh, C.-S., Shiea, J. (2008) Electrospray-assisted Laser Desorption/Ionization Mass

Spectrometry for Continuously Monitoring the States of Ongoing Chemical Reactions in Organic or Aqueous Solution under Ambient Conditions. Anal. Chem. 80: 7699–7705.

144. Peng, I.X., Ogorzalek Loo, R.R., Shiea, J., Loo, J.A. (2008) Reactive-electrospray-assisted Laser Desorption/Ionization for Characterization of Peptides and Proteins. Anal. Chem. 80: 6995–7003.

145. Mühlberger, F., Zimmermann, R., Kettrup, A. (2001) A Mobile Mass Spectrometer for Comprehensive On-line Analysis of Trace and Bulk Components of Complex Gas Mixtures: Parallel Application of the Laser-based Ionization Methods VUV Single-photon Ionization, Resonant Multiphoton Ionization, and Laser-induced Electron Impact Ionization. Anal. Chem. 73: 3590–3604.

146. Rowland, A., Housh, T.B., Holcombe, J.A. (2008) Use of Electrothermal Vaporization for Volatility-based Separation of Rb-Sr Isobars for Determination of Isotopic Ratios by ICP-MS. J. Anal. At. Spectrom. 23: 167–172.

147. Cao, G.X., Jimenez, O., Zhou, F. (2006) Nafion-coated Bismuth Film and Nafion-coated Mercury Film Electrodes for Anodic Stripping Voltammetry Combined On-line with ICP-Mass Spectrometry. J. Am. Soc. Mass Spectrom. 17: 945–952.

148. Hu, S., Liu, R., Zhang, S., Huang, Z., Xing, Z., Zhang, X. (2009) A New Strategy for Highly Sensitive Immunoassay Based on Single-particle Mode Detection by Inductively Coupled Plasma Mass Spectrometry. J. Am. Soc. Mass Spectrom. 20: 1096–1103.

149. Bendall, S.C., Simonds, E.F., Qiu, P., Amir, E.D., Krutzik, P.O., Finck, R., Bruggner, R.V., Melamed, R., Trejo, A., Ornatsky, O.I., Balderas, R.S., Plevritis, S.K., Sachs, K., Pe'er, D., Tanner, S.D., Nolan, G.P. (2011) Single-cell Mass Cytometry of Differential Immune and Drug Responses Across a Human Hematopoietic Continuum. Science 332: 687–696.

150. Zhu, Z., Bartmess, J.E., McNally, M.E., Hoffman, R.M., Cook, K.D., Song, L. (2012) Quantitative Real-time Monitoring of Chemical Reactions by Autosampling Flow Injection Analysis Coupled with Atmospheric Pressure Chemical Ionization Mass Spectrometry. Anal. Chem. 84: 7547–7554.

151. Ma, X., Zhang, S., Lin, Z., Liu, Y., Xing, Z., Yang, C., Zhang, X. (2009) Real-time Monitoring of Chemical Reactions by Mass Spectrometry Utilizing a Low-temperature Plasma Probe. Analyst 134: 1863–1867.

152. Van Berkel, G.J., Kertesz, V. (2010) Electrochemistry of the Electrospray Ion Source. In: Cole, R.B. (ed.) Electrospray and MALDI Mass Spectrometry: Fundamentals, Instrumentation, Practicalities, and Biological Applications, 2nd Edition. John Wiley & Sons, Inc., Hoboken.

153. Pozniak, B.P., Cole, R.B. (2015) Perspective on Electrospray Ionization and Its Relation to Electrochemistry. J. Am. Soc. Mass Spectrom. 26: 369–385.

154. Ochran, R.A., Konermann, L. (2004) Effects of Ground Loop Currents on Signal Intensities in Electrospray Mass Spectrometry. J. Am. Soc. Mass Spectrom. 15: 1748–1754.

155. Corkery, L.J., Pang, H., Schneider, B.B., Covey, T.R., Siu, K.W.M. (2005) Automated Nanospray Using Chip-based Emitters for the Quantitative Analysis of Pharmaceutical Compounds. J. Am. Soc. Mass Spectrom. 16: 363–369.

156. Seiwert, B., Henneken, H., Karst, U. (2004) Ferrocenoyl Piperazide as Derivatizing Agent for the Analysis of Isocyanates and Related Compounds Using Liquid Chromatography/Electrochemistry/Mass Spectrometry (LC/EC/MS). J. Am. Soc. Mass Spectrom. 15: 1727–1736.

157. Diehl, G., Karst, U. (2002) On-line Electrochemistry – MS and Related Techniques. Anal. Bioanal. Chem. 373: 390–398.

158. Bruckenstein, S., Gadde, R.R. (1971) Use of a Porous Electrode for *In Situ* Mass Spectrometric Determination of Volatile Electrode Reaction Products. J. Am. Chem. Soc. 93: 793–794.

159. Kertesz, V., Van Berkel, G.J. (2005) Monitoring Ionic Adducts to Elucidate Reaction Mechanisms: Reduction of Tetracyanoquinodimethane and Oxidation of Triphenylamine Investigated Using On-line Electrochemistry/Electrospray Mass Spectrometry. J. Solid State Electrochem. 9: 390–397.

160. Liu, P., Lu, M., Zheng, Q., Zhang, Y., Dewald, H.D., Chen, H. (2013) Recent Advances of Electrochemical Mass Spectrometry. Analyst 138: 5519–5539.

161. Baltruschat, H. (2004) Differential Electrochemical Mass Spectrometry. J. Am. Soc. Mass Spectrom. 15: 1693–1706.

162. de Groot, M.T., Koper, M.T.M. (2004) The Influence of Nitrate Concentration and Acidity on the Electrocatalytic Reduction of Nitrate on Platinum. J. Electroanal. Chem. 562: 81–94.

163. Hambitzer, G., Heitbaum, J. (1986) Electrochemical Thermospray Mass Spectrometry. Anal. Chem. 58: 1067–1070.

164. Volk, K.J., Yost, R.A., Brajter-Toth, A. (1989) On-line Electrochemistry/Thermospray/Tandem Mass Spectrometry as a New Approach to the Study of Redox Reactions: the Oxidation of Uric Acid. Anal. Chem. 61: 1709–1717.

165. Volk, K.J., Yost, R.A., Brajter-Toth, A. (1990) On-line Mass Spectrometric Investigation of the Peroxidase-catalysed Oxidation of Uric Acid. J. Pharm. Biomed. Anal. 8: 205–215.

166. Zhou, F., Van Berkel, G.J. (1995) Electrochemistry Combined On-line with Electrospray Mass Spectrometry. Anal. Chem. 67: 3643–3649.

167. Bond, A.M., Colton, R., D'Agostino, A., Downard, A.J., Traeger, J.C. (1995) A Role for Electrospray Mass Spectrometry in Electrochemical Studies. Anal. Chem. 67: 1691–1695.

168. Roeser, J., Permentier, H.P., Bruins, A.P., Bischoff, R. (2010) Electrochemical Oxidation and Cleavage of Tyrosine- and Tryptophan-containing Tripeptides. Anal. Chem. 82: 7556–7565.

169. Modestov, A.D., Srebnik, S., Lev, O., Gun, J. (2001) Scanning Capillary Microscopy/Mass Spectrometry for Mapping Spatial Electrochemical Activity of Electrodes. Anal. Chem. 73: 4229–4240.

170. Odijk, M., Olthuis, W., van den Berg, A. (2012) Improved Conversion Rates in Drug Screening Applications Using Miniaturized Electrochemical Cells with Frit Channels. Anal. Chem. 84: 9176–9183.

171. Karst, U. (2004) Electrochemistry/Mass Spectrometry (EC/MS) – A New Tool to Study Drug Metabolism and Reaction Mechanisms. Angew. Chem. Int. Ed. 43: 2476–2478.

172. Jahn, S., Lohmann, W., Bomke, S., Baumann, A., Karst, U. (2012) A Ferrocene-based Reagent for the Conjugation and Quantification of Reactive Metabolites. Anal. Bioanal. Chem. 402: 461–471.

173. Odijk, M., Baumann, A., Lohmann, W., van den Brink, F.T.G., Olthuis, W., Karst, U., van den Berg, A. (2009) A Microfluidic Chip for Electrochemical Conversions in Drug Metabolism Studies. Lab Chip 9: 1687–1693.

174. Brown, T.A., Chen, H., Zare, R.N. (2015) Identification of Fleeting Electrochemical Reaction Intermediates Using Desorption Electrospray Ionization Mass Spectrometry. J. Am. Chem. Soc. 137: 7274–7277.

175. Liu, P., Lanekoff, I.T., Laskin, J., Dewald, H.D., Chen, H. (2012) Study of Electrochemical Reactions Using Nanospray Desorption Electrospray Ionization Mass Spectrometry. Anal. Chem. 84: 5737–5743.

176. Aerodyne Research. Aerosol Mass Spectrometer, http://www.aerodyne.com/products/acrosol-mass-spectrometer (accessed March 26, 2015).

177. EMSL. Mass Spectromer: Aerosol, Time-of-flight, High Resolution, http://www.emsl.pnl.gov/emslweb/instruments/mass-spectrometer-aerosol-time-flight-high-resolution (accessed March 27, 2015).

178. Drewnick, F., Hings, S.S., DeCarlo, P., Jayne, J.T., Gonin, M., Fuhrer, K., Weimer, S., Jimenez, J.L., Demerjian, K.L., Borrmann, S., Worsnop, D.R. (2005) A New Time-of-flight Aerosol Mass Spectrometer (TOF-AMS) – Instrument Description and First Field Deployment. Aerosol Sci. Technol. 39: 637–658.

179. Nölting, B. (2006) Methods in Modern Biophysics, 2nd Edition. Springer, Berlin.

180. Jayne, J.T., Leard, D.C., Zhang, X., Davidovits, P., Smith, K.A., Kolb, C.E., Worsnop, D.R. (2000) Development of an Aerosol Mass Spectrometer for Size and Composition Analysis of Submicron Particles. Aerosol Sci. Technol. 33: 49–70.

181. Elsasser, M., Crippa, M., Orasche, J., DeCarlo, P.F., Oster, M., Pitz, M., Cyrys, J., Gustafson, T.L., Pettersson, J.B.C., Schnelle-Kreis, J., Prévôt, A.S.H., Zimmermann, R. (2012) Organic Molecular Markers and Signature from Wood Combustion Particles in Winter Ambient Aerosols: Aerosol Mass Spectrometer (AMS) and High Time-resolved GC-MS Measurements in Augsburg, Germany. Atmos. Chem. Phys. 12: 6113–6128.

182. Vogel, A.L., Äijälä, M., Brüggemann, M., Ehn, M., Junninen, H. Petäjä, T., Worsnop, D.R., Kulmala, M., Williams, J., Hoffmann, T. (2013) Online Atmospheric Pressure Chemical Ionization Ion Trap Mass Spectrometry (APCI-IT-MS^n) for Measuring Organic Acids in Concentrated Bulk Aerosol – a Laboratory and Field Study. Atmos. Meas. Tech. 6: 431–443.

183. Voisin, D., Smith, J.N., Sakurai, H., McMurry, P.H., Eisele, F.L. (2003) Thermal Desorption Chemical Ionization Mass Spectrometer for Ultrafine Particle Chemical Composition. Aerosol Sci. Technol. 37: 471–475.

184. Pratt, K.A., Prather, K.A. (2012) Mass Spectrometry of Atmospheric Aerosols – Recent Developments and Applications. Part II: On-line Mass Spectrometry Techniques. Mass Spectrom. Rev. 31: 17–48.

185. Partridge, W.P., Storey, J.M.E., Lewis, S.A., Smithwick, R.W., Devault, G.L., Cunningham, M.J., Currier, N.W., Yonushonis, T.M. (2000) Time-resolved Measurements of Emission Transients By Mass Spectrometry. SAE Technical Paper 2000-01-2952, DOI: 10.4271/2000-01-2952.

186. Ellis, A.M., Mayhew, C.A. (2013) Proton Transfer Reaction Mass Spectrometry: Principles and Applications. John Wiley & Sons, Ltd, Chichester.

187. Sánchez-López, J.A., Zimmermann, R., Yeretzian, C. (2014) Insight into the Time-resolved Extraction of Aroma Compounds during Espresso Coffee Preparation: Online Monitoring by PTR-ToF-MS. Anal. Chem. 86: 11696–11704.

188. Adams, N.G., Smith, D. (1976) The Selected Ion Flow Tube (SIFT); A Technique for Studying Ion-neutral Reactions. In. J. Mass Spectrom. Ion Phys. 21: 349–359.

189. Smith, D., Sovová, K., Španěl, P. (2012) A Selected Ion Flow Tube Study of the Reactions of H_3O^+, NO^+ and $O_2^{+\bullet}$ with Seven Isomers of Hexanol in Support of SIFT-MS. Int. J. Mass Spectrom. 319-320: 25–30.

190. Smith, D., Španěl, P. (2015) SIFT-MS and FA-MS Methods for Ambient Gas Phase Analysis: Developments and Applications in the UK. Analyst 140: 2573–2591.

191. Španěl, P., Smith, D. (1996) Selected Ion Flow Tube: A Technique for Quantitative Trace Gas Analysis of Air and Breath. Med. Biol. Eng. Comput. 34: 409–419.

192. Senthilmohan, S.T., McEwan, M.J., Wilson, P.F., Milligan, D.B., Freeman, C.G. (2001) Real Time Analysis of Breath Volatiles Using SIFT-MS in Cigarette Smoking. Redox Rep. 6: 185–187.

193. Milligan, D.B., Wilson, P.F., Mautner, M.N., Freeman, C.G., McEwan, M.J., Clough, T.J., Sherlock, R.R. (2002) Real-time, High-resolution Quantitative Measurement of Multiple Soil Gas Emissions: Selected Ion Flow Tube Mass Spectrometry. J. Environ. Qual. 31: 515–524.

194. Francis, G.J., Langford, V.S., Milligan, D.B., McEwan, M.J. (2009) Real-time Monitoring of Hazardous Air Pollutants. Anal. Chem. 81: 1595–1599.

195. Hastie, D.R., Gray, J., Langford, V.S., Maclagan, R.G.A.R., Milligan, D.B., McEwan, M.J. (2010) Real-time Measurement of Peroxyacetyl Nitrate Using Selected Ion Flow Tube Mass Spectrometry. Rapid Commun. Mass Spectrom. 24: 343–348.

196. Owen, A.W., Nordon, A., Littlejohn, D., Lynch, T.P., Lancaster, J.S., Wright, R.G. (2014) On-line Detection and Quantification of Trace Impurities in Vaporisable Samples by Direct Liquid Introduction Process Mass Spectrometry. Anal. Meth. 6: 8148–8153.

197. Owen, A.W., McAulay, E.A.J., Nordon, A., Littlejohn, D., Lynch, T.P., Lancaster, J.S., Wright, R.G. (2014) Monitoring of an Esterification Reaction by On-line Direct Liquid Sampling Mass Spectrometry and In-line Mid Infrared Spectrometry with an Attenuated Total Reflectance Probe. Anal. Chim. Acta 849: 12–18.

198. Laine, J.E., Auriola, S., Pasanen, M., Juvonen, R.O. (2011) D-isomer of *gly-tyr-pro-cys-pro-his-pro* Peptide: A Novel and Sensitive *In Vitro* Trapping Agent to Detect Reactive Metabolites by Electrospray Mass Spectrometry. Toxicol. in Vitro 25: 411–425.

199. Soglia, J.R., Contillo, L.G., Kalgutkar, A.S., Zhao, S., Hop, C.E.C.A., Boyd, J.G., Cole, M.J. (2006) A Semiquantitative Method for the Determination of Reactive Metabolite Conjugate Levels In Vitro Utilizing Liquid Chromatography-Tandem Mass Spectrometry and Novel Quaternary Ammonium Glutathione Analogues. Chem. Res. Toxicol. 19: 480–490.

200. Harada, H., Endo, T., Momose, Y., Kusama, H. (2009) A Liquid Chromatography/Tandem Mass Spectrometry Method for Detecting UGT-mediated Bioactivation

of Drugs as Their *N*-acetylcysteine Adducts in Human Liver Microsomes. Rapid Commun. Mass Spectrom. 23: 564–570.

201. Motwani, H.V., Fred, C., Haglund, J., Golding, B.T., Törnqvist, M. (2009) Cob(I)alamin for Trapping Butadiene Epoxides in Metabolism with Rat S9 and for Determining Associated Kinetic Parameters. Chem. Res. Toxicol. 22: 1509–1516.

202. Rousu, T., Pelkonen, O., Tolonen, A. (2009) Rapid Detection and Characterization of Reactive Drug Metabolites *In Vitro* Using Several Isotope-labeled Trapping Agents and Ultra-performance Liquid Chromatography/Time-of-flight Mass Spectrometry. Rapid Commun. Mass Spectrom. 23: 843–855.

203. LeBlanc, A., Shiao, T.C., Roy, R., Sleno, L. (2010) Improved Detection of Reactive Metabolites with a Bromine-containing Glutathione Analog Using Mass Defect and Isotope Pattern Matching. Rapid Commun. Mass Spectrom. 24: 1241–1250.

204. Defoy, D., Dansette, P.M., Neugebauer, W., Wagner, J.R., Klarskov, K. (2011) Evaluation of Deuterium Labeled and Unlabeled Bis-methyl Glutathione Combined with Nanoliquid Chromatography-Mass Spectrometry to Screen and Characterize Reactive Drug Metabolites. Chem. Res. Toxicol. 24: 412–417.

205. Rousu, T., Tolonen, A. (2011) Characterization of Cyanide-trapped Methylated Metabonates Formed during Reactive Drug Metabolite Screening *In Vitro*. Rapid Commun. Mass Spectrom. 25: 1382–1390.

206. Creaser, C.S., West, S.K., Wilkins, J.P.G. (2000) Reactions of Perfluorotri-*n*-butylamine Fragment Ions in the Quadrupole Ion Trap: the Origin of Artefacts in the Perfluorotri-*n*-butylamine Calibration Spectrum. Rapid Commun. Mass Spectrom. 14: 538–540.

207. Sloane, T.M., Ratcliffe, J.W. (1983) Time-resolved Mass Spectrometry of a Propagating Methane-Oxygen-Argon Flame. Combust. Sci. Technol. 33: 65–74.

208. Taatjes, C.A., Hansen, N., Osborn, D.L., Kohse-Höinghaus, K., Cool, T.A., Westmoreland, P.R. (2008) "Imaging" Combustion Chemistry via Multiplexed Synchrotron-Photoionization Mass Spectrometry. Phys. Chem. Chem. Phys. 10: 20–34.

209. GSI. The SHIP Setup, https://www.gsi.de/en/work/research/nustarenna/nustarenna_divisions/she_physik/experimentel_setup/ship.htm?nr=%2Fproc%2Fself%2Fenviron%27 (accessed April 9, 2015).

210. The Transuranium Elements, http://oregonstate.edu/instruct/ch374/ch418518/Chapter%2015%20%20The%20Transuranium%20Elements-rev.pdf (accessed April 9, 2015).

211. Franzke, B., Beckert, K., Eickhoff, H., Nolden, F., Reich, H., Schaaf, U., Schlitt, B., Schwinn, A., Steck, M., Winkler, T. (1995) Schottky Mass Spectrometry at the Experimental Storage Ring ESR. Physica Scripta T59: 176–178.

212. Hausmann, M., Attallah, F., Beckert, K., Bosch, F., Dolinskiy, A., Eickhoff, H., Falch, M., Franczak, B., Franzke, B., Geissel, H., Kerscher, T., Klepper, O., Kluge, H.-J., Kozhuharov, C., Löbner, K.E.G., Münzenberg, G., Nolden, F., Novikov, Y.N., Radon, T., Schatz, H., Scheidenberger, C., Stadlmann, J., Steck, M., Winkler, T., Wollnik, H. (2000) First Isochronous Mass Spectrometry at the Experimental Storage Ring ESR. Nucl. Instr. Meth. Phys. Res. A 446: 569–580.

213. Katta, V., Chait, B.T. (1991) Conformational Changes in Proteins Probed by Hydrogen-exchange Electrospray-ionization Mass Spectrometry. Rapid Commun. Mass Spectrom. 5: 214–217.

214. Winger, B.E., Light-Wahl, K.J., Rockwood, A.L., Smith, R.D. (1992) Probing Qualitative Conformation Differences of Multiply Protonated Gas-phase Proteins via H/D Isotopic Exchange with D_2O. J. Am. Chem. Soc. 114: 5898–5900.

215. Katta, V., Chait, B.T. (1993) Hydrogen/Deuterium Exchange Electrospray Ionization Mass Spectrometry: A Method for Probing Protein Conformational Changes in Solution. J. Am. Chem. Soc. 115: 6317–6321.

216. Miranker, A., Robinson, C.V., Radford, S.E., Aplin, R.T., Dobson, C.M. (1993) Detection of Transient Protein Folding Populations by Mass Spectrometry. Science 262: 896–900.

217. Suckau, D., Shi, Y., Beu, S.C., Senko, M.W., Quinn, J.P., Wampler III, F.M., McLafferty, F.W. (1993) Coexisting Stable Conformations of Gaseous Protein Ions. Proc. Natl. Acad. Sci. USA 90: 790–793.

218. Sam, J.W., Tang, X.-J., Peisach, J. (1994) Electrospray Mass Spectrometry of Iron Bleomycin: Demonstration that Activated Bleomycin Is a Ferric Peroxide Complex. J. Am. Chem. Soc. 116: 5250–5256

219. Arakawa, R., Jian, L., Yoshimura, A., Nozaki, K., Ohno, T., Doe, H., Matsuo, T. (1995) On-Line Mass Analysis of Reaction Products by Electrospray Ionization. Photosubstitution of Ruthenium(II) Diimine Complexes. Inorg. Chem. 34: 3874–3878.

220. Hooke, S.D., Eyles, S.J., Miranker, A., Radford, S.E., Robinson, C.V., Dobson, C.M. (1995) Cooperative Elements in Protein Folding Monitored by Electrospray Ionization Mass Spectrometry. J. Am. Chem. Soc. 117: 1548–1549.

221. Arakawa, R., Mimura, S., Matsubayashi, G., Matsuo, T. (1996) Photolysis of (Diamine)bis(2,2′-bipyridine)ruthenium(II) Complexes Using On-line Electrospray Mass Spectrometry. Inorg. Chem. 35: 5725–5729.

222. Gross, D.S., Schnier, P.D., Rodriguez-Cruz, S.E., Fagerquist, C.K., Williams, E.R. (1996) Conformations and Folding of Lysozyme Ions *in Vacuo*. Proc. Natl. Acad. Sci. USA 93: 3143–3148.

223. Ripa, L., Hallberg, A. (1996) Controlled Double-bond Migration in Palladium-catalyzed Intramolecular Arylation of Enamidines. J. Org. Chem. 61: 7147–7155.

224. Schuster, D.I., Cao, J., Kaprinidis, N., Wu, Y., Jensen, A.W., Lu, Q., Wang, H., Wilson, S.R. (1996) [2 + 2] Photocycloaddition of Cyclic Enones to C_{60}. J. Am. Chem. Soc. 118: 5639–5647.

225. Arakawa, R., Lu, J., Mizuno, K., Inoue, H., Doe, H., Matsuo, T. (1997) On-line Electrospray Mass Analysis of Photoallylation Reactions of Dicyanobenzenes by Allylic Silanes via Photoinduced Electron Transfer. Int. J. Mass Spectrom. Ion Proc. 160: 371–376.

226. Arakawa, R., Matsuda, F., Matsubayashi, G., Matsuo, T. (1997) Structural Analysis of Photo-oxidized (Ethylenediamine)bis(2,2'-bipyridine)ruthenium(II) Complexes by Using On-line Electrospray Mass Spectrometry of Labeled Compounds. J. Am. Soc. Mass Spectrom. 8: 713–717.

227. Hinderling, C., Feichtinger, D., Plattner, D.A., Chen, P. (1997) A Combined Gas-phase, Solution-phase, and Computational Study of C-H Activation by Cationic Iridium(III) Complexes. J. Am. Chem. Soc. 119: 10793–10804.

228. Kimura, K., Mizutani, R., Yokoyama, M., Arakawa, R., Matsubayashi, G., Okamoto, M., Doe, H. (1997) All-or-none Type Photochemical Switching of Cation Binding with Malachite Green Carrying a Bis(monoazacrown ether) Moiety. J. Am. Chem. Soc. 119: 2062–2063.

229. Konermann, L., Rosell, F.I., Mauk, A.G., Douglas, D.J. (1997) Acid-induced Denaturation of Myoglobin Studied by Time-resolved Electrospray Ionization Mass Spectrometry. Biochemistry 36: 6448–6454.
230. Konermann, L., Collings, B.A., Douglas, D.J. (1997) Cytochrome *c* Folding Kinetics Studied by Time-resolved Electrospray Ionization Mass Spectrometry. Biochemistry 36: 5554–5559.
231. Yang, H., Smith, D.L. (1997) Kinetics of Cytochrome *c* Folding Examined by Hydrogen Exchange and Mass Spectrometry. Biochemistry 36: 14992–14999.
232. Brum, J., Dell'Orco, P. (1998) On-line Mass Spectrometry: Real-time Monitoring and Kinetics Analysis for the Photolysis of Idoxifene. Rapid Commun. Mass Spectrom. 12: 741–745.
233. Wang, C.-H., Huang, M.-W., Lee, C.-Y., Chei, H.-L., Huang, J.-P., Shiea, J. (1998) Detection of a Thermally Unstable Intermediate in the Wittig Reaction Using Low-temperature Liquid Secondary Ion and Atmospheric Pressure Ionization Mass Spectrometry. J. Am. Soc. Mass Spectrom. 9: 1168–1174.
234. Zechel, D.L., Konermann, L., Withers, S.G., Douglas, D.J. (1998) Pre-steady State Kinetic Analysis of an Enzymatic Reaction Monitored by Time-resolved Electrospray Ionization Mass Spectrometry. Biochemistry 37: 7664–7669.
235. Aramendía, M.A., Lafont, F. (1999) Electrospray Ionization Mass Spectrometry Detection of Intermediates in the Palladium-catalyzed Oxidative Self-coupling of Areneboronic Acids. J. Org. Chem. 64: 3592–3594.
236. Dell'Orco, P., Brum, J., Matsuoka, R., Badlani, M., Muske, K. (1999) Monitoring Process-scale Reactions Using API Mass Spectrometry. Anal. Chem. 71: 5165–5170.
237. Freitas, M.A., Hendrickson, C.L., Emmett, M.R., Marshall, A.G. (1999) Gas-phase Bovine Ubiquitin Cation Conformations Resolved by Gas-phase Hydrogen/Deuterium Exchange Rate and Extent. Int. J. Mass Spectrom. 185/186/187: 565–575.
238. Fujii, A., Hagiwara, E., Sodeoka, M. (1999) Mechanism of Palladium Complex-catalyzed Enantioselective Mannich-type Reaction: Characterization of a Novel Binuclear Palladium Enolate Complex. J. Am. Chem. Soc. 121: 5450–5458.
239. Hinderling, C., Chen, P. (1999) Rapid Screening of Olefin Polymerization Catalyst Libraries by Electrospray Ionization Tandem Mass Spectrometry. Angew. Chem. Int. Ed. 38: 2253–2256.
240. Hofstadler, S.A., Sannes-Lowery, K.A., Griffey, R.H. (1999) A Gated-beam Electrospray Ionization Source with an External Ion Reservoir. A New Tool for the Characterization of Biomolecules Using Electrospray Ionization Mass Spectrometry. Rapid Commun. Mass Spectrom. 13: 1971–1979.
241. Lee, V.W.S., Chen, Y.-L., Konermann, L. (1999) Reconstitution of Acid-denatured Holomyoglobin Studied by Time-resolved Electrospray Ionization Mass Spectrometry. Anal. Chem. 71: 4154–4159.
242. Tsui, V., Garcia, C., Cavagnero, S., Siuzdak, G., Dyson, H.J., Wright, P.E. (1999) Quench-flow Experiments Combined with Mass Spectrometry Show Apomyoglobin Folds through an Obligatory Intermediate. Protein Sci. 8: 45–49.
243. Adam, W., Mock-Knoblauch, C., Saha-Möller, C.R., Herderich, M. (2000) Are Mn^{IV} Species Involved in Mn(Salen)-catalyzed Jacobsen–Katsuki Epoxidations? A Mechanistic Elucidation of Their Formation and Reaction Modes by EPR Spectroscopy,

Mass-spectral Analysis, and Product Studies: Chlorination versus Oxygen Transfer. J. Am. Chem. Soc. 122: 9685–9691.

244. Hinderling, C., Chen P. (2000) Mass Spectrometric Assay of Polymerization Catalysts for Combinational Screening. Int. J. Mass Spectrom. 195/196: 377–383.

245. Hong, C.-M., Tsai, F.-C., Shiea, J. (2000) A Multiple Channel Electrospray Source Used to Detect Highly Reactive Ketenes from a Flow Pyrolyzer. Anal. Chem. 72: 1175–1178.

246. Kotiaho, T., Eberlin, M.N., Vainiotalo, P., Kostiainen, R. (2000) Electrospray Mass and Tandem Mass Spectrometry Identification of Ozone Oxidation Products of Amino Acids and Small Peptides. J. Am. Soc. Mass Spectrom. 11: 526–535.

247. Nishimura, C., Prytulla, S., Dyson, H.J., Wright, P.E. (2000) Conservation of Folding Pathways in Evolutionarily Distant Globin Sequences. Nature Struct. Biol. 7: 679–686.

248. Sogbein, O.O., Simmons, D.A., Konermann, L. (2000) Effects of pH on the Kinetic Reaction Mechanism of Myoglobin Unfolding Studied by Time-resolved Electrospray Ionization Mass Spectrometry. J. Am. Soc. Mass Spectrom. 11: 312–319.

249. Waters, T., O'Hair, R.A.J., Wedd, A.G. (2000) Probing the Catalytic Oxidation of Alcohols *via* an Anionic Dimolybdate Centre Using Multistage Mass Spectrometry. Chem. Commun. 225–226.

250. Badman, E.R., Hoaglund-Hyzer, C.S., Clemmer, D.E. (2001) Monitoring Structural Changes of Proteins in an Ion Trap over ~10–200 ms: Unfolding Transitions in Cytochrome *c* Ions. Anal. Chem. 73: 6000–6007.

251. Ding, W., Johnson, K.A., Amster, I.J., Kutal, C. (2001) Identification of Photogenerated Intermediates by Electrospray Ionization Mass Spectrometry. Inorg. Chem. 40: 6865–6866.

252. Feichtinger, D., Plattner, D.A. (2001) Probing the Reactivity of Oxomanganese-Salen Complexes: An Electrospray Tandem Mass Spectrometric Study of Highly Reactive Intermediates. Chem. Eur. J. 7: 591–599.

253. Ge, X., Sirich, T.L., Beyer, M.K., Desaire, H., Leary, J.A. (2001) A Strategy for the Determination of Enzyme Kinetics Using Electrospray Ionization with an Ion Trap Mass Spectrometer. Anal. Chem. 73: 5078–5082.

254. Breuker, K., Oh, H., Horn, D.M., Cerda, B.A., McLafferty, F.W. (2002) Detailed Unfolding and Folding of Gaseous Ubiquitin Ions Characterized by Electron Capture Dissociation. J. Am. Chem. Soc. 124: 6407–6420.

255. Griep-Raming, J., Meyer, S., Bruhn, T., Metzger, J.O. (2002) Investigation of Reactive Intermediates of Chemical Reactions in Solution by Electrospray Ionization Mass Spectrometry: Radical Chain Reactions. Angew. Chem. Int. Ed. 41: 2738–2742.

256. Simmons, D.A., Konermann, L. (2002) Characterization of Transient Protein Folding Intermediates during Myoglobin Reconstitution by Time-resolved Electrospray Mass Spectrometry with On-line Isotopic Pulse Labeling. Biochemistry 41: 1906–1914.

257. Zhang, Y.H., Yan, X., Maier, C.S., Schimerlik, M.I., Deinzer, M.L. (2002) Conformational Analysis of Intermediates Involved in the In Vitro Folding Pathways of Recombinant Human Macrophage Colony Stimulating Factor β by Sulfhydryl Group Trapping and Hydrogen/Deuterium Pulsed Labeling. Biochemistry 41: 15495–15504.

258. Ding, W., Johnson, K.A., Kutal, C., Amster, I.J. (2003) Mechanistic Studies of Pho-
    tochemical Reactions with Millisecond Time Resolution by Electrospray Ionization
    Mass Spectrometry. Anal. Chem. 75: 4624–4630.
259. Gerdes, G., Chen, P. (2003) Comparative Gas-phase and Solution-phase Investiga-
    tions of the Mechanism of C-H Activation by [(N-N)Pt(CH₃)(L)]⁺. Organometallics
    22: 2217–2225.
260. Li, Z., Sau, A.K., Shen, S., Whitehouse, C., Baasov, T., Anderson, K.S. (2003) A
    Snapshot of Enzyme Catalysis Using Electrospray Ionization Mass Spectrometry. J.
    Am. Chem. Soc. 125: 9938–9939.
261. Masllorens, J., Moreno-Mañas, M., Pla-Quintana, A., Roglans, A. (2003) First Heck
    Reaction with Arenediazonium Cations with Recovery of Pd-triolefinic Macrocyclic
    Catalyst. Org. Lett. 5: 1559–1561.
262. Meyer, S., Koch, R., Metzger, J.O. (2003) Investigation of Reactive Intermediates
    of Chemical Reactions in Solution by Electrospray Ionization Mass Spectrometry:
    Radical Cation Chain Reactions. Angew. Chem. Int. Ed. 42: 4700–4703.
263. Nishimura, C., Wright, P.E., Dyson, H.J. (2003) Role of the B Helix in Early Folding
    Events in Apomyoglobin: Evidence from Site-directed Mutagenesis for Native-like
    Long Range Interactions. J. Mol. Biol. 334: 293–307.
264. Simmons, D.A., Dunn, S.D., Konermann, L. (2003) Conformational Dynamics
    of Partially Denatured Myoglobin Studied by Time-resolved Electrospray Mass
    Spectrometry with Online Hydrogen–Deuterium Exchange. Biochemistry 42:
    5896–5905.
265. Chevrin, C., Le Bras, J., Hénin, F., Muzart, J. (2004) Allylic Substitution Mediated by
    Water and Palladium: Unusual Role of a Palladium(II) Catalyst and ESI-MS Analysis.
    Organometallics 23: 4796–4799.
266. Comelles, J., Moreno-Mañas, M., Pérez, E., Roglans, A., Sebastián, R.M., Vallribera,
    A. (2004) Ionic and Covalent Copper(II)-based Catalysts for Michael Additions. The
    Mechanism. J. Org. Chem. 69: 6834–6842.
267. Dietiker, R., Chen, P. (2004) Gas-phase Reactions of the [(PHOX)IrL₂]⁺ Ion Olefin-
    hydrogenation Catalyst Support an Irᴵ/Irᴵᴵᴵ Cycle. Angew. Chem. Int. Ed. 43:
    5513–5516.
268. Domingos, J.B., Longhinotti, E., Brandão, T.A.S., Bunton, C.A., Santos, L.S., Eber-
    lin, M.N., Nome, F. (2004) Mechanisms of Nucleophilic Substitution Reactions of
    Methylated Hydroxylamines with Bis(2,4-dinitrophenyl)phosphate. Mass Spectro-
    metric Identification of Key Intermediates. J. Org. Chem. 69: 6024–6033.
269. Domingos, J.B., Longhinotti, E., Brandão, T.A.S., Santos, L.S., Eberlin, M.N.,
    Bunton, C.A., Nome, F. (2004) Reaction of Bis(2,4-dinitrophenyl) Phosphate with
    Hydrazine and Hydrogen Peroxide. Comparison of O- and N- Phosphorylation. J.
    Org. Chem. 69: 7898–7905.
270. Eyles, S.J., Kaltashov, I.A. (2004) Methods to Study Protein Dynamics and Folding
    by Mass Spectrometry. Methods 34: 88–99.
271. Gerdes, G., Chen, P. (2004) Cationic Platinum(II) Carboxylato Complexes Are Com-
    petent in Catalytic Arene C-H Activation under Mild Conditions. Organometallics 23:
    3031–3036.
272. Markert, C., Pfaltz, A. (2004) Screening of Chiral Catalysts and Catalyst Mixtures by
    Mass Spectrometric Monitoring of Catalytic Intermediates. Angew. Chem. Int. Ed.
    43: 2498–2500.

273. Modestov, A.D., Gun, J., Savotine, I., Lev, O. (2004) On-line Electrochemical-Mass Spectrometry Study of the Mechanism of Oxidation of *N,N*-dimethyl-*p*-phenylenediamine in Aqueous Electrolytes. J. Electroanal. Chem. 565: 7–19.

274. Pi, N., Leary, J.A. (2004) Determination of Enzyme/Substrate Specificity Constants Using a Multiple Substrate ESI-MS Assay. J. Am. Soc. Mass Spectrom. 15: 233–243.

275. Pi, N., Yu, Y., Mougous, J.D., Leary, J.A. (2004) Observation of a Hybrid Random Ping-Pong Mechanism of Catalysis for NodST: A Mass Spectrometry Approach. Protein Sci. 13: 903–912.

276. Raminelli, C., Prechtl, M.H.G., Santos, L.S., Eberlin, M.N., Comasseto, J.V. (2004) Coupling of Vinylic Tellurides with Alkynes Catalyzed by Palladium Dichloride: Evaluation of Synthetic and Mechanistic Details. Organometallics 23: 3990–3996.

277. Sabino, A.A., Machado, A.H.L., Correia, C.R.D., Eberlin, M.N. (2004) Probing the Mechanism of the Heck Reaction with Arene Diazonium Salts by Electrospray Mass and Tandem Mass Spectrometry. Angew. Chem. Int. Ed. 43: 2514–2518.

278. Santos, L.S., Pavam, C.H., Almeida, W.P., Coelho, F., Eberlin, M.N. (2004) Probing the Mechanism of the Baylis–Hillman Reaction by Electrospray Ionization Mass and Tandem Mass Spectrometry. Angew. Chem. Int. Ed. 43: 4330–4333.

279. Wilson, D.J., Konermann, L. (2004) Mechanistic Studies on Enzymatic Reactions by Electrospray Ionization MS Using a Capillary Mixer with Adjustable Reaction Chamber Volume for Time-resolved Measurements. Anal. Chem. 76: 2537–2543.

280. Yu, Y., Kirkup, C.E., Pi, N., Leary, J.A. (2004) Characterization of Noncovalent Protein-Ligand Complexes and Associated Enzyme Intermediates of GlcNAc-6-O-Sulfotransferase by Electrospray Ionization FT-ICR Mass Spectrometry. J. Am. Soc. Mass Spectrom. 15: 1400–1407.

281. Ferraz, H.M.C., Pereira, F.L.C., Gonçalo, É.R.S., Santos, L.S., Eberlin, M.N. (2005) Unexpected Synthesis of Conformationally Restricted Analogues of γ-Amino Butyric Acid (GABA): Mechanism Elucidation by Electrospray Ionization Mass Spectrometry. J. Org. Chem. 70: 110–114.

282. Gao, H., Yu, Y., Leary, J.A. (2005) Mechanism and Kinetics of Metalloenzyme Phosphomannose Isomerase: Measurement of Dissociation Constants and Effect of Zinc Binding Using ESI-FTICR Mass Spectrometry. Anal. Chem. 77: 5596–5603.

283. Pi, N., Hoang, M.B., Gao, H., Mougous, J.D., Bertozzi, C.R., Leary, J.A. (2005) Kinetic Measurements and Mechanism Determination of Stf0 Sulfotransferase Using Mass Spectrometry. Anal. Biochem. 341: 94–104.

284. Qian, R., Guo, H., Liao, Y., Guo, Y., Ma, S. (2005) Probing the Mechanism of the Palladium-catalyzed Addition of Organoboronic Acids to Allenes in the Presence of AcOH by ESI-FTMS. Angew. Chem. 117: 4849–4852.

285. Trage, C., Schröder, D., Schwarz, H. (2005) Coordination of Iron(III) Cations to β-Keto Esters as Studied by Electrospray Mass Spectrometry: Implications for Iron-catalyzed Michael Addition Reactions. Chem. Eur. J. 11: 619–627.

286. van den Heuvel, R.H.H., Gato, S., Versluis, C., Gerbaux, P., Kleanthous, C., Heck, A.J.R. (2005) Real-time Monitoring of Enzymatic DNA Hydrolysis by Electrospray Ionization Mass Spectrometry. Nucleic Acids Res. 33: e96.

287. Wilson, D.J., Rafferty, S.P., Konermann, L. (2005) Kinetic Unfolding Mechanism of the Inducible Nitric Oxide Synthase Oxygenase Domain Determined by Time-resolved Electrospray Mass Spectrometry. Biochemistry 44: 2276–2283.

288. Enquist, P.-A., Nilsson, P., Sjöberg, P., Larhed, M. (2006) ESI-MS Detection of Proposed Reaction Intermediates in the Air-promoted and Ligand-modulated Oxidative Heck Reaction. J. Org. Chem. 71: 8779–8786.

289. Pan, J., Rintala-Dempsey, A.C., Li, Y., Shaw, G.S., Konermann, L. (2006) Folding Kinetics of the S100A11 Protein Dimer Studied by Time-resolved Electrospray Mass Spectrometry and Pulsed Hydrogen-Deuterium Exchange. Biochemistry 45: 3005–3013.

290. Abella, C.A.M., Benassi, M., Santos, L.S., Eberlin, M.N., Coelho, F. (2007) The Mechanism of Tröger's Base Formation Probed by Electrospray Ionization Mass Spectrometry. J. Org. Chem. 72: 4048–4054.

291. Jha, S.K., Udgaonkar, J.B. (2007) Exploring the Cooperativity of the Fast Folding Reaction of a Small Protein Using Pulsed Thiol Labeling and Mass Spectrometry. J. Biol. Chem. 282: 37479–37491.

292. Marquez, C.A., Fabbretti, F., Metzger, J.O. (2007) Electrospray Ionization Mass Spectrometric Study on the Direct Organocatalytic α-Halogenation of Aldehydes. Angew. Chem. Int. Ed. 46: 6915–6917.

293. Milagre, C.D.F., Milagre, H.M.S., Santos, L.S., Lopes, M.L.A., Moran, P.J.S., Eberlin, M.N., Rodrigues, J.A.R. (2007) Probing the Mechanism of Direct Mannich-type α-Methylenation of Ketoesters via Electrospray Ionization Mass Spectrometry. J. Mass Spectrom. 42: 1287–1293.

294. Wu, Z.-J., Luo, S.-W., Xie, J.-W., Xu, X.-Y., Fang, D.-M., Zhang, G.-L. (2007) Study of Michael–Michael–Retro Michael Addition Catalyzed by 9-Amino-9-Deoxyepiquinine Using ESI-MS. J. Am. Soc. Mass Spectrom. 18: 2074–2080.

295. Breuker, K., McLafferty, F.W. (2008) Stepwise Evolution of Protein Native Structure with Electrospray into the Gas Phase, 10^{-12} to 10^2 s. Proc. Natl. Acad. Sci. USA 105: 18145–18152.

296. Pan, J., Han, J., Borchers, C.H., Konermann, L. (2008) Electron Capture Dissociation of Electrosprayed Protein Ions for Spatially Resolved Hydrogen Exchange Measurements. J. Am. Chem. Soc. 130: 11574–11575.

297. Santos, L.S., Metzger, J.O. (2008) On-line Monitoring of Brookhart Polymerization by Electrospray Ionization Mass Spectrometry. Rapid Commun. Mass Spectrom. 22: 898–904.

298. Wang, H., Metzger, J.O. (2008) ESI-MS Study on First-generation Ruthenium Olefin Metathesis Catalysts in Solution: Direct Detection of the Catalytically Active 14-Electron Ruthenium Intermediate. Organometallics 27: 2761–2766.

299. Gau, B.C., Sharp, J.S., Rempel, D.L., Gross, M.L. (2009) Fast Photochemical Oxidation of Protein Footprints Faster than Protein Unfolding. Anal. Chem. 81: 6563–6571.

300. Li, Z., Song, F., Zhuang, Z., Dunaway-Mariano, D., Anderson, K.S. (2009) Monitoring Enzyme Catalysis in the Multimeric State: Direct Observation of *Arthrobacter* 4-Hydroxybenzoyl-Coenzyme A Thioesterase Catalytic Complexes Using Time-resolved Electrospray Ionization Mass Spectrometry. Anal. Biochem. 394: 209–216.

301. Pan, J., Han, J., Borchers, C.H., Konermann, L. (2009) Hydrogen/Deuterium Exchange Mass Spectrometry with Top-down Electron Capture Dissociation for Characterizing Structural Transitions of a 17 kDa Protein. J. Am. Chem. Soc. 131: 12801–12808.

302. Rand, K.D., Pringle, S.D., Murphy III, J.P., Fadgen, K.E., Brown, J., Engen, J.R. (2009) Gas-Phase Hydrogen/Deuterium Exchange in a Traveling Wave Ion Guide for the Examination of Protein Conformations. Anal. Chem. 81: 10019–10028.

303. Rob, T., Wilson, D.J. (2009) A Versatile Microfluidic Chip for Millisecond Time-scale Kinetic Studies by Electrospray Mass Spectrometry. J. Am. Soc. Mass Spectrom. 20: 124–130.

304. Weimann, D.P., Winkler, H.D.F., Falenski, J.A., Koksch, B., Schalley, C.A. (2009) Highly Dynamic Motion of Crown Ethers along Oligolysine Peptide Chains. Nature Chem. 1: 573–577.

305. Beierlein, C.H., Breit, B. (2010) Online Monitoring of Hydroformylation Intermediates by ESI-MS. Organometallics 29: 2521–2532.

306. Clarke, D.J., Stokes, A.A., Langridge-Smith, P., Mackay, C.L. (2010) Online Quench-flow Electrospray Ionization Fourier Transform Ion Cyclotron Resonance Mass Spectrometry for Elucidating Kinetic and Chemical Enzymatic Reaction Mechanisms. Anal. Chem. 82: 1897–1904.

307. Gau, B.C., Chen, H., Zhang, Y., Gross, M.L. (2010) Sulfate Radical Anion as a New Reagent for Fast Photochemical Oxidation of Proteins. Anal. Chem. 82: 7821–7827.

308. Liuni, P., Rob, T., Wilson, D.J. (2010) A Microfluidic Reactor for Rapid, Low-pressure Proteolysis with On-chip Electrospray Ionization. Rapid Commun. Mass Spectrom. 24: 315–320.

309. Pan, J., Han, J., Borchers, C.H., Konermann, L. (2010) Characterizing Short-lived Protein Folding Intermediates by Top-down Hydrogen Exchange Mass Spectrometry. Anal. Chem. 82: 8591–8597.

310. Reynolds, J.C., Blackburn, G.J., Guallar-Hoyas, C., Moll, V.H., Bocos-Bintintan, V., Kaur-Atwal, G., Howdle, M.D., Harry, E.L., Brown, L.J., Creaser, C.S., Thomas, C.L. (2010) Detection of Volatile Organic Compounds in Breath Using Thermal Desorption Electrospray Ionization-Ion Mobility-Mass Spectrometry. Anal Chem. 82: 2139–2144.

311. Roberts, A., Furdui, C., Anderson, K.S. (2010) Observation of a Chemically Labile, Noncovalent Enzyme Intermediate in the Reaction of Metal-dependent *Aquifex pyrophilus* KDO8PS by Time-resolved Mass Spectrometry. Rapid Commun. Mass Spectrom. 24: 1919–1924.

312. Schade, M.A., Fleckenstein, J.E., Knochel, P., Koszinowski, K. (2010) Charged Tags as Probes for Analyzing Organometallic Intermediates and Monitoring Cross-coupling Reactions by Electrospray-Ionization Mass Spectrometry. J. Org. Chem. 75: 6848–6857.

313. Zhu, Y., Fang, Q. (2010) Integrated Droplet Analysis System with Electrospray Ionization-Mass Spectrometry Using a Hydrophilic Tongue-based Droplet Extraction Interface. Anal. Chem. 82: 8361–8366.

314. Bujara, M., Schümperli, M., Pellaux, R., Heinemann, M., Panke, S. (2011) Optimization of a Blueprint for *In Vitro* Glycolysis by Metabolic Real-time Analysis. Nature Chem. Biol. 7: 271–277.

315. Chevreux, G., Atmanene, C., Lopez, P., Ouazzani, J., Van Dorsselaer, A., Badet, B., Badet-Denisot, M.-A., Sanglier-Cianférani, S. (2011) Monitoring the Dynamics of Monomer Exchange Using Electrospray Mass Spectrometry: The Case of the

Dimeric Glucosamine-6-phosphate Synthase. J. Am. Soc. Mass Spectrom. 22: 431–439.

316. Oliveira, F.F.D., dos Santos, M.R., Lalli, P.M., Schmidt, E.M., Bakuzis, P., Lapis, A.A.M., Monteiro, A.L., Eberlin, M.N., Neto, B.A.D. (2011) Charge-tagged Acetate Ligands As Mass Spectrometry Probes for Metal Complexes Investigations: Applications in Suzuki and Heck Phosphine-free Reactions. J. Org. Chem. 76: 10140–10147.

317. Regiani, T., Santos, V.G., Godoi, M.N., Vaz, B.G., Eberlin, M.N., Coelho, F. (2011) On the Mechanism of the *aza*-Morita–Baylis–Hillman Reaction: ESI-MS Interception of a Unique New Intermediate. Chem. Commun. 47: 6593–6595.

318. Silva, B.V., Violante, F.A., Pinto, A.C., Santos, L.S. (2011) The Mechanism of Sandmeyer's Cyclization Reaction by Electrospray Ionization Mass Spectrometry. Rapid Commun. Mass Spectrom. 25: 423–428.

319. Wilson, E.F., Miras, H.N., Rosnes, M.H., Cronin, L. (2011) Real-time Observation of the Self-assembly of Hybrid Polyoxometalates Using Mass Spectrometry. Angew. Chem. Int. Ed. 50: 3720–3724.

320. Zhao, Z.-X., Wang, H.-Y., Guo, Y.-L. (2011) ESI-MS Study on Transient Intermediates in the Fast Cyclopropenium-activated Chlorination Reaction of Alcohols. J. Mass. Spectrom. 46: 856–858.

321. Liuni, P., Jeganathan, A., Wilson, D.J. (2012) Conformer Selection and Intensified Dynamics during Catalytic Turnover in Chymotrypsin. Angew. Chem. Int. Ed. 51: 9666–9669.

322. Rob, T., Liuni, P., Gill, P.K., Zhu, S., Balachandran, N., Berti, P.J., Wilson, D.J. (2012) Measuring Dynamics in Weakly Structured Regions of Proteins Using Microfluidics-enabled Subsecond H/D Exchange Mass Spectrometry. Anal. Chem. 84: 3771–3779.

323. Song, P., Hershey, N.D., Mabrouk, O.S., Slaney, T.R., Kennedy, R.T. (2012) A Mass Spectrometry "Sensor" for In Vivo Acetylcholine Monitoring. Anal. Chem. 84: 4659–4664.

324. Verano-Braga, T., Schwämmle, V., Sylvester, M., Passos-Silva, D.G., Peluso, A.A.B., Etelvino, G.M., Santos, R.A.S., Roepstorff, P. (2012) Time-resolved Quantitative Phosphoproteomics: New Insights into Angiotensin-(1–7) Signaling Networks in Human Endothelial Cells. J. Proteome Res. 11: 3370–3381.

325. Ahmadi, Z., McIndoe, J.S. (2013) A Mechanistic Investigation of Hydrodehalogenation Using ESI-MS. Chem. Commun. 49: 11488–11490.

326. Roscioli, K.M., Zhang, X., Li, S.X., Goetz, G.H., Cheng, G., Zhang, Z., Siems, W.F., Hill Jr, H.H. (2013) Real Time Pharmaceutical Reaction Monitoring by Electrospray Ion Mobility-Mass Spectrometry. Int. J. Mass Spectrom. 336: 27–36.

327. Vahidi, S., Stocks, B.B., Liaghati-Mobarhan, Y., Konermann, L. (2013) Submillisecond Protein Folding Events Monitored by Rapid Mixing and Mass Spectrometry-based Oxidative Labeling. Anal. Chem. 85: 8618–8625.

328. Hsu, F.-J., Liu, T.-L., Laskar, A.H., Shiea, J., Huang, M.-Z. (2014) Gravitational Sampling Electrospray Ionization Mass Spectrometry for Real-time Reaction Monitoring. Rapid Commun. Mass Spectrom. 28: 1979–1986.

329. Jarrell, T., Riedeman, J., Carlsen, M., Replogle, R., Selby, T., Kenttämaa, H. (2014) Multiported Pulsed Valve Interface for a Linear Quadrupole Ion Trap Mass Spectrometer to Enable Rapid Screening of Multiple Functional-group Selective Ion–Molecule Reactions. Anal. Chem. 86: 6533–6539.

330. dos Santos, M.R., Coriolano, R., Godoi, M.N., Monteiro, A.L., de Oliveira, H.C.B., Eberlin, M.N., Neto, B.A.D. (2014) Phosphine-free Heck Reaction: Mechanistic Insights and Catalysis "On Water" Using a Charge-tagged Palladium Complex. New J. Chem. 38: 2958–2963.

331. Yan, X., Sokol, E., Li, X., Li, G., Xu, S., Cooks, R.G. (2014) On-line Reaction Monitoring and Mechanistic Studies by Mass Spectrometry: Negishi Cross-coupling, Hydrogenolysis, and Reductive Amination. Angew. Chem. Int. Ed. 53: 5931–5935.

332. Bain, R.M., Pulliam, C.J., Cooks, R.G. (2015) Accelerated Hantzsch Electrospray Synthesis with Temporal Control of Reaction Intermediates. Chem. Sci. 6: 397–401.

333. Donohoe, G.C., Khakinejad, M., Valentine, S.J. (2015) Ion Mobility Spectrometry-Hydrogen Deuterium Exchange Mass Spectrometry of Anions: Part 1. Peptides to Proteins. J. Am. Soc. Mass Spectrom. 26: 564–576.

334. Limberger, J., Leal, B.C., Monteiro, A.L., Dupont, J. (2015) Charge-tagged Ligands: Useful Tools for Immobilising Complexes and Detecting Reaction Species during Catalysis. Chem. Sci. 6: 77–94.

335. Rajabi, K. (2015) Time-resolved Pulsed Hydrogen/Deuterium Exchange Mass Spectrometry Probes Gaseous Proteins Structural Kinetics. J. Am. Soc. Mass Spectrom. 26: 71–82.

336. Schenk, E.R., Almeida, R., Miksovska, J., Ridgeway, M.E., Park, M.A., Fernandez-Lima, F. (2015) Kinetic Intermediates of Holo- and Apo-myoglobin Studied Using HDX-TIMS-MS and Molecular Dynamic Simulations. J. Am. Soc. Mass Spectrom. 26: 555–563.

337. Simithy, J., Gill, G., Wang, Y., Goodwin, D.C., Calderón, A.I. (2015) Development of an ESI-LC-MS-based Assay for Kinetic Evaluation of *Mycobacterium tuberculosis* Shikimate Kinase Activity and Inhibition. Anal. Chem. 87: 2129–2136.

338. Fändrich, M., Tito, M.A., Leroux, M.R., Rostom, A.A., Hartl, F.U., Dobson, C.M., Robinson, C.V. (2000) Observation of the Noncovalent Assembly and Disassembly Pathways of the Chaperone Complex MtGimC by Mass Spectrometry. Proc. Natl. Acad. Sci. USA 97: 14151–14155.

339. Brivio, M., Liesener, A., Oosterbroek, R.E., Verboom, W., Karst, U., van den Berg, A., Reinhoudt, D.N. (2005) Chip-based On-line Nanospray MS Method Enabling Study of the Kinetics of Isocyanate Derivatization Reactions. Anal. Chem. 77: 6852–6856.

340. Painter, A.J., Jaya, N., Basha, E., Vierling, E., Robinson, C.V., Benesch, L.P. (2008) Real-time Monitoring of Protein Complexes Reveals Their Quaternary Organization and Dynamics. Chem. Biol. 15: 246–253.

341. Stengel, F., Baldwin, A.J., Painter, A.J., Jaya, N., Basha, E., Kay, L.E., Vierling, E., Robinson, C.V., Benesch, J.L.P. (2010) Quaternary Dynamics and Plasticity Underlie Small Heat Shock Protein Chaperone Function. Proc. Natl. Acad. Sci. USA 107: 2007–2012.

342. Smith, D.P., Radford, S.E., Ashcroft, A.E. (2010) Elongated Oligomers in β_2-Microglobulin Amyloid Assembly Revealed by Ion Mobility Spectrometry-Mass Spectrometry. Proc. Natl. Acad. Sci. USA 107: 6794–6798.

343. Fritzsche, S., Ohla, S., Glaser, P., Giera, D.S., Sickert, M., Schneider, C., Belder D. (2011) Asymmetric Organocatalysis and Analysis on a Single Microfluidic Nanospray Chip. Angew. Chem. Int. Ed. 50: 9467–9470.

344. Li, P.-H., Ting, H., Chen, Y.-C., Urban, P.L. (2012) Recording Temporal Characteristics of Convection Currents by Continuous and Segmented-flow Sampling. RSC Adv. 2: 12431–12437.
345. Olivero, D., LaPlaca, M., Kottke, P.A. (2012) Ambient Nanoelectrospray Ionization with In-Line Microdialysis for Spatially Resolved Transient Biochemical Monitoring within Cell Culture Environments. Anal. Chem. 84: 2072–2075.
346. Browne, D.L., Wright, S., Deadman, B.J., Dunnage, S., Baxendale, I.R., Turner, R.M., Ley, S.V. (2012) Continuous Flow Reaction Monitoring Using an On-line Miniature Mass Spectrometer. Rapid Commun. Mass Spectrom. 26: 1999–2010.
347. Sakamoto, S., Yamaguchi, K. (2003) Hyperstranded DNA Architectures Observed by Cold-spray Ionization Mass Spectrometry. Angew. Chem. Int. Ed. 42: 905–908.
348. Chen, H., Cotte-Rodríguez, I., Cooks, R.G. (2006) cis-Diol Functional Group Recognition by Reactive Desorption Electrospray Ionization (DESI). Chem. Commun. 597–599.
349. Zhang, T., Zhou, W., Jin, W., Jin, Q., Chen, H. (2013) Direct Detection of Aromatic Amines and Observation of Intermediates of Schiff-base Reactions by Reactive Desorption Electrospray Ionization Mass Spectrometry. Microchem. J. 108: 18–23.
350. Xu, G., Chen, B., Guo, B., He, D., Yao, S. (2011) Detection of Intermediates for the Eschweiler–Clarke Reaction by Liquid-phase Reactive Desorption Electrospray Ionization Mass Spectrometry. Analyst 136: 2385–2390.
351. Perry, R.H., Cahill III, T.J., Roizen, J.L., Du Bois, J., Zare, R.N. (2012) Capturing Fleeting Intermediates in a Catalytic C-H Amination Reaction Cycle. Proc. Natl. Acad. Sci. USA 109: 18295–18299.
352. Boeser, C.L., Holder, J.C., Taylor, B.L.H., Houk, K.N., Stoltz, B.M., Zare, R.N. (2015) Mechanistic Analysis of an Asymmetric Palladium-catalyzed Conjugate Addition of Arylboronic Acids to β-Substituted Cyclic Enones. Chem. Sci. 6: 1917–1922.
353. Liu, W., Wang, N., Lin, X., Ma, Y., Lin, J.-M. (2014) Interfacing Microsampling Droplets and Mass Spectrometry by Paper Spray Ionization for Online Chemical Monitoring of Cell Culture. Anal. Chem. 86: 7128–7134.
354. Touboul, D., Jecklin, M.C., Zenobi, R. (2007) Rapid and Precise Measurements of Gas-phase Basicity of Peptides and Proteins at Atmospheric Pressure by Electrosonic Spray Ionization-Mass Spectrometry. J. Phys. Chem. B 111: 11629–11631.
355. Pan, N., Rao, W., Kothapalli, N.R., Liu, R., Burgett, A.W.G., Yang, Z. (2014) The Single-probe: A Miniaturized Multifunctional Device for Single Cell Mass Spectrometry Analysis. Anal. Chem. 86: 9376–9380.
356. Cheah, M.H., Millar, A.H., Myers, R.C., Day, D.A., Roth, J., Hillier, W., Badger, M.R. (2014) Online Oxygen Kinetic Isotope Effects Using Membrane Inlet Mass Spectrometry Can Differentiate between Oxidases for Mechanistic Studies and Calculation of Their Contributions to Oxygen Consumption in Whole Tissues. Anal. Chem. 86: 5171–5178.
357. Creaser, C., Freitas dos Santos, L., Gómez Lamarca, D., New, A., Wolff, J.-C. (2002) Biodegradation Studies of 4-Fluorobenzoic Acid and 4-Fluorocinnamic Acid: an Evaluation of Membrane Inlet Mass Spectrometry as an Alternative to High Performance Liquid Chromatography and Ion Chromatography. Anal. Chim. Acta 454: 137–145.

358. Roussel, M.R., Lloyd, D. (2007) Observation of a Chaotic Multioscillatory Metabolic Attractor by Real-time Monitoring of a Yeast Continuous Culture. FEBS J. 274: 1011–1018.

359. Petucci, C., Diffendal, J., Kaufman, D., Mekonnen, B., Terefenko, G., Musselman, B. (2007) Direct Analysis in Real Time for Reaction Monitoring in Drug Discovery. Anal. Chem. 79: 5064–5070.

360. Smith, N.J., Domin, M.A., Scott, L.T. (2008) HRMS Directly From TLC Slides. A Powerful Tool for Rapid Analysis of Organic Mixtures. Org. Lett. 10: 3493–3496.

361. Cho, D.S., Gibson, S.C., Bhandari, D., McNally, M.E., Hoffman, R.M., Cook, K.D., Song, L. (2011) Evaluation of Direct Analysis in Real Time Mass Spectrometry for Onsite Monitoring of Batch Slurry Reactions. Rapid Commun. Mass Spectrom. 25: 3575–3580.

362. Sanchez, L.M., Curtis, M.E., Bracamonte, B.E., Kurita, K.L., Navarro, G., Sparkman, O.D., Linington, R.G. (2011) Versatile Method for the Detection of Covalently Bound Substrates on Solid Supports by DART Mass Spectrometry. Org. Lett. 13: 3770–3773.

363. Oh, J.-S., Aranda-Gonzalvo, Y., Bradley, J.W. (2011) Time-resolved Mass Spectroscopic Studies of an Atmospheric-pressure Helium Microplasma Jet. J. Phys. D: Appl. Phys. 44: 365202.

364. Slagle, I.R., Yamada, F., Gutman, D. (1981) Kinetics of Free Radicals Produced by Infrared Multiphoton-induced Decompositions. 1. Reactions of Allyl Radicals with Nitrogen Dioxide and Bromine. J. Am. Chem. Soc. 103: 149–153.

365. Slagle, I.R., Gutman, D. (1982) Kinetics of Free Radicals Produced by Infrared Multiphoton-induced Decomposition. 2. Formation of Acetyl and Dichlorofluoromethyl Radicals and Their Reactions with Nitrogen Dioxide. J. Am. Chem. Soc. 104: 4741–4748.

366. Slagle, I.R., Feng, Q., Gutman, D. (1984) Kinetics of the Reaction of Ethyl Radicals with Molecular Oxygen from 294 to 1002 K. J. Phys. Chem. 88: 3648–3653.

367. Fockenberg, C., Bernstein, H.J., Hall, G.E., Muckerman, J.T., Preses, J.M., Sears, T.J., Weston Jr, R.E. (1999) Repetitively Sampled Time-of-flight Mass Spectrometry for Gas-phase Kinetics Studies. Rev. Sci. Instrum. 70: 3259–3264.

368. Qi, F., Yang, R., Yang, B., Huang, C., Wei, L., Wang, J., Sheng, L., Zhang, Y. (2006) Isomeric Identification of Polycyclic Aromatic Hydrocarbons Formed in Combustion with Tunable Vacuum Ultraviolet Photoionization. Rev. Sci. Instrum. 77: 084101.

369. Osborn, D.L., Zou, P., Johnsen, H., Hayden, C.C., Taatjes, C.A., Knyazev, V.D., North, S.W., Peterka, D.S., Ahmed, M., Leone, S.R. (2008) The Multiplexed Chemical Kinetic Photoionization Mass Spectrometer: A New Approach to Isomer-resolved Chemical Kinetics. Rev. Sci. Instr. 79: 104103.

370. Selby, T.M., Meloni, G., Goulay, F., Leone, S.R., Fahr, A., Taatjes, C.A., Osborn, D.L. (2008) Synchrotron Photoionization Mass Spectrometry Measurements of Kinetics and Product Formation in the Allyl Radical (H_2CCHCH_2) Self-reaction. J. Phys. Chem. A 112: 9366–9373.

371. Genbai, C., Jun, C., Fuyi, L., Liusi, S. (2012) Kinetics of Gas-phase Radical Reactions Using Photoionization Mass Spectrometry with Synchrotron Source. Progr. Chem. 24: 2097–2105.

372. Savee, J.D., Welz, O., Taatjes, C.A., Osborn, D.L. (2012) New Mechanistic Insights to the O(^{3}P) + Propene Reaction from Multiplexed Photoionization Mass Spectrometry. Phys. Chem. Chem. Phys. 14: 10410–10423.

373. Zhou, Z.Y., Wang, Y., Tang, X.F., Wu, W.H., Qi, F. (2013) A New Apparatus for Study of Pressure-dependent Laminar Premixed Flames with Vacuum Ultraviolet Photoionization Mass Spectrometry. Rev. Sci. Instrum. 84: 014101.

374. Middaugh, J.E. (2014) The Study of Bimolecular Radical Reactions Using a Novel Time-resolved Photoionization Time-of-flight Mass Spectrometry and Laser Absorption Spectrometry Apparatus. PhD Thesis, Massachusetts Institute of Technology.

375. Lynch, P.T., Troy, T.P., Ahmed, M., Tranter, R.S. (2015) Probing Combustion Chemistry in a Miniature Shock Tube with Synchrotron VUV Photo Ionization Mass Spectrometry. Anal. Chem. 87: 2345–2352.

376. Hua, L., Wu, Q., Hou, K., Cui, H., Chen, P., Wang, W., Li, J., Li, H. (2011) Single Photon Ionization and Chemical Ionization Combined Ion Source Based on a Vacuum Ultraviolet Lamp for Orthogonal Acceleration Time-of-flight Mass Spectrometry. Anal. Chem. 83: 5309–5316.

377. Rusteika, N., Brogaard, R.Y., Sølling, T.I., Rudakov, F.M., Weber, P.M. (2009) Excited-state Ions in Femtosecond Time-resolved Mass Spectrometry: An Investigation of Highly Excited Chloroamines. J. Phys. Chem. A 113: 40–43.

378. Studzinski, H., Zhang, S., Wang, Y., Temps, F. (2008) Ultrafast Nonradiative Dynamics in Electronically Excited Hexafluorobenzene by Femtosecond Time-resolved Mass Spectrometry. J. Chem. Phys. 128: 164314.

379. Cool, T.A., McIlroy, A., Qi, F., Westmoreland, P.R., Poisson, L., Peterka, D.S., Ahmed, M. (2005) Photoionization Mass Spectrometer for Studies of Flame Chemistry with a Synchrotron Light Source. Rev. Sci. Instrum. 76: 094102.

380. Litvinov, Y.A., Bosch, F., Geissel, H., Weick, H., Beckert, K., Beller, P., Boutin, D., Brandau, C., Chen, L., Klepper, O., Knöbel, R., Kozhuharov, C., Kurcewicz, J., Litvinov, S.A., Mazzocco, M., Münzenberg, G., Nociforo, C., Nolden, F., Plaß, W., Scheidenberger, C., Steck, M., Sun, B., Winkler, M. (2005) New Mass and Lifetime Measurements of ^{152}Sm Projectile Fragments with Time-resolved Schottky Mass Spectrometry, http://arxiv.org/abs/nucl-ex/0509019v1 (accessed September 11, 2015).

381. Fitzgerald, M.C., Harris, K., Shevlin, C.G., Siuzdak, G. (1996) Direct Characterization of Solid Phase Resin-bound Molecules by Mass Spectrometry. Bioorg. Med. Chem. Lett. 6: 979–982.

382. Carrasco, M.R., Fitzgerald, M.C., Oda, Y., Kent, S.B.H. (1997) Direct Monitoring of Organic Reactions on Polymeric Supports. Tetrahedron Lett. 38: 6331–6334.

383. Apuy, J.L., Park, Z.-Y., Swartz, P.D., Dangott, L.J., Russell, D.H., Baldwin, T.O. (2001) Pulsed-alkylation Mass Spectrometry for the Study of Protein Folding and Dynamics: Development and Application to the Study of a Folding/Unfolding Intermediate of Bacterial Luciferase. Biochemistry 40: 15153–15163.

384. Apuy, J.L., Chen, X., Russell, D.H., Baldwin, T.O., Giedroc, D.P. (2001) Ratiometric Pulsed Alkylation/Mass Spectrometry of the Cysteine Pairs in Individual Zinc Fingers of MRE-binding Transcription Factor-1 (MTF-1) as a Probe of Zinc Chelate Stability. Biochemistry 40: 15164–15175.

385. Apuy, J.L., Busenlehner, L.S., Russell, D.H., Giedroc, D.P. (2004) Ratiometric Pulsed Alkylation Mass Spectrometry as a Probe of Thiolate Reactivity in Different Metalloderivatives of *Staphylococcus aureus* pI258 CadC. Biochemistry 43: 3824–3834.

386. Min, D.-H., Yeo, W.-S., Mrksich, M. (2004) A Method for Connecting Solution-phase Enzyme Activity Assays with Immobilized Format Analysis by Mass Spectrometry. Anal. Chem. 76: 3923–3929.

387. Brivio, M., Tas, N.R., Goedbloed, M.H., Gardeniers, H.J., Verboom, W., van den Berg, A., Reinhoudt, D.N. (2005) A MALDI-chip Integrated System with a Monitoring Window. Lab Chip 5: 378–381.

388. Nichols, K.P., Gardeniers, H.J.G.E. (2007) A Digital Microfluidic System for the Investigation of Pre-steady-state Enzyme Kinetics Using Rapid Quenching with MALDI-TOF Mass Spectrometry. Anal. Chem. 79: 8699–8704.

389. Yoon, S.H., Moon, J.H., Kim, M.S. (2010) Dissociation Mechanisms and Implication for the Presence of Multiple Conformations for Peptide Ions with Arginine at the C-Terminus: Time-resolved Photodissociation Study. J. Mass. Spectrom. 45: 806–814.

390. Minegishi, Y., Morimoto, D., Matsumoto, J., Shiromaru, H., Hashimoto, K., Fujino, T. (2012) Desorption Dynamics of Tetracene Ion from Tetracene-doped Anthracene Crystals Studied by Femtosecond Time-resolved Mass Spectrometry. J. Phys. Chem. C 116: 3059–3064.

391. Chen, C.-C., Yang, Y.-L., Ou, C.-L., Chou, C.-H., Liaw, C.-C., Lin, P.-C. (2013) Direct Monitoring of Chemical Transformations by Combining Thin Layer Chromatography with Nanoparticle-assisted Laser Desorption/Ionization Mass Spectrometry. Analyst 138: 1379–1385.

392. Nichols, K.P., Azoz, S., Gardeniers, H.J.G.E. (2008) Enzyme Kinetics by Directly Imaging a Porous Silicon Microfluidic Reactor Using Desorption/Ionization on Silicon Mass Spectrometry. Anal. Chem. 80: 8314–8319.

393. Chuang, T.J., Hussla, I. (1984) Time-resolved Mass-spectrometric Study on Infrared Laser Photodesorption of Ammonia from Cu(100). Phys. Rev. Lett. 52: 2045.

394. Pei, J., Kang, Y., Huang, G. (2014) Reactive Intermediate Detection in Real Time via Paper Assisted Thermal Ionization Mass Spectrometry. Analyst 139: 5354–5357.

395. Smith, L.A., Caprioli, R.M. (1983) Following Enzyme Catalysis in Real-time Inside a Fast Atom Bombardment Mass Spectrometer. Biol. Mass Spectrom. 10: 98–102.

396. Smith, L.A., Caprioli, R.M. (1984) Enzyme Reaction Rates Determined by Fast Atom Bombardment Mass Spectrometry. Biol. Mass Spectrom. 11: 392–395.

397. Heidmann, M., Fonrobert, P., Przybylski, M., Platt, K.L. Seidel, A., Oesch, F. (1988) Conjugation Reactions of Polyaromatic Quinones to Mono- and Bisglutathionyl Adducts: Direct Analysis by Fast Atom Bombardment Mass Spectrometry. Biomed. Environ. Mass Spectrom. 15: 329–332.

398. Lossing, F.P., Tickner, A.W. (1952) Free Radicals by Mass Spectrometry. I. The Measurement of Methyl Radical Concentrations. J. Chem. Phys. 20: 907.

399. Lossing, F.P., Ingold, K.U., Tickner, A.W. (1953) Free Radicals by Mass Spectrometry. Part II. – The Thermal Decomposition of Ethylene Oxide, Propyline Oxide, Dimethyl Ether, and Dioxane. Discuss. Faraday Soc. 14: 34–44.

400. Brodbelt, J.S., Cooks, R.G., Tou, J.C., Kallos, G.J., Dryzga, M.D. (1987) In Vivo Mass Spectrometric Determination of Organic Compounds in Blood with a Membrane Probe. Anal. Chem. 59: 454–458.
401. Chakravarty, T., Windig, W., Hill, G.R., Meuzelaar, H.L.C. (1988) Time-resolved Pyrolysis Mass Spectrometry of Coal: A New Tool for Mechanistic and Kinetic Studies. Energy Fuels 2: 400–405.
402. Savickas, P.J., LaPack, M.A., Tou, J.C. (1989) Hollow Fiber Membrane Probes for the In Situ Mass Spectrometric Monitoring of Nitrogen Trichloride Formation during Wastewater Treatment. Anal. Chem. 61: 2332–2336.
403. Kotiaho, T., Lister, A.K., Hayward, M.J., Cooks, R.G. (1991) On-line Monitoring of Chloramine Reactions by Membrane Introduction Mass Spectrometry. Talanta 38: 195–200.
404. Lauritsen, F.R., Gylling, S. (1995) On-line Monitoring of Biological Reactions at Low Parts-per-trillion Levels by Membrane inlet Mass Spectrometry. Anal. Chem. 67: 1418–1420.
405. Tedder, L.L., Rubloff, G.W., Shareef, I., Anderle, M., Kim, D.-H., Parsons, G.N. (1995) Real-Time Process and Product Diagnostics in Rapid Thermal Chemical Vapor Deposition Using *In Situ* Mass Spectrometric Sampling. J. Vac. Sci. Technol., B 13: 1924–1927.
406. Alberici, R.M., Mendes, M.A., Jardim, W.F., Eberlin, M.N. (1998) Mass Spectrometry On-Line Monitoring and MS^2 Product Characterization of TiO_2/UV Photocatalytic Degradation of Chlorinated Volatile Organic Compounds. J. Am. Soc. Mass Spectrom. 9: 1321–1327.
407. Rios, R.V.R.A., da Rocha, L.L., Vieira, T.G., Lago, R.M., Augusti, R. (2000) On-line Monitoring by Membrane Introduction Mass Spectrometry of Chlorination of Organics in Water. Mechanistic and Kinetic Aspects of Chloroform Formation. J. Mass Spectrom. 35: 618–624.
408. Somboon, P., Kinoshita, M., Wyszynski, B., Nakamoto, T. (2009) Development of Odor Recorder with Enhanced Recording Capabilities Based on Real-time Mass Spectrometry. Sens. Actuat. B 141: 141–146.
409. Zhou, L., Piekiel, N., Chowdhury, S., Zachariah, M.R. (2009) T-jump/Time-of-flight Mass Spectrometry for Time-resolved Analysis of Energetic Materials. Rapid Commun. Mass Spectrom. 23: 194–202.
410. Kotiaho, T., Hayward, M.J., Cooks, R.G. (1991) Direct Determination of Chlorination Products of Organic Amines Using Membrane Introduction Mass Spectrometry. Anal. Chem. 63: 1794–1801.
411. Johnson, R.C., Koch, K., Cooks, R.G. (1999) On-line Monitoring of Reactions of Epichlorohydrin in Water Using Liquid Membrane Introduction Mass Spectrometry. Ind. Eng. Chem. Res. 38: 343–351.
412. Hillier, W., Messinger, J., Wydrzynski, T. (1998) Kinetic Determination of the Fast Exchanging Substrate Water Molecule in the S_3 State of Photosystem II. Biochemistry 37: 16908–16914.
413. Messinger, J., Badger, M., Wydrzynski, T. (1995) Detection of *One* Slowly Exchanging Substrate Water Molecule in the S_3 State of Photosystem II. Proc. Natl. Acad. Sci. USA 92: 3209–3213.

414. Hillier, W., Wydrzynski, T. (2000) The Affinities for the Two Substrate Water Binding Sites in the O_2 Evolving Complex of Photosystem II Vary Independently during S-State Turnover. Biochemistry 39: 4399–4405.

415. Hendry, G., Wydrzynski, T. (2003) ^{18}O Isotope Exchange Measurements Reveal that Calcium Is Involved in the Binding of One Substrate-Water Molecule to the Oxygen-evolving Complex in Photosystem II. Biochemistry 42: 6209–6217.

416. Nilsson, H., Krupnik, T., Kargul, J., Messinger, J. (2014) Substrate Water Exchange in Photosystem II Core Complexes of the Extremophilic Red Alga *Cyanidioschyzon merolae*. Biochim. Biophys. Acta 1837: 1257–1262.

417. Harren, F.J.M., Cristescu, S.M. (2013) Online, Real-time Detection of Volatile Emissions from Plant Tissue. AoB Plants 5: plt003.

418. Xie, Y.; Hua, L.; Hou, K.; Chen, P.; Zhao, W.; Chen, W.; Ju, B.; Li, H. (2014) Long-Term Real-time Monitoring Catalytic Synthesis of Ammonia in a Microreactor by VUV-Lamp-based Charge-transfer Ionization Time-of-flight Mass Spectrometry. Anal. Chem. 86: 7681–7687.

419. Cho, Y.S., Kim, M.S., Choe, J.C. (1995) Reinvestigation of the Photodissociation Kinetics of *m*-Iodotoluene Molecular Ion. Int. J. Mass Spectrom. Ion Proc. 145: 187–195.

5

Balancing Acquisition Speed and Analytical Performance of Mass Spectrometry

5.1 Overview

Mass resolving power, *mass accuracy*, and *sensitivity* are three critical parameters determining analytical performance in conventional mass spectrometry (MS). On the other hand, *spectrum acquisition speed* is the most important factor influencing time-resolved mass spectrometry (TRMS) studies. An instrument with higher spectrum acquisition speed can certainly facilitate more TRMS applications, especially for real-rime monitoring of chemical and physical processes. Unfortunately, high performance mass analyses typically require longer *spectrum acquisition times* (SATs) than regular operation modes. Obtaining better resolved and accurate mass spectra normally relies on slower spectrum scanning speeds (and consequently longer SATs), but a slower spectrum acquisition speed normally results in lower sensitivity and duty cycles. Since the ideal mass spectrometer (i.e., one that incorporates the fastest acquisition speeds and highest mass resolving power and sensitivity) is beyond reach, mass spectrometer users must find a compromise among these performance features. This chapter discusses the relationship between spectrum acquisition speeds, mass resolving power, mass accuracy, and sensitivity. It also contains practical considerations for performing real-time monitoring by MS, which is aimed at assisting mass spectrometrists to find optimal instrument and operating conditions for their TRMS experiments.

5.2 Spectrum Acquisition Speed

Before discussing relationships among various parameters in MS analysis, it is important to understand the concept of SAT. SAT indicates the time required for a complete scan of

Time-Resolved Mass Spectrometry: From Concept to Applications, First Edition.
Pawel Lukasz Urban, Yu-Chie Chen and Yi-Sheng Wang.
© 2016 John Wiley & Sons, Ltd. Published 2016 by John Wiley & Sons, Ltd.

a mass spectrum from the instant of ion generation till the completion of ion detection. It is convenient to use SAT rather than speed, because SAT enables direct comparisons of different instruments in acquiring a complete spectrum without complex calculations. For real-world applications, effective SAT should take into account the so-called "dead time", which is defined as the time interval between two adjacent scanning events. Based on effective SAT, we can deduce *duty cycle* – another important parameter affecting the speed of MS analysis. Unlike SAT, the duty cycle does not correlate noticeably with mass resolving power and sensitivity, but it is as equally important as SAT when it comes to the performance of an instrument in TRMS applications.

5.2.1   Spectrum Acquisition Time

As mentioned above, SAT is a measure of the time interval between the instant of sample ionization and the completion of data acquisition. The total time it takes for ions to travel through a mass spectrometer is typically orders of magnitude greater than the time needed for ion production. Hence, SAT closely relates with the operation mode of static-field mass analyzers, such as time-of-flight (TOF) and Fourier transform (FT) mass analyzers, and also with *scanning speed* across a mass range (u s^{-1}) in dynamic-field mass analyzers, such as quadrupole mass filters (QMFs) and ion trap (IT) mass analyzers. Other parameters affecting SAT include the physical dimensions of ion transmission regions (ion optics, guiding devices, etc.) as well as the kinetic energy of ions traversing these regions.

The SATs of mass spectrometers vary considerably, ranging from nanoseconds to seconds. In trapping-type mass spectrometers (e.g., IT and FT), the SAT can be adjusted by trapping ions over a controllable time interval. The trapping time depends on the experiment being conducted. However, as a rule of thumb, increasing the SAT reduces overall ion abundance due to limited ion lifetimes as well as ion losses during transmission or trapping. In contrast to the trapping-type instruments, beam-type mass spectrometers [e.g., TOF, sector (S), QMF, triple quadrupole (QqQ), and quadrupole (Q)-TOF] have shorter SATs because they disallow ion beams to be stored in any region. Therefore, knowing the practical range of the SAT of important mass spectrometers can be useful when planning TRMS studies. Figure 5.1 compares the typical SATs of popular mass spectrometers. Only the practical SAT region of trapping-type instruments is displayed, although such instruments can control ion trapping times beyond the ranges indicated in this figure.

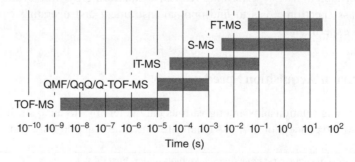

Figure 5.1   *The typical spectrum acquisition times of important types of mass spectrometers*

Among all mass spectrometers, the TOF-MS devices exhibit the shortest SATs because ions travel along a simple straight or round-trip path with the highest velocity, even though TOF analyzers are normally longer than other instruments. In matrix-assisted laser desorption/ionization (MALDI)-TOF-MS, ions are normally accelerated to 20 000–25 000 eV, resulting in an ion velocity of roughly 10^3–10^5 m s^{-1}. If the flight tube is 1–2 m long, the flight time is in the sub-microsecond and hundreds of microseconds range in the case of small organic and large biological molecules, respectively.

The SAT of a quadrupole MS system and for hybrid instruments containing quadrupole mass analyzers (QMF, QqQ-MS, and Q-TOF-MS) is longer than that of TOF-MS devices. In the instruments containing QMF, ions are typically accelerated to tens or hundreds of electronvolts when entering the QMF, or roughly 10^2–10^3 m s^{-1}. Because a QMF is normally 10–20 cm long, ions spend from a few tens to hundreds of microseconds in this region. The scan rate of a QMF and a QqQ-MS device is roughly 10^3–10^4 u s^{-1}, so their SATs range from tens of microseconds to hundreds of milliseconds. The SAT of a Q-TOF-MS system is still in the same range as that of a QMF because the SAT of TOF is much shorter than that of QMF.

The SAT of trapping-type instruments is generally longer than that of beam-type instruments because trapping-type instruments scan mass spectra after enough ions have accumulated. In IT-MS, an ion trap normally spends sub-millisecond to hundreds of milliseconds to accumulate ions. The long ion accumulation times make the SAT of IT-MS longer than that of a QMF and other types of quadrupole mass spectrometers. However, the advantage of IT-MS is that ion trapping times can be adjusted in a wide range for various experimental purposes. The high flexibility of ion trapping times in IT-MS, in conjunction with good tandem MS capability, make IT-MS devices one of the most popular type of mass spectrometers in the market, and also good choices for TRMS.

The S-MS device scans mass spectra with low speed due to the slow scanning rate of the magnetic field. The scanning rate is typically 10^2–10^3 u s^{-1}. However, ions in sector instruments travel at high velocities because the ion beam in such instruments is typically accelerated to a kinetic energy of a few kiloelectronvolts, corresponding to an ion velocity of roughly 10^4 m s^{-1}. Commercial S-MS devices typically consist of two sector regions and a long transmission path, so ions just spend few microseconds to hundreds of microseconds on the way from the ion source to the detector. These times are only slightly longer than those in TOF-MS.

FT-MS is the slowest among all MS techniques. The long SAT of FT-MS is due to an extended time for acquiring the transient, which is essential for the accurate determination of frequencies of periodic motions of ions. Moreover, FT mass analyzers need to be installed in ultrahigh vacuum environments, and such environments require long ion transmission paths and several differentially pumped vacuum stages. Because commercial FT mass analyzers are installed downstream of IT mass analyzers, the typical SATs of FT-MS range from hundreds of milliseconds to tens of seconds.

5.2.2 Duty Cycle

The duty cycle is an essential temporal characteristic of mass spectrometers [1]. It is defined as the ratio of the time spent on the collection of ions for analysis to the operation time of the ion source within a complete scan. It is typically expressed as a percentage. This parameter is important when combining a mass analyzer with continuous ion sources because discrete mass spectra are recorded by modulating/gating the transmission of ions, which results in ion losses within this time interval (i.e., before and after the ion acceptance by the mass analyzer). The duty cycle should be taken into account while selecting instruments for TRMS, in particular, for real-time monitoring applications.

The duty cycle of a mass spectrometer varies with the type of ion source, mass analyzer, operational mode, and control electronics. It can be as high as 100% when using pulsed ion sources since the time interval for ion collection of most mass spectrometers is longer than the typical time spread of ion packages ejected from pulsed ion sources. In contrast, when continuous ion sources are used, the duty cycle goes down to a few tens of percent because ion collection by the mass analyzer needs to be temporally stopped for spectrum recording. Before the start of the next event, a short period or dead time elapses (i.e., a few microseconds) to rest and stabilize the electronics, to recover the ion detector, and to record the spectrum. The short recovery time for the ion detector is necessary to restore the detector to its optimal working conditions because static charges on the surface of microchannel plates (MCPs) or electron multipliers (EMs) have been discharged by the previous ion beam or ion package. In addition, data storage time from the random-access memory (RAM) of digitizers to the hard drive typically has a speed of roughly several tens to hundreds of megabits per second (Mb s^{-1}), so the time necessary for data storage is data-size dependent. In fact, the major limitation of data storage time is the speed of the hard drive. Figure 5.2 shows typical sequence of mass analysis with a continuous ion source. For trapping-type instruments, such as in IT-MS and FT-MS, one analysis event starts from the input of ions into the mass analyzer and it lasts for a period, denoted by t_i. Once the ion input event stops, mass analysis starts. Mass analysis includes ion fragmentation, separation, and detection. This period is denoted by t_a. For beam-type instruments, such as in a QMF and S-MS, t_i is unnecessary because ions enter the mass analyzer during mass analysis. Thus, the

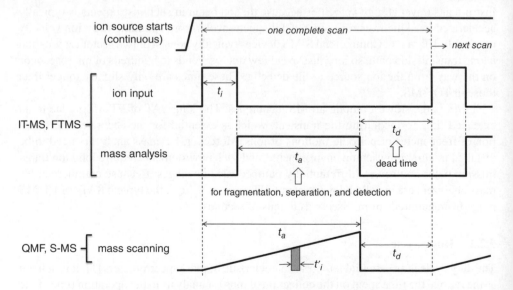

Figure 5.2   *The sequence of mass scans in different mass spectrometers*

SAT for the trapping-type instruments is the sum of ion input and analysis (i.e., the upper seqnece in Figure 5.2); that is $SAT = t_i + t_a$, and for beam-type instruments it is the time required for mass analysis, or $SAT = t_a$ (i.e., the bottom sequence in Figure 5.2). After detection, the experiment stops for a dead time (t_d) before the next scanning cycle starts. The t_d is optimized by the manufacturer and cannot be adjusted by the user. Since the t_d is important for continuous or real-time measurements, the effective SAT, or the exact time interval between adjacent scans, needs to take into account the t_d. That is, the effective $SAT = t_i + t_a + t_d$.

Because the useful portion of ion streams flowing into ion traps or the FT-MS device is t_i, the duty cycle of such instruments is $[t_i/(t_i + t_a + t_d)] \times 100\%$. Although extending the collection window results in a higher duty cycle, it reduces the quality of mass spectra due to the space-charge effect (too many ions loaded into the mass analyzer). For the QMF, on the other hand, the duty cycle is defined as the portion of time the mass analyzer spends passing through ions at one m/z during a complete scan. Assuming this time is t_i', the duty cycle of the QMF is then $[t_i'/(t_a + t_d)] \times 100\%$. However, a duty cycle as high as 100% is achievable with continuous ion sources when monitoring ions at a specific m/z (without scanning) using quadrupole mass filters. This operation mode is referred to as the selected ion monitoring (SIM) or selected reaction monitoring (SRM) mode. Notably, if multiple precursor ions are selected, the duty cycle will be significantly reduced.

The duty cycle of important mass spectrometers varies significantly. Under general working conditions, the duty cycle is 100% for MALDI-TOF-MS, ~30% for Q-TOF-MS, ~10% for IT-MS, ~0.1% for S-MS and QMF, and ~0.01% for FT-MS.

5.3 Relationship between Spectrum Acquisition Time and Mass Spectrometer Performance

5.3.1 Mass Resolving Power

Mass resolving power is one of the most important factors characterizing the analytical performance of a mass spectrometer. It is defined as the ratio of the ion m/z to the full width at half maximum (FWHM) of its spectral feature, or $m/\Delta m$ in the case of singly charged species [2]. It represents the sharpness of the spectral feature. That is, a higher mass resolving power indicates a better capability of resolving a spectral feature from other nearby features. Figure 5.3 shows important characteristics of spectral features used to calculate the mass resolving power. The mass resolving power for ions with mass m_1 and m_2 is $m_1/\Delta m_1$ and $m_2/\Delta m_2$, respectively. Therefore, the mass resolving power is not necessarily a constant value within a mass spectrum. Another similar but less commonly used factor is *mass resolution*, which defines the maximum mass range at which a mass analyzer can well resolve two spectral features from ions with mass difference of one mass unit (i.e., m vs. $m + 1$ or $m - 1$). One definition of well-resolved features is having two adjacent peaks separated by a valley of <10% of the height of any of the two peaks.

Although there is no clear definition of a high-resolution mass spectrometer, it is generally accepted that S-MS, FT-MS, and TOF-MS devices and the related hybrid mass spectrometers are high-resolution instruments. For example, the maximum mass resolving power of a double-focusing S-MS system is roughly 60 000, and in a TOF-MS

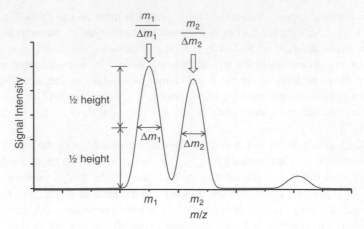

Figure 5.3 *Spectral characteristics used to calculate mass resolving power and mass resolution. The mass resolving power of an individual peak is indicated at the top of the peak*

device and related hybrid instruments is 5000–80 000. The highest mass resolving power in an FT-MS device depends on the mass analyzer; it is roughly 400 000 for orbitrap and 3 000 000 for Fourier transform ion cyclotron resonance (FT-ICR) mass analyzers. Notably, an ultrahigh mass resolving power FT-ICR mass spectrometer with an optimized electric potential inside its ICR cell has recently became commercially available, providing a mass resolving power of above 10 000 000 [3]. The remaining mass spectrometers are considered low or of medium resolution. For example, members of the quadrupole mass spectrometer family are considered low resolution instruments because the practical mass resolving powers of such instruments are 1000–3000.

The advantage of high mass resolving power is that the maximum number of resolvable ions in a spectrum is relatively large. This condition is highly advantageous for analyzing complex samples because multiple ions may present in a narrow mass range. Highly complex biological or organic samples may contain analytes with almost identical molecular weights or isobaric ions (distinct ions with the same nominal mass). This complexity is especially apparent when analyzing multiply charged molecules produced by electrospray ionization (ESI). In such a spectrum, multiply charged spectral patterns of every component in a complicated mixture may overlap. In addition, every charge state of an analyte produced by ESI shows isotopical patterns. Thus, the higher the mass resolving power, the higher the ability of mass spectrometers to resolve complicated spectral features and to extract useful information. Methods of structural identification and charge deconvolution are detailed in Chapter 9.

Although high mass resolving power is advantageous, it generally requires long SATs. As mentioned in Section 3.2.4, a longer transient in FT-MS results in higher confidence of periodic frequency determination [4, 5]. An example illustrating the relationship between the spectral width in the frequency domain and numbers of sinusoidal cycles in the time domain is shown in Figure 5.4. For sinusoidal waves with a frequency of 1000 Hz, a transient of 0.1 s contains 100 cycles. Such a transient results in a feature in the frequency domain with an FWHM of 19.5 Hz. Increasing the transient to 1 s (or 1000 cycles) results in

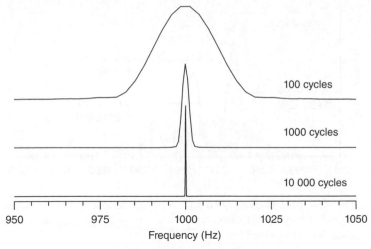

Figure 5.4 *Change of frequency-domain spectral bandwidth with the number of sinusoidal cycles recorded in the time domain by FT-MS. This result indicates that the spectral linewidth is smaller at a larger number of cycles*

an improvement of FWHM in the frequency domain to 2.2 Hz. With 10 s transient (10 000 cycles), the FWHM in the frequency domain is 0.2 Hz.

A similar effect can also be observed in quadrupole MS, in which a higher resolving power can be achieved when ions are subjected to more oscillations in the mass analyzer. The probability of ejecting ions right at the boundary of the unstable region is higher if the preset electrical condition is applied for a longer time before switching to another condition. The reason for the lower ejection probability is due to the energy and position spreads of ions within the mass analyzers. Thus, if an ion is at the boundary of the unstable region but is not ejected in the first oscillation, it can be ejected when subjected to more oscillations under the same electrical conditions. In IT devices, for instance, spectra with a higher mass resolving power can be recorded by reducing the scanning speed of the electric potentials [6]. In a QMF device, ions can follow only one trajectory along the mass analyzer. Commercial QMF devices typically have a length of 10–20 cm, within which ions experience a few tens of electrical sinusoidal waves. In this case, reducing ion speed or increasing rod length (to allow ions to experience more electric cycles) can enhance mass resolving power. However, reducing ion speed and increasing QMF length are unfavorable because they reduce ion transmission efficiency and increase the dimensions of the vacuum system, respectively.

5.3.2 Mass Accuracy

Mass accuracy describes correctness of the observed mass. It is defined as the ratio of mass error to the theoretical mass of an ion, typically expressed in parts per million (ppm). Mass accuracy and mass resolving power are equally important parameters when implementing MS in molecular identification because well-resolved spectral features are useful only if the mass is correct. However, mass resolving power is mainly determined by the mass analyzer, whereas mass accuracy is affected by both the mass analyzer and mass calibration.

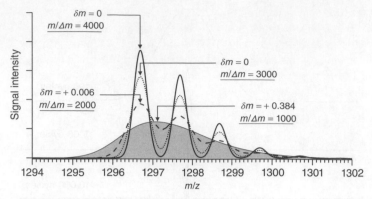

Figure 5.5 *Shift of the peak of protonated angiotensin I due to a decrease of the mass resolving power. The peak shift results in a decrease in mass accuracy*

Because mass accuracy and mass resolving power are correlated, high mass accuracy also relies on a long SAT. This dependence results from the uncertainty in determining the position of broad peaks, which are more difficult to characterize than sharp peaks. For example, the difficulty in peak determination in Figure 5.4 increases from the bottom to the top spectrum. In addition to peak width, the observed mass may shift as peak shape changes. This effect can be due to the presence of isotopes or contaminants. The change in spectral shape is unimportant when analyzing small ions with high-resolution mass spectrometers. In the mass spectra obtained with low-resolution mass spectrometers, however, the isotopic variants of an ion may merge and become indistinguishable. In this case, determination of accurate mass becomes highly challenging and unreliable. Figure 5.5 illustrates the predicted shift of the peak of protonated angiotensin I ($[C_{62}H_{89}N_{17}O_{14} +$ $H]^+$, $m/z = 1296.685$) as mass resolving power decreases. The shift is negligible when the mass resolving power is above 3000, in which the valleys between the isotope features are roughly 12% of peak amplitude. The mass shift becomes noticeable when the valleys become 50% of peak amplitude, corresponding to a mass resolving power of roughly 2000. The mass difference in this case is 0.006 u, resulting in a mass error of ~5 ppm. When mass resolving power is 1000, no isotope pattern can be resolved and the observed mass is $m/z = 1297.069$. The observed mass is close to the average mass (1297.484) of protonated angiotensin I. The mass difference in this case is 0.384 u, corresponding to an error of 300 ppm. This case demonstrates that the observed peak position depends on the mass resolving power of the instrument.

Another important factor determining mass accuracy is the stability of the electric fields of mass analyzers. This stability issue includes the ripple and drift of electric potentials during measurements. Such a change in electric potential affects the kinetic energy and hence the motion of ions in mass analyzers. In a TOF mass analyzer, for example, even a 0.01% drift in the acceleration voltage causes an error of roughly 100 ppm. To achieve a mass accuracy of 10 ppm, power sources with ultralow ripple (<0.001% ripple) are necessary. Such a specification requirement approaches the limits of commercial DC power supplies. In order to facilitate high mass accuracy measurements, most commercial mass spectrometers use high precision power supplies and electronic components.

With good instrument conditions and mass calibration, mass accuracy of different mass analyzers follows the same trend as mass resolving power. For quadrupole MS, the typical mass accuracy is roughly 50–100 ppm. Without performing tandem MS, such mass accuracy is suitable only for molecule identification below m/z 100. The typical mass accuracy ranges from 1 to 50 ppm for TOF-MS and S-MS. These instruments can be used for identification of small molecules or peptides without performing tandem MS below m/z 500. FT-MS exhibits the highest mass accuracy, typically within 5 ppm to sub-parts per million range. Therefore, FT-MS provides the best analytical performance for peptide and protein sequencing.

5.3.3   Sensitivity and Detection Limit

Sensitivity and *detection limit* [or *limit of detection* (LOD)] represent other important characteristics of MS. Both are essential when defining the analytical performance of mass spectrometers; however, the two terms are often confused and need to be clarified. The sensitivity of a mass spectrometer is its capability to respond to changes in analyte quantity. A more specific definition of sensitivity is the electric charge recorded by the ion detector of the mass spectrometer per unit quantity of analyte, normally with units of coulomb per microgram (C μg^{-1}) for condensed phase analytes, or coulomb per pascal (C Pa^{-1}) for gaseous analytes. In contrast, the LOD is the lowest analyte quantity a mass spectrometer can successfully detect. The calculation of LOD is based on a predefined signal intensity, normally with a signal-to-noise (S/N) ratio of 3. The LOD is usually reported in absolute amounts of analytes. Thus, a mass spectrometer with a higher sensitivity or lower LOD has better analytical performance. The LOD can also be expressed in concentration units. The use of analyte quantity or concentration units depends on the technique used. For example, MALDI-MS is mass (or quantity) sensitive, and the LOD of such methods should be reported in quantity units. In contrast, ESI-MS may be concentration- or mass-sensitive, depending on the flow rate used in the measurement. Thus, the LOD of ESI-MS may be expressed in quantity or concentration units, and the experimental conditions used to achieve the reported LODs should be clarified.

Sensitivity is one of the first parameters considered when selecting instruments for TRMS measurements. In such measurements, monitoring chemical or biochemical reactions usually involves the detection of reactants and products with different concentrations. In some cases, the concentration of reactants is normally high initially and reduces with time as a reaction proceeds. The pattern of products reversed. In order to detect the extent of a reaction with the highest precision, the sensitivity of a mass spectrometer needs to be high enough to detect small concentration changes in reaction products at the beginning of a reaction. Without a sensitive mass spectrometer, the early stages of chemical reactions cannot be precisely characterized. On the other hand, LOD should also be low enough to detect small amounts of reactants at the end of chemical reactions. Thus, it is important to ensure that the sensitivity and LOD of the mass spectrometer are optimized for both reactants and products. If the instrumemts are not sensitive to pronounced differences in mass, structure, or charge states then the obtained data are unreliable.

The sensitivity and LOD of mass spectrometers depend on several factors. The first and most important factor is the ionization efficiency of analytes. Ionization efficiency is analyte-dependent because it is normally higher for analytes of higher polarity, higher volatility, and lower mass (see Chapter 2). The second critical factor is the ionization technique. This factor includes the environment under which ionization occurs and the transmission efficiency of ions from the ion source to the mass analyzer. In the case of atmospheric-pressure ionization techniques, the interface bridging ambient and vacuum regions causes serious ion losses. Experimental observations suggest that the typical transmission efficiency through a vacuum interface is 0.1% or less [7]. Although transmission efficiency can be improved by ion guiding devices, the sensitivity of ambient ion sources is still one to two orders of magnitude lower than that of vacuum ionization methods. The third critical factor is the presence of background or "unwanted" species affecting the appearance of analyte ion peaks. Background species may compete with analytes for charges (electrons or protons). Other abundant ions also suppress analyte ions. This common interference is referred to as the "matrix effect" [8, 9]. If the charge density in an ion source or mass analyzer is too high, the sensitivity of a mass spectrometer declines due to space-charge effects.

Beside the above determinants, temporal properties of mass spectrometers also critically affect sensitivity and LOD. The most serious problem is the combination of the ion source and mass analyzer. Combining pulsed ion sources with TOF or trapping-type mass analyzers provides superior sensitivity. In this combination, every component of an ion packets can eventually reach the ion detector. The best example is a TOF mass analyzer equipped with an LDI or MALDI ion source, which offers a duty cycle as high as 100%. The LOD of commercial MALDI-TOF mass spectrometers is in the femtomole range. In contrast, most beam-type mass analyzers are unsuitable for pulsed ion sources because the duration of the typical ion packets is too short for beam-type mass analyzers to complete a mass spectral scan. For example, it is not practical to integrate a MALDI ion source into a QMF because the ion package produced from every laser shot only lasts for nanoseconds. Even with the highest scanning speed of a QMF (e.g., $10000\,u\,s^{-1}$), scanning 100 ns in MALDI-MS only cover a mass range of 0.001 u! Therefore, such a combination leads to extremely low efficiency. On the other hand, combining pulsed ion sources with trapping-type mass analyzers provides high efficiency because an ion packet lasting for a short period of time can be captured by an IT device and analyzed. In contrast, continuous ion sources offer lower sensitivity. The major reason is the duty cycle problem (as mentioned in Section 5.2.2). That is, neither beam-type nor trapping-type mass analyzers are able to process all ions produced in continuous ion sources.

Because SATs rely on all the experimental details given above, changing the SAT can also affect the sensitivity and LOD of a mass spectrometer. As a rule of thumb, the analytical performances of a mass spectrometer with a pulsed ion source decreases as the SAT increases. The reason for this is that the lifetime of an ion is limited and ion abundance decreases with time. Thus, a lower spectrum scanning speed results in lower signal intensity; for example, when analyzing MALDI generated ions with trapping-type mass spectrometers (e.g., IT-MS and FT-MS devices). Since the SATs of IT-MS and FT-MS devices are in the range of milliseconds to seconds, these instruments usually record stable ions. In contrast, when continuous ion sources are coupled with trapping-type mass analyzers, sensitivity is affected by a competition between ion accumulation and ion loss in the mass analyzer. That is, in the mass analyzer, ion losses increase as scanning time increases, but the number of ions can be increased by extending the ion accumulation time. However, a very long accumulation time is normally not helpful, because ion losses dominate in the case of low ion flux. In contrast, charge-competition and space-charge problems occur in the case of high ion flux. In order to control the number of ions present in mass analyzers, commercial instruments are able to adjust ion accumulation times automatically according to the observed total ion signal of the previous scanning event. This functionality is called *auto-gain control* or *ion-current control*.

A decrease in S/N ratio with increasing SAT is very pronounced when comparing the results of low- and high-resolution spectra obtained from the same trapping-type mass spectrometer. This characteristic is especially apparent in IT-MS and FT-MS devices because SATs of such instruments are relatively long. Thus, high-resolution scans take longer time than low-resolution scans and normally exhibit lower S/N ratios. Hence, in many cases, it is necessary to find a compromise between scanning speed, sensitivity, and resolving power.

References

1. Gross, J.H. (2004) Mass Spectrometry: A Textbook, 1st Edition. Springer, Berlin.
2. Marshall, A.G., Hendrickson, C.L., Shi, S.D.H. (2002) Scaling MS Plateaus with High-resolution FT-ICRMS. Anal. Chem. 74: 253a–259a.
3. Nikolaev, E.N., Boldin, I.A., Jertz, R., Baykut, G. (2011) Initial Experimental Characterization of a New Ultra-high Resolution FTICR Cell with Dynamic Harmonization. J. Am. Soc. Mass Spectrom. 22: 1125–1133.
4. Li, Y.Z., Hunter, R.L., McIver, R.T. (1996) Ultrahigh-resolution Fourier Transform Mass Spectrometry of Biomolecules Above m/z 5000. Int. J. Mass Spectrom. Ion Proc. 157: 175–188.
5. Marshall, A.G., Verdun, F.R. (1990) Fourier Transforms in NMR, Optical, and Mass Spectrometry: A User's Handbook. Elsevier, Amsterdam.
6. Williams, J.D., Cox, K.A., Cooks, R.G., Kaiser, R.E., Schwartz, J.C. (1991) High Mass-resolution Using a Quadrupole Ion-trap Mass-spectrometer. Rapid Commun. Mass Spectrom. 5: 327–329.
7. Kelly, R.T., Tolmachev, A.V., Page, J.S., Tang, K.Q., Smith, R.D. (2010) The Ion Funnel: Theory, Implementations, and Applications. Mass Spectrom. Rev. 29: 294–312.
8. Matuszewski, B.K., Constanzer, M.L., Chavez-Eng, C.M. (2003) Strategies for the Assessment of Matrix Effect in Quantitative Bioanalytical Methods Based on HPLC-MS/MS. Anal. Chem. 75: 3019–3030.
9. Niessen, W.M.A., Manini, P., Andreoli, R. (2006) Matrix Effects in Quantitative Pesticide Analysis Using Liquid Chromatography–Mass Spectrometry. Mass Spectrom. Rev. 25: 881–899.

6

Hyphenated Mass Spectrometric Techniques

6.1 Introduction

The so-called hyphenated mass spectrometric techniques take advantage of temporal ion recording enabled by mass spectrometry (MS) instruments. In principle, all types of mass spectrometers can be coupled on-line or off-line with flow-based techniques to conduct analysis of effluents from separation or sample preparation devices. For a few decades now MS has been operated in conjunction with separation techniques, in particular liquid chromatography (LC; for a review see [1]), gas chromatography (GC; for a reference textbook see [2–4]) and capillary electrophoresis (CE; for a review see [5]). The purpose of coupling separation techniques with MS is to reduce the effects of matrix interferences (ion suppression), increase analytical selectivity, and improve quantitative capabilities of the assays (see also Chapter 8). Coupling separation stages with MS relies on the high frequency data acquisition enabled by MS instruments. Effluents of chromatographic or electrophoretic columns can be regarded as dynamic samples because identities and concentrations of the eluting compounds change over time. Alternative separation and sample treatment schemes have also been coupled with MS to enable seamless analysis of dynamic systems, increase sensitivity, or to reduce human involvement. Certainly, the goal of MS detection is to obtain molecular information on the eluted compounds with high analytical sensitivity and without losing temporal properties. Thus, before we move on to the discussion of hyphenated systems, it is necessary to revisit a few basic terms and dependencies which characterize the performance of common liquid- and gas-phase separation techniques.

6.1.1 Chromatography

In typical chromatographic separations, the sample is loaded into a column filled with stationary phase (solid or liquid supported on a solid). The movement of the sample

---

Time-Resolved Mass Spectrometry: From Concept to Applications, First Edition.
Pawel Lukasz Urban, Yu-Chie Chen and Yi-Sheng Wang.
© 2016 John Wiley & Sons, Ltd. Published 2016 by John Wiley & Sons, Ltd.

components along the column is facilitated by the flow of a mobile phase (liquid, supercritical fluid, or gas). Various analytes present in the sample interact with the stationary phase to a different extent. Thus, some of them can move faster while the others traverse the column with a lower speed. If the affinities of given sample components to the stationary phase are significantly different, complete separation can be achieved.

The concept of a *theoretical plate* is introduced to characterize separation of compounds in chromatographic columns. A theoretical plate is a virtual discrete section of the chromatographic column in which equilibrium between the fraction of the analyte in the mobile phase and the stationary phase can occur at the provided conditions. The number of theoretical plates (N) is a measure of column efficiency. It depends on the length of the column (L) and the *height equivalent to a theoretical plate* (HETP):

$$N = \frac{L}{HETP} \tag{6.1}$$

The HETP value is given by the *Van Deemter equation* [6]:

$$HETP = A + \frac{B}{u} + Cu \tag{6.2}$$

where A, B, C are the factors describing various processes taking place in the column (eddy diffusion, longitudinal diffusion, and resistance to mass transfer, respectively) while u is the average velocity of the mobile phase. This dependence is illustrated in a Van Deemter plot (Figure 6.1).

From the Van Deemter plot, it is clear that there exists an optimum velocity of mobile phase at which the HETP function attains its minimum. The low HETP value is favorable because it will imply a large number of theoretical plates in the column of a given length (cf. Equation 6.1). Thus, interactions between the sample and the stationary phase will favor differentiation of *retention times* of various analytes traversing the chromatographic column.

According to the Van Deemter equation, eddy diffusion is not affected by the velocity of the mobile phase. Longitudinal diffusion has the largest effect on HETP at low velocities. Therefore, it is preferable to pump mobile phases with relatively high flow rates. However, the resistance to the flow exerted by the column imposes the requirement for pumping the mobile phase with a high pressure. When using conventional LC columns, the mass transfer term of the Van Deemter equation becomes significant at high velocities,

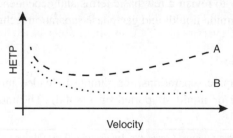

Figure 6.1   *Van Deemter plots: (A) high contribution of resistance to mass transfer; (B) low contribution of resistance to mass transfer*

thus increasing the HETP value (Figure 6.1A). This unfavorable influence of the resistance to mass transfer can be limited by decreasing the size of the beads of the stationary phase (Figure 6.1B). Columns packed with very small beads <3 µm are used in *ultra-high-performance liquid chromatography* (UHPLC). In this case, the backpressure produced in the column is particularly high. Thus, UHPLC pumps are more expensive than the conventional *high-performance liquid chromatography* (HPLC) pumps. They can handle high pressures (~1000 bar or ~100 MPa) and overcome the backpressure that occurs during the flow of the mobile phase through the voids among small beads of the stationary phase.

Most of the modern GC systems use open tubular capillary columns. Here, the shape of the Van Deemter plot is strongly influenced by the type of the mobile phase gas used. Helium is a commonly used mobile phase gas in GC. The HETP value can be brought down significantly by using hydrogen gas as the mobile phase. Hydrogen has particularly low viscosity while the diffusivity of volatile analytes in hydrogen is high [7]. However, due to safety considerations, hydrogen has not become popular in GC applications.

For a given pair of analytes, one can define the *chromatographic resolution (R)*:

$$R = 1.18 \left(\frac{t_{R(2)} - t_{R(1)}}{W_{1/2(1)} + W_{1/2(2)}} \right) \tag{6.3}$$

where $t_{R(1)}$ and $t_{R(2)}$ are the retention times of peaks 1 and 2, respectively, while $W_{1/2(1)}$ and $W_{1/2(2)}$ are the *full-width-at-half-maximum* (FWHM) values of these peaks. This equation assumes that the chromatographic peaks are Gaussian, which is correct for many optimized separations. The R value is unitless. If it is greater than ~1.5, one can conclude that the two peaks are *baseline-separated*.

Chromatographic techniques can be divided according to multiple criteria. It is helpful to consider the specifics of the common modes of LC used on-line with MS, taking into account the choice of stationary and mobile phases:

- *Size-exclusion chromatography* (SEC) uses porous particles as stationary phases. Small molecules enter the pores, following a complex movement path; thus, their migration along the column is slow. Large species (e.g., macromolecules) cannot readily diffuse into the pores; thus, they move faster.
- *Ion-exchange chromatography* separates ionic species based on the ionic interactions between them and the charged groups immobilized on the surface of stationary phase particles.
- *Normal-phase liquid chromatography* (NP-LC) separates analytes based on the differences in their interactions with the polar stationary phase. In this mode, the stationary phase is more polar than the mobile phase. It normally uses organic solvent as mobile phase but there exists a variant of NP-LC in which water-based mobile phases are implemented (*aqueous normal-phase liquid chromatography*, ANP-LC).
- *Reversed-phase liquid chromatography* (RP-LC) separates analytes based on the differences in their interactions with the non-polar stationary phase. In this mode, the stationary phase is less polar than the mobile phase. The reversed-phase stationary phases often contain alkyl groups (C_2, C_8, or C_{18}).
- *Hydrophilic interaction chromatography* (HILIC) uses polar stationary phases. Thus, polar analytes can be separated with high efficiency. It is sometimes regarded as a type of *partition chromatography* – one of the early chromatographic modes.

• *Affinity chromatography* takes advantage of affinity interactions between analytes and molecules (e.g., proteins, chelating agents) immobilized on the surface of the solid support.

6.1.2   Electrophoresis

Charged particles – for example, ions – can be separated in an electrolyte medium in the presence of an electric field. In biochemistry, it is common to separate large biomolecules in gel matrices (agarose, polyacrylamide), soaked with buffered electrolyte solutions. In a modern version of the technique, hair-thin capillaries are used as electrophoretic columns. Differences in the electromigration of charged species in a capillary filled with an electrolyte lead to separation. Conventional capillary electrophoresis (CE) apparatus incorporates fused silica capillaries with diameters of 20–100 μm and lengths of 20–100 cm. However, there exist miniaturized CE devices which implement microchips with much smaller and shorter separation channels (see Chapter 7). Typically, the separation capillary is filled with an electrolyte, and its two ends are dipped into reservoirs filled with electrolyte. Electrodes are also placed in the electrolyte reservoirs to create an electric field inside the capillary. The potential difference between the positive electrode (anode) and the negative electrode (cathode) can be as high as 30 kV. Higher voltages are not normally used – mainly because of the occurrence of *Joule heating* inside the capillary.

The simplest mode of CE is referred to as *capillary zone electrophoresis* (CZE). In most separations, a short plug of sample is introduced to the capillary column at its anode end. Subsequently, it is dipped in the electrolyte, and the electric potential is applied. Charged species (e.g., ions) are separated according to the ratios of their electric charges and hydrodynamic diameters. Positively charged species are attracted by the negatively polarized electrode (cathode) and move fastest toward the outlet end of the capillary (opposite to the injection end). Negatively charged species are attracted by the positively polarized electrode (anode) and tend to move toward the inlet end (where the injection took place). However, both negatively charged and neutral species can still move toward the outlet end of the column. Their migration toward the capillary outlet is sustained due to the existence of an *electroosmotic flow* (EOF).

The EOF is produced in the CE column because of the presence of electric charges at its wall. If the electrolyte is buffered at a pH higher than 3, the surface of silica is covered with a multitude of negative electric charges. They are formed due to dissociation of protons from silanol groups ($\equiv$Si-OH). The negatively charged residues attract positively charged cations present in the electrolyte (including H^+). Some of those cations firmly associate with the negatively charged surface but others weakly interact with it while maintaining mobility. In the presence of an electric field, some of the positively charged species move toward the cathode end of the capillary exerting a drag on all the contents of the capillary – including solvent molecules and ions which are far away from the capillary wall. The EOF is most effective in narrow capillaries, where wall-to-wall distances are small. EOF affects separation speed and (indirectly) its efficiency. Most importantly, it supports the movement of neutral and negatively charged species toward the cathode. Therefore, ions can be separated from other ions with the same and different charge, and detected in a single CE run. Neutral species are also detected; however, they are not separated from other neutral species in the CZE mode.

The electrophoretic velocity (v) of a charged particle in CZE is directly proportional to the electric field strength (E):

$$v = \mu_{ep}E \qquad (6.4)$$

The proportionality factor – electrophoretic mobility (μ_{ep}) – is given by:

$$\mu_{ep} = \frac{q}{6\pi\eta R_h} \qquad (6.5)$$

where q is the electric charge of the particle, η is the dynamic viscosity of the electrolyte, and R_h is the hydrodynamic radius of the particle. The mobility of the EOF can be expressed as:

$$\mu_{EOF} = \frac{\varepsilon\zeta E}{4\pi\eta} \qquad (6.6)$$

where ε is the dielectric constant of the electrolyte and ζ denotes the so-called *zeta potential* of the capillary surface. The zeta potential is a measure of the electric charge of the surface where it is hard to estimate the exact number of individual electric charges per area unit. It depends on the overall electric charge of the surface of silica. Thus, it is affected by the acidity of the electrolyte solution. The final electrophoretic mobility (μ_{tot}) of a particle can be expressed as the sum of the electrophoretic mobility and the EOF mobility:

$$\mu_{tot} = \mu_{ep} + \mu_{EOF} \qquad (6.7)$$

One can realize that the velocity of separated particles (v) is a non-negative variable. However, electrophoretic mobilities can have negative, positive or null values. In the normal polarity mode (anode at the inlet), if the final electrophoretic mobility (μ_{tot}) is positive, the particles will move towards the outlet (typically, cathode) and will possibly be detected at the end of the capillary column.

The above considerations assume the absence of hydrodynamic flow. However, hydrodynamic flow may contribute to the migration of solutes. Hydrodynamic flow can be introduced unintentionally (e.g., by erroneous positioning of inlet and outlet reservoirs) or intentionally (e.g., by applying pressurized gas to the inlet or underpressurized atmosphere to the outlet) to improve flow stability, repeatability, and transfer to an off-capillary detector. The existence of additional hydrodynamic flow in the CE column cannot be ignored in some CE-MS systems.

The number of theoretical plates in CE is given by:

$$N = \frac{\mu_{tot}U}{2D_c} \qquad (6.8)$$

where U is the applied voltage and D_c is the diffusion coefficient of the analyte. The above formula considers electromigration and diffusion as the main factors that influence separation efficiency. However, in reality, the widths of peaks recorded in the electropherograms are affected by various other factors. The variance of signal distribution within a given electrophoretic peak (σ_{tot}^2) can be described by [8]:

$$\sigma_{tot}^2 = \sigma_{diff}^2 + \sigma_{inj}^2 + \sigma_{det}^2 + \sigma_{wall}^2 + \sigma_{Joule}^2 + \sigma_{ED}^2 + \sigma_{dest}^2 + \sigma_{others}^2 \qquad (6.9)$$

where σ_{diff}^2 is the variance contribution of diffusion, σ_{inj}^2 of sample injection, σ_{det}^2 of detection, σ_{wall}^2 of wall effects, σ_{Joule}^2 of Joule heating, σ_{ED}^2 of electrodispersion, σ_{dest}^2 of destacking, and σ_{others}^2 describes contributions of any other factors, some of which are hard to predict (e.g., breakage of capillary, electrolyte siphoning). Some of the factors affecting shapes of electrophoretic peaks can be controlled to a certain extent while others are hard to control. For example, convective dispersion of analyte zones due to Joule heating can be minimized by using low conductivity electrolytes, moderate electric field strength, and thermostatting the capillary column. Geometry of the detector can also contribute to the broadening of recorded signals. If the dead volume of the interface/detection zone is in the order of a few tens to a few hundred nanoliters, this can already lead to deterioration of peak shapes. Moreover, a high time constant of the detector (or the smoothing algorithm applied during detection) can lead to peak tailing and anomalous asymmetry [8]. Thus, optimization of the CE-MS interface is important for the preservation of the high resolving power which is intrinsic to CE.

Various modes of CE exist, which facilitate separation of the ions and molecules that cannot be resolved by CZE. They include:

- *Micellar electrokinetic chromatography* (MEKC). The electrolyte contains a surfactant [e.g., sodium dodecyl sulfate (SDS)] above its critical micelle concentration. Micelles present in the capillary interact with the sample components. Differences in the interactions between micelles and various analytes give rise to separation. Some non-ionogenic species can also interact with the surfactant micelles, thus promoting separation. Due to its resemblance to chromatography, the micellar solution in MEKC is sometimes referred to as *pseudostationary phase*.
- *Electrokinetic chromatography* (EKC) is a more general term that encompasses CE methods using buffer additives, for example cyclodextrins, which interact with analyte species.
- *Capillary gel electrophoresis* (CGE) uses capillaries filled with gel matrix. Penetration of gel pores by charged particles provides additional selectivity since small species encounter less resistance than large particles.
- *Capillary isotachophoresis* (cITP) uses capillaries with pH gradients for separation of macromolecular species with different isoelectric points (pI). On application of an electric field, ions move until they reach the zone with the pH corresponding to their pI value, at which point their overall electric charge is null. This mode is particularly useful in the characterization of protein samples.
- *Capillary electrochromatography* (CEC) is a combination of capillary (miniaturized) chromatography with CE. Here, selectivity is both due to interactions of solutes with the stationary phase as well as different electrophoretic mobilities.

6.2   Separation Techniques Coupled with Mass Spectrometry

Modern chromatographic and electrophoretic techniques (e.g., UHPLC, GC, and CE) are characterized with high apparent numbers of theoretical plates (typically, 10^4-10^6). While chromatograms and electropherograms obtained with these separation techniques can accommodate a few tens up to a few hundred peaks (for certain optimized separations), any of these platforms cannot be considered as a replacement for MS. Chromatography, electrophoresis, and MS are orthogonal approaches. That means that the separations are performed due to differences in distinct physicochemical properties of analytes. For example, RP-LC can separate non-polar compounds, GC can separate volatile compounds based on the differences in their boiling points and polarity, CZE can separate liquid-phase ions based on their charges and hydrodynamic radii while MS separates gas-phase ions according to their mass-to-charge (m/z) ratios. Orthogonal separations can also be achieved by combining different modes of certain separation techniques (e.g., reversed-phase and ion-exchange chromatography). Such multi-dimensional analytic systems provide superior selectivity thus enabling qualitative and quantitative characterization of complex samples, especially those that cannot be fully resolved taking into account only one physicochemical criterion (e.g., only polarity or only m/z ratio).

Robust coupling between GC and MS was achieved already in the 1960s (Figure 6.2; [9]). In GC, the dynamic sample reaching the ionization region is already in the gas phase which greatly simplifies operation of the interface. However, there persists the requirement to balance two different pressure regimes of the GC column and the mass analyzer. In the early days, packed GC columns were in common use. Their efficiency was far behind that of the modern version of GC using hair-thin capillary columns. Capillary GC systems are directly joined with ion sources, so that the whole volume of the mobile phase can be used

Figure 6.2 *An early time-of-flight mass spectrometer and gas chromatograph (ca. 1960s). Courtesy of the Chemical Heritage Foundation.*

in the ionization step. The implementation of computers in the GC-MS platforms greatly facilitated data acquisition at high speed in real time.

Chromatographic systems with multiple columns are devised to enhance analytical selectivity. For example, in one study, Van Geem *et al.* [10] presented a setup for two-dimensional (2D) gas chromatography (GC×GC) that allows switching between flame ionization detection (FID) and a time-of-flight (TOF) mass spectrometer. It allows automatic sampling of the hot reactor effluent gases and immediate injection of the sample to the GC×GC. While the use of multi-column chromatography enhances selectivity, it also limits the temporal resolution of analysis in the case of monitoring dynamic chemical systems.

Please note that analytes introduced to GC columns need to be volatile and amenable to detection using the detector at the column outlet. To render less volatile compounds amenable to analysis by GC-MS, derivatization is often conducted by reacting sample components with a generic or specific reagent. While derivatization is normally conducted off-line, in principle, it can be done on-line. In one study, a system was developed in which analytes were first separated in an LC column followed by on-line derivatization, and introduction of the resulting mixture to a GC column for further separation and detection by MS using an electron ion source and a quadrupole ion trap analyzer [11].

In the case of LC, samples are in the liquid phase, and they need to be transferred into the gas phase, which was a huge challenge at the beginning. The first attempts of interfacing LC with MS date back to the late 1960s [12, 13]. Liquid was sprayed into an electron ion source of a mass spectrometer. However, the electron ion source can only work at very low pressures which limited the volume of liquid which could be introduced. This obstacle was mitigated when electron ionization was replaced with chemical ionization in the direct liquid introduction probe [14]. The technique was improved in the 1970s [15, 16]. A *moving belt interface* was also proposed to transmit dry samples to the ion source [17]; however, it did not attract the attention of the scientific community for a long time.

A chief development in the coupling of LC separations with MS was the invention of atmospheric pressure ion sources [18]. To this end, the development of the atmospheric pressure chemical ionization (APCI) ion source was important because it enabled the generation of ions at atmospheric pressure (Figure 6.3; [19]). Thus, vaporized solvents did not compromise the vacuum of the mass spectrometer anymore. Hyphenation of LC with tandem MS with the APCI source provided superior analytical sensitivity [20–22]. Another significant milestone was the development of thermospray interface [23, 24], which became quite successful [13]. However, the revolution in LC-MS interfacing came soon after the introduction of electrospray ionization (ESI)-MS by John Fenn ([25]; see also Chapter 2). In the 1980s, due to the improvements in liquid-/gas-phase interfaces and ion sources, the sensitivity of MS analyses was boosted (cf. [26]), and the newly produced LC-MS sets became user-friendly [13]. These features greatly enhanced pickup of LC-MS by industry. Atmospheric pressure ion sources enable analysis of compounds with varied polarity, lability and molecular weights; thus, they were widely implemented in LC-MS separation systems [18]. Already in the 1980s, it became clear that LC-MS, conducted using an on-line source (APCI), can warrant robust and fast analysis: 60 samples of crude equine urine and plasma extracts could be analyzed in 1 h using this approach [21].

An important issue to address is matching the flow rates of separation systems and ion sources. Conventional LC systems operate at high flow rates (milliliters per minute)

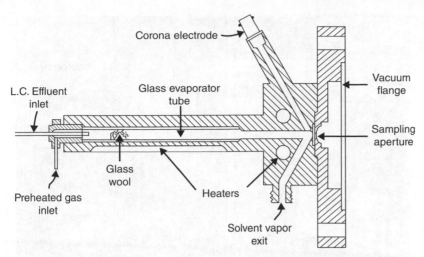

Figure 6.3 *Schematic diagram of the corona source [19]. Reprinted with permission from Carroll, D.I., Dzidic, I., Stillwell, R.N., Haegele, K.D., Horning, E. (1975) Atmospheric Pressure Ionization Mass Spectrometry. Corona Discharge Ion Source for Use in a Liquid Chromatograph-Mass Spectrometer-Computer Analytical System. Anal. Chem. 47: 2369–2373. Copyright (1975) American Chemical Society*

which are incompatible with several ion sources – including the standard ESI source. In these cases, the flow of column effluent may be split, and only a fraction of it directed to the ion source. Microscale LC columns operate at lower flow rates (microliters per minute), and can directly be coupled with standard ESI sources (Figure 6.4). On the other hand, capillary LC columns operate at extremely low flow rates (nanoliters per minute); therefore, they are compatible with nanospray electrospray ionization (nanoESI) ion sources which can handle similar flow rates. Capillary LC columns have become very popular, and are widely used in shotgun proteomic analyses [27]. They are coupled directly to the nanoESI-MS, and the whole amount of eluent is used for ionization. The use of microscale LC in conjunction with nanoESI-MS provides good mass sensitivity. In the case of separations, it is vital that the separated zones are transferred to MS without losing peak resolution. This condition is achieved by reducing the dead volume between the column outlet and the ionization region. There exist capillary columns with a variety of stationary phases – in the form of particles (beads) or monoliths. In some cases, the stationary phase reaches the end of the ion source emitter which helps to limit band broadening due to the dead volume section, and decrease the analysis time.

As noted above, LC can be conducted in different modes. The reversed-phase chromatography takes advantage of non-polar stationary phases and polar mobile phases. Most importantly, it does not usually require addition of salts to the mobile phase. Mobile phases in the reversed-phase LC involve common organic solvents (e.g., methanol, ethanol, acetonitrile) and water. Therefore, they are highly compatible with MS detection. On the other hand, the methods which use non-volatile mobile phases (e.g., some of the electrolyte solutions used in ion-exchange chromatography) are much less compatible with MS systems. For example, infusing salt solutions through the ESI needle may lead

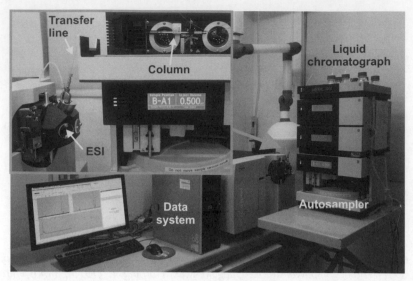

Figure 6.4 *A modern liquid chromatograph hyphenated with an ion-trap mass spectrometer using an ESI interface. Courtesy of E.P. Dutkiewicz. See colour plate section for colour figure*

to clogging, electric discharges, contamination of the ion source and the mass analyzer, and – in a longer term – damage of the ion source.

Modern chromatographic systems are equipped with autosamplers. Samples can be injected directly into the separation column, or they can be passed through enrichment (trapping) columns. The concentrated analytes can subsequently be directed to the separation columns. This step extends analysis time; however, it may be indispensable in the case of analysis of low abundance components of matrix-rich samples.

Metabolomic studies are often conducted using LC-MS methods [28]. Here, various problems are normally faced: Ionization efficiency differs among metabolites. It can substantially vary even for related structures. There exist no universal ion sources which can provide satisfactory ionization of compounds representing different groups of biomolecules present in cells and biofluids. Fast chromatographic separations (e.g., using HILIC columns), combined with high-resolution mass spectrometers, are also useful in metabolomic analysis, for example, of thousands of metabolites in mammalian cells [29]. Identification of unknown metabolites normally requires coupling separations with high-resolution Fourier transform (FT)-MS analyzers [28], often at the expense of temporal resolution.

Proteomic analysis frequently involves comparison of protein levels in experimental and control samples, which is linked with the quantitative analysis (see Chapter 8). In some cases, it is necessary to follow protein abundances over time. Changes in protein levels in cells or biofluids may be monitored over a few hours or days, which does not put much pressure on the temporal resolution of the analytical procedure [at least not to the extent observed in ultrafast time-resolved mass spectrometry (TRMS) studies introduced in other chapters of this book]. LC-MS is a simple approach to obtain relative abundances of proteins in highly complex samples [30]. It is possible to carry out quantitative comparisons of tens of thousands of ions from control and experimental samples. Relative changes of abundances in the two conditions can be assessed by looking at accurate masses and retention times [30]. Thus, in proteomic analyses, it is common to carry out LC-MS separations of complex mixtures of peptides. For instance, Batycka *et al.* [31] showed that polymeric monolithic columns can be used to separate peptide mixtures within 9 min. Comparable results were obtained on standard C_{18} columns within 30 min. In this case, the separation efficiency was the limiting factor of the analytical throughput.

Off-line detection by MS of chromatographic eluents can also be accomplished (Figure 6.5; [32]). However, in this case, the temporal resolution of chromatographic analysis is compromised due to undersampling the eluted analyte zones. This issue may affect quantitative analysis based on the off-line chromatographic results. Auxiliary

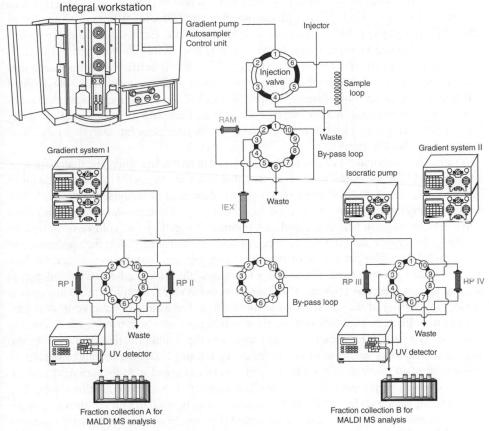

Figure 6.5 *Schematic representation of the on-line comprehensive two-dimensional HPLC system including an integrated sample preparation step [32]. Reprinted from Journal of Chromatography B, 803, Machtejevas, E., John, H., Wagner, K., Standker, L., Marko-Varga, G., Forssmann, W.G., Bischoff, R., Unger, K.K., Automated Multi-Dimensional Liquid Chromatography: Sample Preparation and Identification of Peptides from Human Blood Filtrate, 121–130. Copyright (2004), with permission from Elsevier*

detectors (e.g., ultraviolet–visible absorption) are occasionally used to assist quantitative analysis, and to control the quality of separation and for trouble-shooting purposes.

Two-dimensional LC separations become increasingly popular [33]. They provide additional selectivity, as required in analysis of complex samples, for example in protein studies. Compared with the standard methods, 2D LC-MS has a high resolving power, and is particularly useful in the analysis of complex proteome samples [34]. While this methodology enables separation of multiple analytes with diverse physicochemical properties, it is usually time-consuming. Thus, it is not particularly suitable for the analysis of highly labile samples. Multi-dimensional analysis involving ion mobility separations (see Section 6.3) is more adequate for real-time monitoring of dynamic systems because the ion mobility separation is very fast compared with chromatography and electrophoresis.

Various unconventional interfaces have also been devised to facilitate on-line introduction of samples present in different forms to LC-MS systems. For instance, surface sampling and analysis can be conducted by means of a liquid microjunction (LMJ) interface coupled with LC-ESI-MS [35]. A similar approach is applicable to off-line coupling of thin-layer chromatography (TLC) separations with MS [36]. TLC can also be coupled with MS via various other interfaces, including matrix-assisted laser desorption/ionization (MALDI) [37], desorption electrospray ionization (DESI) [38], and dielectric barrier discharge ionization (DBDI) [39]. Salentijn *et al.* [40] demonstrated a three-dimensional (3D)-printed cartridge which can accommodate paper-spray ionization and rudimentary paper chromatographic separation monitored on-line by MS. Indeed, there is a great potential for using 3D-printing technology to fabricate prototypes of MS interfaces for analysis of dynamic samples (see also Chapter 7).

Coupling CE with mass spectrometers is perceived to be more problematic than setting up GC-MS and LC-MS systems. It was first presented in the late 1980s [41]. Since then, a number of CE-MS interfaces have been set up and implemented in various areas. *Sheath-flow* and *sheathless* interfaces are in common use (Figure 6.6). CE separations require application of electric field in hair-thin capillaries (similar to capillary GC columns). In on-line coupling, the electric field in CE has to be applied simultaneously with the application of an electric field to the ion source of the mass spectrometer. In order to mitigate the loss of electrophoretic resolution, it is desirable to position the outlet end of the CE column as close to the ion source as possible. However, this option incurs technical problems related to electrical wiring of the ion source and the column. For example, Fang *et al.* [42] presented an on-line CE-TOF-MS system. Here, a thin gold wire was used to apply electric potential. The ions were extracted at right angles to the initial direction of the ion beam, and a complete mass spectrum was recorded every 100 μs [42]. However, in the sheathless CE-MS interface, the capillary outlet is often directly coupled with the nanoESI emitter. This mode ensures efficient transfer of the CE column effluent to the ionization region. The metal coating on the surface of the nanoESI emitter can be used to provide electric contact for CE separation and ionization of the separated analytes. In some cases the outlet sections of the CE columns are tapered so that they can act as nanoESI emitters and provide high analytical sensitivity. In the sheath-flow coupling, a flow of make-up (sheath) liquid is introduced. It renders the eluting analytes readily ionizable, and warrants stable electric contact with the capillary outlet. However, it also dilutes the analyte zones. Thus, the detection limits are normally lower in the sheath-flow interfaces compared with the sheathless interfaces. Nonetheless, the stability of the sheath-flow interfaces often outweighs their lower

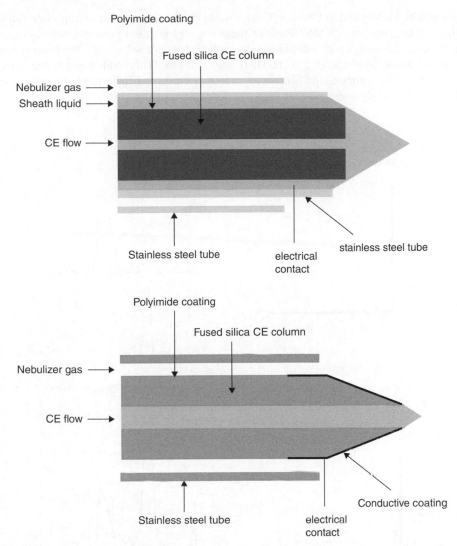

Polyimide coating

Fused silica CE column

Nebulizer gas

Sheath liquid

CE flow

Stainless steel tube   electrical
                            contact

stainless steel tube

Polyimide coating

Fused silica CE column

Nebulizer gas

CE flow

Stainless steel tube   electrical
                            contact

Conductive coating

Figure 6.6 *Schematic drawing of commonly used CE-ESI-MS interfacing: sheath-flow and sheathless coupling [5]. Reproduced from Mischak, H., Coon, J.J., Novak, J., Weissinger, L.M., Schanstra, J.P., Dominiczak, A.F. (2009) Capillary Electrophoresis-Mass Spectrometry as a Powerful Tool in Biomarker Discovery and Clinical Diagnosis: an Update of Recent Developments. Mass Spectrom. Rev. 28: 703–724 with permission from John Wiley and Sons*

analytical sensitivity [5]. Apart from the instrumental obstacles, a major issue to address when coupling CE with MS is electrolyte compatibility. CE separations are frequently conducted using non-volatile buffers. Pumping such solutions toward an MS interface leads to ion suppression, clogging of the thin emitter orifice, and rapid contamination of the ion source region. Thus, volatile or semi-volatile electrolytes are used in CE-MS separations, including aqueous solutions of acetic acid, formic acid, or ammonium acetate.

As will be highlighted in Chapter 7, microscale separations carried out on microfluidic chips can be coupled with MS. Some of them are very fast. For example, separations of pharmaceuticals and peptides could be accomplished within ~2 s while the mass spectra were obtained by TOF operated at 100 Hz (Figure 6.7) [43]. This result suggests that it may be possible to incorporate liquid-phase separations as a fast on-line sample treatment step while monitoring dynamic samples (e.g., reaction mixtures) by MS.

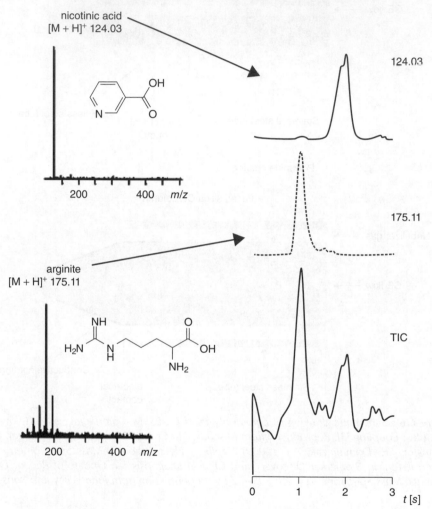

Figure 6.7 *Total and reconstructed selected ion electropherograms with mass spectra of model analytes arginine (50 μg ml⁻¹) and nicotinic acid (200 μg ml⁻¹). Electrolyte: 2.5 mM ammonium acetate containing 12.5% methanol (v/v). Injection time: 10 s. Separation potentials: BI: 17.5 kV; SI: 10 kV; SO: 10 kV [43]. Reproduced from Fritzsche, S., Hoffmann, P., Belder, D. (2010) Chip Electrophoresis with Mass Spectrometric Detection in Record Speed. Lab Chip 10: 1227–1230 with permission of The Royal Society of Chemistry*

6.3   Ion-mobility Spectrometry

Ion-mobility spectrometry (IMS) separates ions based on their size/charge ratios and their interactions with a buffer gas [44]. The shape of ions also has an effect on the separation [45]. Following ionization, the ions are introduced into a chamber filled with a neutral gas at controlled pressure [45]. The separation proceeds in the presence of a relatively weak field. While IMS alone has great importance in national security applications (e.g., detection of explosives), if coupled with MS – it supports analyses of biomolecular species (proteins, lipid isomers) which cannot be fully resolved by MS alone. Both IMS and MS handle gas-phase ions, which makes them particularly compatible with each other. In IMS, the velocity of the ions is proportional to the electric field with the proportionality factor (K) [44, 46]:

$$K = \left(\frac{3q}{16N}\right)\left(\frac{2\pi}{kT}\right)^{1/2}\left(\frac{m+M}{mM}\right)^{1/2}\left(\frac{1}{\Omega}\right) \qquad (6.10)$$

where q represents the charge of the ion, N is the number density of the buffer gas, k is Boltzmann's constant, T is the absolute temperature, m is the mass of the buffer gas, M is the mass of the ion, and Ω is the collision cross-section (CCS) of the ion.

IM-MS data are "three-dimensional", including drift time, m/z, and intensity [45]. The IM stage further enhances the selectivity, sensitivity, and speed of mass spectrometers [47]. Due to the short duration of IM separations, the IM stage can readily be embedded between chromatographic separation and MS analysis [47]. Typically, they occur in millisecond-range intervals although sub-millisecond resolution is often required to detect separated ion zones leaving the drift cell. In certain cases, they can outperform chromatographic methods [47, 48]. The information obtained by IMS and MS is complementary, and the obtained IM-MS data are interpreted simultaneously [45]. Duration of single analysis is not extended significantly because of the incorporation of an IMS stage in front of the mass analyzer. The high operational throughput of IM-MS is important for application in the screening of biochemical samples.

Various ionization techniques can be used in conjunction with IM-MS. Moreover, different types of IMS instruments can be coupled with mass spectrometers, including drift time [49], aspiration [50, 51], differential [52], and traveling wave [53]-based devices. For instance, in the traveling wave ion guides (TWIGs), the ions within the IMS cell are driven away from the potential hills and carried with the waves, decreasing their transit time (Figure 6.8a; [54]). Here, radio-frequency voltages with opposite phases are applied to adjacent electrodes in order to confine the ions in the radial direction [47]. Direct current voltage pulses, applied to each electrode in succession, guide the ions in the axial direction [47]. A TWIG drift cell was implemented in a commercial IM-MS instrument, as illustrated in Figure 6.8b, incorporating quadrupole and TOF mass analyzers. The time-based ion mobility instruments can provide information about cross-sectional area while others are used as filtration devices [47]. Traveling wave IMS can provide CCS values following calibration under defined conditions [47].

IM-MS enabled separation of singly and multiply charged peptides as well as conformers [55]: it improves analytical selectivity thus facilitating identification of ions. However, IM-MS can also be used to obtain kinetic and thermochemical data [56]. The temperature of the drift cell (and the buffer gas) can be controlled over a wide range. In some cases,

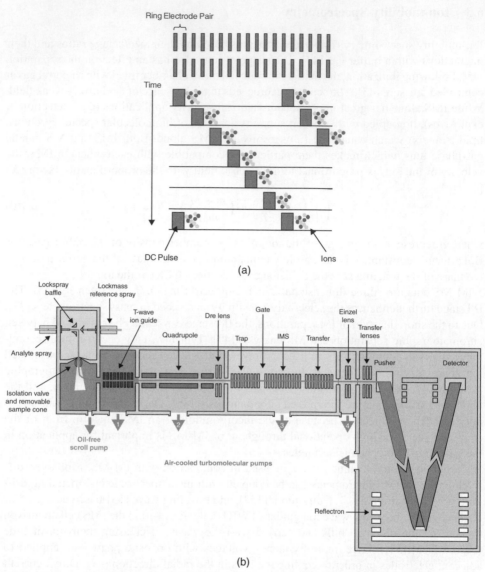

Figure 6.8 *Traveling wave ion guide (TWIG) ion-mobility mass spectrometry. (a) Illustration of the operation of a TWIG for ion propulsion in the presence of background gas. (b) A schematic diagram of the Synapt HDMS system. All of the turbomolecular pumps have $220\,l\,s^{-1}$ pumping speed (model EXT255H, Edwards, Crawley, UK) and are backed by a $35\,m^3\,h^{-1}$ scroll pump (model XDS35i, Edwards, Crawley, UK) [54]. Reprinted from International Journal of Mass Spectrometry, 261, Pringle, S.D., Giles, K., Wildgoose, J.L., Williams, J.P., Slade, S.E., Thalassinos, K., Bateman, R.H., Bowers, M.T., Scrivens, J.H., An Investigation of the Mobility Separation of Some Peptide and Protein Ions Using a New Hybrid Quadrupole/Travelling Wave IMS/oa-ToF Instrument, 1–12. Copyright (2007), with permission from Elsevier*

rate constants of isomerization and dissociation can be obtained. If a species undergoes transformation to another species – with a different cross-section – during the drift time, one will obtain a cross-section distribution indicating such a process [56]. For example, the technique has been applied to follow the dissociation of serine clusters, bradykinin dimers, and interconversion of poly(ethylene terephthalate) (see also [56] and the references cited therein).

IM-MS has been used in the monitoring of organic reactions in the liquid phase. Harry *et al.* [57] applied IM-MS in real-time monitoring of the deprotonation reaction of 7-fluoro-6-hydroxy-2-methylindole in the presence of aqueous sodium hydroxide. They observed enhancement of product ion signal compared with conventional MS analysis. Roscioli *et al.* [58] applied IM-MS in the monitoring of pharmaceutical synthesis reactions. The reaction completion time was determined by monitoring the starting, intermediate and product materials during the process (300 s). The method provided more complete information about the reaction steps including complex formation [58]. Traveling wave IM-MS was used in on-line monitoring of Fab-arm exchange in antibodies. The reaction time was in the order of a few hours while the sampling interval was between a few minutes and a few hours [59]. In an elegant study, Smith *et al.* [60] used ESI-IM-MS to shed new light on the amyloid fibril assembly process. Moreover, thermal desorption and ESI can be coupled with field asymmetric waveform IM-MS to enable direct analysis of toxic compounds. Desorption profiles can be recorded with good temporal resolution preserving information on the characteristics of the sample that is orthogonal to the IM-MS data [61]. Interfacing thermal desorption with ESI-IM-TOF-MS also helps to analyze and identify breath metabolites [62]. Since the separation by IMS is very fast (microseconds or milliseconds), it has almost no effect on the monitoring of the chemical reactions or physical processes that take seconds to hours.

6.4   Other Hyphenated Systems

Certainly, the enabling role of the hyphenation of MS with various separation systems in the analysis of complex samples is behind many applications of MS in general. However, in some cases, time-consuming sample preparation and/or separations are required. Therefore, it is appealing to design TRMS setups which avoid long-term sample processing and possibly replace them with faster and highly automated alternatives.

Solid-phase microextraction (SPME) [63] is normally used for sample collection, preconcentration and desalting before analysis. It is usually used off-line. However, SPME can be implemented in temporal monitoring of biomolecules with low to medium temporal resolution (minutes, hours) [64]. In one method, SPME fibers were used to sample metabolites from live animals followed by analysis using LC-MS [65]. The method enabled extraction of metabolites directly in the tissue of moving animals. It was not necessary to withdraw a representative biological sample for analysis. In this case, the amount of analyte extracted into the SPME fiber was independent of the sample volume [65]. Recently, SPME was also coupled on-line with a MS ion source operated at atmospheric pressure [66].

On-line extraction and preconcentration systems are sometimes coupled directly with MS decreasing sample preparation time [67–69]. C_{18} pipette tips can be used to purify complex samples, and directly introduce them to the mass spectrometer via an ESI interface [70]. This technique is characterized by great simplicity; however, one may expect that it may not be applicable to all complex samples, especially when quantitative results are desired. A microdialysis-paper spray ionization MS system has also been proposed for sampling and droplet analysis [71]. Using this system, hormone regulation of glucose concentration could be investigated.

Flow injection analysis (FIA) often uses microliter-per-minute flow rates which are compatible with many atmospheric pressure ion sources such as ESI. If the flow rate is too high, then only a fraction of the flow is directed to the ion source while the remainder is diverted to waste. Zhu *et al.* [72] introduced an autosampling FIA method coupled with APCI-MS (Figure 6.9). Using this system, quantitative real-time monitoring of a model reaction was performed although the concentrations of reactants were higher than 1 M, which is considered to be an extremely high concentration for MS.

Other on-line sample pretreatment methods can be combined with MS directly or before the separation column. For example, on-line hydrogen/deuterium exchange (HDX) could

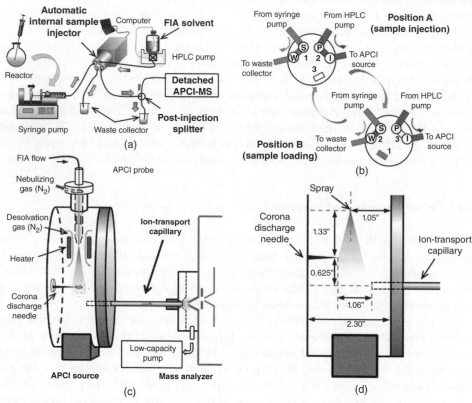

Figure 6.9 (a) Diagram of the autosampling flow injection analysis (FIA)/APCI-MS system; (b) diagram of the automatic internal sample injector; (c) diagram of the detached APCI-MS setup; and (d) diagram of the relative position of the ion-transport capillary, the corona discharge needle, and the APCI probe [72]. Reprinted with permission from Zhu, Z., Bartmess, J.E., McNally, M.E., Hoffman, R.M., Cook, K.D., Song, L. (2012) Quantitative Real-Time Monitoring of Chemical Reactions by Autosampling Flow Injection Analysis Coupled with Atmospheric Pressure Chemical Ionization Mass Spectrometry. Anal. Chem. 84: 7547–7554. Copyright (2012) American Chemical Society

be performed [73]. Determination of the number of exchangeable hydrogen atoms is an important parameter which can facilitate the analysis of constituents of mixtures by GC-MS techniques without previous isolation. On-line digestion is frequently used in shotgun proteomics [27]. The protein digests eluted from immobilized enzyme reactors can be directed to a mass spectrometer for characterization [74]. In other noteworthy work, Van Berkel and Kertesz [75] combined the liquid microjunction surface sampling probe (cf. Section 6.2) with a two-electrode electrochemical cell to enable tagging of analyte thiol functionalities (e.g., peptide cysteine residues) with hydroquinone tags. On the other hand, Hogenboom *et al.* [76] presented a continuous-flow analytical screening system to measure the interaction of biologically active compounds with soluble affinity proteins. It integrates biochemical assay with fluorescence and mass spectrometric detection.

6.5 Influence of Data Acquisition Speed

When recording mass spectra of separation column effluents, it is important that the sampling rate of the mass spectrometer is such that several data points (i.e., mass spectra) are collected for every chromatographic or electrophoretic peak. Data acquisition has to be fast enough to ensure faithful representation of temporal signals. This representation is particularly important in the case of high-performance separation techniques in which the duration of the signal is very short. Too low data acquisition speed leads to deterioration of temporal signals (Figure 6.10). To define a Gaussian peak, ideally at least 20 data points should be recorded. This is possible in the case of relatively wide peaks obtained by HPLC. However, the width of peaks obtained by CE, UHPLC or nanoflow LC can be very small (<10 s; see, e.g., [77]). Some capillary GC separations produce peaks much narrower than 1 s. In these cases, only a high data acquisition rate can assure faithful representation of the peaks. Undersampling narrow peaks can lead to decreased repeatability and sensitivity of the measurements.

LC is occasionally coupled off-line with MALDI-MS. Here, sample collection and discrete scan rate determines the quality of the temporal representation of the chromatographs. In one study, the effluent of the CE column was transferred directly into a matrix-filled groove in a metal plate [78]. This kind of continuous effluent collection enabled chromatograms to be obtained by MALDI-MS. In that case, the deposition of analytes was continuous. However, the resolution of separated analytes deteriorated due to dispersion inside the wetted groove, and the fact that several sub-spectra recorded along the groove were averaged. *Micro-arrays for mass spectrometry* (MAMS) enable discrete aliquoting of particles and liquid into micro-posts which can be individually scanned by MALDI-MS [79]. This technology has also been used to collect micro-fractions of effluents from chromatographic columns [80], providing the opportunity for fast aliquoting and subsequent detection of temporal peaks.

6.6 Concluding Remarks

Temporal resolution is important for obtaining quality output in analyses using hyphenated techniques which combine mass spectrometers with liquid- or gas-phase separation

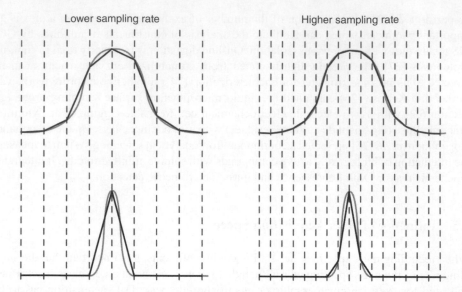

Figure 6.10 *Influence of sampling rate on temporal peak shape. Vertical dashed lines represent time points when signals are logged. The gray line represents the spatial concentration profile of the eluting zones. The black line joins the intensity points recorded by the mass spectrometer in the time domain. Deterioration of the zone/peak shape due to undersampling can be observed*

devices. High data acquisition rates provide adequate representation of transient currents of ions generated in the course of separation. The peaks obtained in HPLC are relatively wide (seconds to tens of seconds): in this case, even "slow" mass spectrometers such as FT-MS instruments have sufficient sampling rate to represent broad chromatographic features. On the other hand, separations conducted using UHPLC, nanoflow LC, CE, and some of the microchip devices lead to particularly short ion currents (fractions of seconds to a few seconds), and require higher data acquisition rates, which can be provided by ion trap or quadrupole mass spectrometers (see also Chapter 5). Whenever hyphenated techniques are implemented in the monitoring of dynamic samples in real time, one needs to assure that the total analysis time (including separation and conditioning of the column) is negligible compared with the total time of the studied process. The typical separation times in chromatographic and electrophoretic methods are in the order of few minutes although some exceptionally fast separations were demonstrated using microfluidic devices. Conversely, ion mobility separations are conducted within a few milliseconds. However, all these techniques separate analyte species based on different principles and cannot – in most cases – be used interchangeably. Oftentimes, a trade-off is made to attain sufficient temporal resolution of the MS record and quality of the analysis output (e.g., qualitative or quantitative information).

References

1. Sharma, K., Mullangi, R. (2013) A Concise Review of HPLC, LC-MS and LC-MS/MS Methods for Determination of Azithromycin in Various Biological Matrices. Biomed. Chromatogr. 27: 1243–1258.
2. McMaster, M., McMaster, C. (1998) GC/MS – A Practical User's Guide. John Wiley & Sons, Inc., New York.
3. Hübschmann, H.-J. (2008) Handbook of GC/MS: Fundamentals and Applications, 2nd Edition. Wiley-VCH Verlag GmbH, Weinheim.
4. Bouchonnet, S. (2013) Introduction to GC-MS Coupling. CRC Press, Boca Raton.
5. Mischak, H., Coon, J.J., Novak, J., Weissinger, E.M., Schanstra, J.P., Dominiczak, A.F. (2009) Capillary Electrophoresis-Mass Spectrometry as a Powerful Tool in Biomarker Discovery and Clinical Diagnosis: an Update of Recent Developments. Mass Spectrom. Rev. 28: 703–724.
6. van Deemter, J.J., Zuiderweg, F.J., Klinkenberg, A. (1956) Longitudinal Diffusion and Resistance to Mass Transfer as Causes of Nonideality in Chromatography. Chem. Eng. Sci. 5: 271–289.
7. Grob, K. (1997) Carrier Gases for GC, http://www.restek.com/Technical-Resources/Technical-Library/Editorial/editorial_A017 (accessed April 8, 2015).
8. Microsolvtech, Achieving Efficiency in HPCE, http://microsolvtech.com/ceget.asp (accessed April 8, 2015).
9. Brock, D.C. (2011) A Measure of Success. Chemical Heritage Magazine, http://www.chemheritage.org/discover/media/magazine/articles/29-1-a-measure-of-success.aspx (accessed September 15, 2015).
10. Van Geem, K.M., Pyl, S.P., Reyniers, M.-F., Vercammen, J., Beens, J., Marin, G.B. (2010) On-line Analysis of Complex Hydrocarbon Mixtures Using Comprehensive Two-dimensional Gas Chromatography. J. Chromatogr. A 1217: 6623–6633.
11. Chappell, C.G., Creaser, C.S., Shepherd, M.J. (1997) On-line Derivatisation in Combined High performance Liquid Chromatography–Gas Chromatography–Mass Spectrometry. Analyst 122: 955–961.
12. Tal 'roze, V.L., Karpov, G.V. (1968) Russ. J. Phys. Chem. 42: 1658–1664.
13. Pullen, F. (2010) The Fascinating History of the Development of LC-MS, a Personal Perspective. Chromatography Today (February/March) 4-6.
14. Baldwin, M.A., McLafferty, F.W. (1973) Liquid Chromatography-Mass Spectrometry Interface–I: The Direct Introduction of Liquid Solutions into a Chemical Ionization Mass Spectrometer. Org. Mass Spectrom. 7: 1111–1112.
15. Henion, J.D. (1978) Drug Analysis by Continuously Monitored Liquid Chromatography/Mass Spectrometry with a Quadrupole Mass Spectrometer. Anal. Chem. 50: 1687–1693.
16. Arpino, P.J., Guiochon, G., Krien, P., Devant, G. (1979) Optimization of the Instrumental Parameters of a Combined Liquid Chromatograph–Mass Spectrometer, Coupled by an Interface for Direct Liquid Introduction: I. Performance of the Vacuum Equipment. J. Chromatogr. 185: 529–547.

17. McFadden, W.H., Schwartz, H.L., Evans, S. (1976) Direct Analysis of Liquid Chromatographic Effluents. J. Chromatogr. 122: 389–396.
18. Linscheid, M., Westmoreland, D.G. (1994) Analytical Techniques for Trace Organic Compounds: VI. Application of Liquid Chromatography–Mass Spectrometry. Pure Appl. Chem. 66: 1913–1930.
19. Carroll, D.I., Dzidic, I., Stillwell, R.N., Haegele, K.D., Horning, E. (1975) Atmospheric Pressure Ionization Mass Spectrometry. Corona Discharge Ion Source for use in a Liquid Chromatograph-Mass Spectrometer-Computer Analytical System. Anal. Chem. 47: 2369–2373.
20. Henion, J.D., Thomson, B.A., Dawson, P.H. (1982) Determination of Sulfa Drugs in Biological Fluids by Liquid Chromatography/Mass Spectrometry/Mass Spectrometry. Anal. Chem. 54: 451–456.
21. Covey, T.R., Lee, E.D., Henion, J.D. (1986) High-speed Liquid Chromatography/Tandem Mass Spectrometry for the Determination of Drugs in Biological Samples. Anal. Chem. 58: 2453–2460.
22. Thomson, B.A. (1998) Atmospheric Pressure Ionization and Liquid Chromatography/Mass Spectrometry – Together at Last. J. Am. Soc. Mass Spectrom. 9: 187–193.
23. Blakley, C.R., McAdams, M.J., Vestal, M.L. (1978) Crossed-beam Liquid Chromatoraph–Mass Spectrometer Combination. J. Chromatogr. 158: 261–276.
24. Blakley, C.R., Carmody, J.C., Vestal, M.L. (1980) Combined Liquid Chromatograph/Mass Spectrometer for Involatile Biological Samples. Clin. Chem. 26: 1467–1473.
25. Yamashita, M., Fenn, J.B. (1984) Negative Ion Production with the Electrospray Ion Source. J. Phys. Chem. 88: 4671–4675.
26. Kandiah, M., Urban, P.L. (2013) Advances in Ultrasensitive Mass Spectrometry of Organic Molecules. Chem. Soc. Rev. 42: 5299–5322.
27. Hu, L., Ye, M., Jiang, X., Feng, S., Zou, H. (2007) Advances in Hyphenated Analytical Techniques for Shotgun Proteome and Peptidome Analysis – a Review. Anal. Chim. Acta 598: 193–204.
28. Hyötyläinen, T., Wiedmer, S. (eds) (2013) Chromatographic Methods in Metabolomics. RSC Publishing, Cambridge.
29. Liu, X., Ser, Z., Locasale, J.W. (2014) Development and Quantitative Evaluation of a High-resolution Metabolomics Technology. Anal. Chem. 86: 2175–2184.
30. Silva, J.C., Denny, R., Dorschel, C.A., Gorenstein, M., Kass, I.J., Li, G.-Z., McKenna, T., Nold, M.J., Richardson, K., Young, P., Geromanos, S. (2005) Quantitative Proteomic Analysis by Accurate Mass Retention Time Pairs. Anal. Chem. 77: 2187–2200.
31. Batycka, M., Inglis, N.F., Cook, K., Adam, A., Fraser-Pitt, D., Smith, D.G.E., Main, L., Lubben, A., Kessler, B.M. (2006) Ultra-fast Tandem Mass Spectrometry Scanning Combined with Monolithic Column Liquid Chromatography Increases Throughput in Proteomic Analysis. Rapid Commun. Mass Spectrom. 20: 2074–2080.
32. Machtejevas, E., John, H., Wagner, K., Standker, L., Marko-Varga, G., Forssmann, W.G., Bischoff, R., Unger, K.K. (2004) Automated Multi-dimensional Liquid Chromatography: Sample Preparation and Identification of Peptides from Human Blood Filtrate. J. Chromatogr. B 803: 121–130.
33. Nägele, E., Vollmer, M., Hörth, P., Vad, C. (2004) 2D-LC/MS Techniques for the Identification of Proteins in Highly Complex Mixtures. Expert Rev. Proteomics 1: 37–46.

34. Vollmer, M., Hörth, P., Vad, C., Nägele, E. (2004) Multidimensional LC-MS for Proteomics – Present and Future. LC GC Eur. 17: 14–20.
35. Kertesz, V., Van Berkel, G.J. (2010) Liquid Microjunction Surface Sampling Coupled with High-Pressure Liquid Chromatography-Electrospray Ionization-Mass Spectrometry for Analysis of Drugs and Metabolites in Whole-body Thin Tissue Sections. Anal. Chem. 82: 5917–5921.
36. Kertesz, V., Ford, M.J., Van Berkel, G.J. (2005) Automation of a Surface Sampling Probe/Electrospray Mass Spectrometry System. Anal. Chem. 77: 7183–7189.
37. Crecelius, A., Clench, M.R., Richards, D.S. (2003) TLC-MALDI in Pharmaceutical Analysis. LC GC Europe (April) 2-6.
38. Paglia, G., Ifa, D.R., Wu, C., Corso, G., Cooks, R.G. (2010) Desorption Electrospray Ionization Mass Spectrometry Analysis of Lipids after Two-dimensional High-Performance Thin-Layer Chromatography Partial Separation. Anal. Chem. 82: 1744–1750.
39. Cegłowski, M., Smoluch, M., Babij, M., Gotszalk, T., Silberring, J., Schroeder, G. (2014) Dielectric Barrier Discharge Ionization in Characterization of Organic Compounds Separated on Thin-Layer Chromatography Plates. PLoS ONE 9: e106088.
40. Salentijn, G.I.J., Permentier, H.P., Verpoorte, E. (2014) 3D-Printed Paper Spray Ionization Cartridge with Fast Wetting and Continuous Solvent Supply Features. Anal. Chem. 86: 11657–11665.
41. Olivares, J.A., Nguyen, N.T., Yonker, C.R., Smith, R.D. (1987) On-line Mass Spectrometric Detection for Capillary Zone Electrophoresis. Anal. Chem. 59: 1230–1232.
42. Fang, L., Zhang, R., Williams, E.R., Zare, R.N. (1994) On-line Time-of-flight Mass Spectrometric Analysis of Peptides Separated by Capillary Electrophoresis. Anal. Chem. 66: 3696–3701.
43. Fritzsche, S., Hoffmann, P., Belder, D. (2010) Chip Electrophoresis with Mass Spectrometric Detection in Record Speed. Lab Chip 10: 1227–1230.
44. Kanu, A.B., Dwivedi, P., Tam, M., Matz, L., Hill Jr, H.H. (2008) Ion Mobility–Mass Spectrometry. J. Mass Spectrom. 43: 1–22.
45. Ruotolo, B.T., Benesch, J.L.P., Sandercock, A.M., Hyung, S.-J., Robinson, C.V. (2008) Ion Mobility-Mass Spectrometry Analysis of Large Protein Complexes. Nature Protoc. 3: 1139–1152.
46. Revercomb, H.E., Mason, E.A. (1975) Theory of Plasma Chromatography/Gaseous Electrophoresis – A Review. Anal. Chem. 47: 970–983.
47. Lanucara, F., Holman, S.W., Gray, C.J., Eyers, C.E. (2014) The power of ion mobility-mass spectrometry for structural characterization and the study of conformational dynamics. Nature Chem. 6: 281–294.
48. Parson, W.B., Schneider, B.B., Kertesz, V., Corr, J.J., Covey, T.R., Van Berkel, G.J. (2011) Rapid Analysis of Isomeric Exogenous Metabolites by Differential Mobility Spectrometry-Mass Spectrometry. Rapid. Commun. Mass Spectrom. 25: 3382–3386.
49. Cohen, M.J., Karasek, F.W. (1970) Plasma Chromatography™ – A New Dimension for Gas Chromatography and Mass Spectrometry. J. Chromatogr. Sci. 8: 330–337.
50. Sacristan E., Solis A.A. (1998) A Swept-field Aspiration Condenser as an Ion-mobility Spectrometer. IEEE Trans. Instrum. Meas. 47: 769–775.
51. Solis, A.A., Sacristan, E. (2006) Designing the Measurement Cell of a Swept-field Differential Aspiration Condenser. Revista Mexicana De Fisica 52: 322–328.

52. Buryakov, I.A., Krylov, E.V., Nazarov, E.G., Rasulev, U.K. (1993) A New Method of Separation of Multi-atomic Ions by Mobility at Atmospheric Pressure Using a High-frequency Amplitude-asymmetric Strong Electric Field. Int. J. Mass Spectrom. Ion Proc. 128: 143–148.

53. Giles, K., Pringle, S.D., Worthington, K.R., Little, D., Wildgoose, J.L., Bateman, R.H. (2004) Applications of a Travelling Wave-based Radio-frequency-only Stacked Ring Ion Guide. Rapid Commun. Mass Spectrom. 18: 2401–2414.

54. Pringle, S.D., Giles, K., Wildgoose, J.L., Williams, J.P., Slade, S.E., Thalassinos, K., Bateman, R.H., Bowers, M.T., Scrivens, J.H. (2007) An Investigation of the Mobility Separation of Some Peptide and Protein Ions Using a New Hybrid Quadrupole/Travelling Wave IMS/oa-ToF Instrument. Int. J. Mass Spectrom. 261: 1–12.

55. Brown, L.J., Toutoungi, D.E., Devenport, N.A., Reynolds, J.C., Kaur-Atwal, G., Boyle, P., Creaser, C.S. (2010) Miniaturized Ultra High Field Asymmetric Waveform Ion Mobility Spectrometry Combined with Mass Spectrometry for Peptide Analysis. Anal. Chem. 82: 9827–9834.

56. Wyttenbach, T., Bowers, M.T. (2003) Gas-phase Conformations: The Ion Mobility/Ion Chromatography Method. Top. Curr. Chem. 225: 207–232.

57. Harry, E.L., Bristow, A.W.T., Wilson, I.D., Creaser, C.S. (2011) Real-time Reaction Monitoring Using Ion Mobility-Mass Spectrometry. Analyst 136: 1728–1732.

58. Roscioli, K.M., Zhang, X., Li, S.X., Goetz, G.H., Cheng, G., Zhang, Z., Siems, W.F., Hill Jr, H.H. (2013) Real Time Pharmaceutical Reaction Monitoring by Electrospray Ion Mobility-Mass Spectrometry. Int. J. Mass Spectrom. 336: 27- 36.

59. Debaene, F., Wagner-Rousset, E., Colas, O., Ayoub, D., Corvaïa, N., Van Dorsselaer, A., Beck, A., Cianférani, S. (2013) Time Resolved Native Ion-Mobility Mass Spectrometry to Monitor Dynamics of IgG4 Fab Arm Exchange and "Bispecific" Monoclonal Antibody Formation. Anal. Chem. 85: 9785–9792.

60. Smith, D.P., Radford, S.E., Ashcroft, A.E. (2010) Elongated Oligomers in β_2-Microglobulin Amyloid Assembly Revealed by Ion Mobility Spectrometry-Mass Spectrometry. Proc. Natl. Acad. Sci. USA 107: 6794–6798.

61. Smith, R.W., Reynolds, J.C., Lee, S.-L., Creaser, C.S. (2013) Direct Analysis of Potentially Genotoxic Impurities by Thermal Desorption-Field Asymmetric Waveform Ion Mobility Spectrometry-Mass Spectrometry. Anal. Meth. 5: 3799–3802.

62. Reynolds, J.C., Blackburn, G.J., Guallar-Hoyas, C., Moll, V.H., Bocos-Bintintan, V., Kaur-Atwal, G., Howdle, M.D., Harry, E.L., Brown, L.J., Creaser, C.S., Thomas, C.L.P. (2010) Detection of Volatile Organic Compounds in Breath Using Thermal Desorption Electrospray Ionization-Ion Mobility-Mass Spectrometry. Anal. Chem. 82: 2139–2144.

63. Pawliszyn, J. (2012) Handbook of Solid Phase Microextraction. Elsevier, Waltham.

64. Vuckovic, D., Risticevic, S., Pawliszyn, J. (2011) *In Vivo* Solid-phase Microextraction in Metabolomics: Opportunities for the Direct Investigation of Biological Systems. Angew. Chem. Int. Ed. 50: 5618–5628.

65. Vuckovic, D., de Lannoy, I., Gien, B., Shirey, R.E., Sidisky, L.M., Dutta, S., Pawliszyn, J. (2011) *In Vivo* Solid-phase Microextraction: Capturing the Elusive Portion of Metabolome. Angew. Chem. Int. Ed. 50: 5344–5348.

66. Ahmad, S., Tucker, M., Spooner, N., Murnane, D., Gerhard, U. (2015) Direct Ionization of Solid-phase Microextraction Fibers for Quantitative Drug Bioanalysis: from Peripheral Circulation to Mass Spectrometry Detection. Anal. Chem. 87: 754–759.

67. Hu, J.-B., Chen, S.-Y., Wu, J.-T., Chen, Y.-C., Urban, P.L. (2014) Automated System for Extraction and Instantaneous Analysis of Millimeter-sized Samples. RSC Adv. 4: 10693–10701.

68. See, H.H., Hauser, P.C. (2014) Automated Electric-field-driven Membrane Extraction System Coupled to Liquid Chromatography–Mass Spectrometry. Anal. Chem. 86: 8665–8670.

69. Chen, S.-Y., Urban, P.L. (2015) On-line Monitoring of Soxhlet Extraction by Chromatography and Mass Spectrometry to Reveal Temporal Extract Profiles. Anal. Chim. Acta 881: 74–81.

70. Wang, H., So, P.-K., Ng, T.-T., Yao, Z.-P. (2014) Rapid Analysis of Raw Solution Samples by C_{18} Pipette-tip Electrospray Ionization Mass Spectrometry. Anal. Chim. Acta 844: 1–7.

71. Liu, W., Wang, N., Lin, X., Ma, Y., Lin, J.-M. (2014) Interfacing Microsampling Droplets and Mass Spectrometry by Paper Spray Ionization for Online Chemical Monitoring of Cell Culture. Anal Chem. 86: 7128–7134.

72. Zhu, Z., Bartmess, J.E., McNally, M.E., Hoffman, R.M., Cook, K.D., Song, L. (2012) Quantitative Real-time Monitoring of Chemical Reactions by Autosampling Flow Injection Analysis Coupled with Atmospheric Pressure Chemical Ionization Mass Spectrometry. Anal. Chem. 84: 7547–7554.

73. Blum, W., Schlumpf, E., Liehr, J.G., Richter, W.J. (1976) On-line Hydrogen/Deuterium Exchange in Capillary Gas Chromatography–Chemical Ionization Mass Spectrometry (GC-CIMS) as a Means of Structure Analysis in Complex Mixtures. Tetrahedron Lett. 7: 565–568.

74. Urban, P.L., Goodall, D.M., Bruce, N.C. (2006) Enzymatic Microreactors in Chemical Analysis and Kinetic Studies. Biotechnol. Adv. 24: 42–57.

75. Van Berkel, G.J., Kertesz, V. (2009) Electrochemically Initiated Tagging of Thiols Using an Electrospray Ionization Based Liquid Microjunction Surface Sampling Probe Two-electrode Cell. Rapid Commun. Mass Spectrom. 23: 1380–1386.

76. Hogenboom, A.C., de Boer, A.R., Derks, R.J.E., Irth, H. (2001) Continuous-flow, On-line Monitoring of Biospecific Interactions Using Electrospray Mass Spectrometry. Anal. Chem. 73: 3816–3823.

77. Martin, S.E., Shabanowitz, J., Hunt, D.F., Marto, J.A. (2000) Subfemtomole MS and MS/MS Peptide Sequence Analysis Using Nano-HPLC Micro-ESI Fourier Transform Ion Cyclotron Resonance Mass Spectrometry. Anal. Chem. 72: 4266–4274.

78. Amantonico, A., Urban, P.L., Zenobi, R. (2009) Facile Analysis of Metabolites by Capillary Electrophoresis Coupled to Matrix-assisted Laser Desorption/Ionization Mass Spectrometry Using Target Plates with Polysilazane Nanocoating and Grooves. Analyst 134: 1536–1540.

79. Urban, P.L., Jefimovs, K., Amantonico, A., Fagerer, S.R., Schmid, T., Mädler, S., Puigmarti-Luis, J., Goedecke, N., Zenobi, R. (2010) High-density Micro-arrays for Mass Spectrometry. Lab Chip 10: 3206–3209.

80. Küster, S.K., Pabst, M., Jefimovs, K., Zenobi, R., Dittrich, P.S. (2014) High-resolution Droplet-based Fractionation of Nano-LC Separations onto Microarrays for MALDI-MS Analysis. Anal. Chem. 86: 4848–4855.

7

Microfluidics for Time-resolved Mass Spectrometry

7.1 Overview

In recent years there has been substantial progress in the development of microfabricated systems for the use in chemical and biological sciences [1]. This chapter will discuss the usefulness of such microscale devices in time-resolved and high-throughput mass spectrometric analysis.

Microfluidics is the area of microtechnology which deals with fabrication and utilization of miniature systems for handling minute volumes of fluids. Microfluidic systems are applied in high-throughput and high-information-content chemical and biological analyses [2]. They often integrate multiple functions [3]. Apart from pre-processed chemical samples, living specimens can also be investigated using microfluidic chips [4]. Some physical and chemical phenomena only emerge at the micrometer scale. For instance, reactions are completed in short times, and flow often remains laminar. The surface-to-volume ratios in the microscale are high which favors interfacial interactions [5]. Microfluidic systems can be portable. They provide increased safety and reduce reagent consumption thus being more environmentally friendly and economical [6]. They accelerate chemical assays. The field of microfluidics is still at an early stage of development. However, it is anticipated that microfluidics will contribute to discoveries and inventions in chemical synthesis, biological analysis, optics, and information technology [7]. Although – arguably – no single "killer application" has been found for microfluidic chips yet, such miniature devices already enhance experimental protocols in the fields of analytical chemistry, materials science, and biomedicine [8].

7.2 Fabrication

A wide range of materials are utilized to fabricate microfluidic units [5]. On account of their rigidity as well as mechanical and chemical resistance, glass and silicon are versatile

Time-Resolved Mass Spectrometry: From Concept to Applications, First Edition.
Pawel Lukasz Urban, Yu-Chie Chen and Yi-Sheng Wang.
© 2016 John Wiley & Sons, Ltd. Published 2016 by John Wiley & Sons, Ltd.

substrates for microfabrication. Glass chips are compatible with optical detectors – at least in the visible part of the electromagnetic spectrum. The use of short-wavelength ultraviolet (UV) detectors requires implementation of chips made of fused silica glass, which makes them more expensive. Glass and silicon are ideal for on-chip capillary electrophoresis (CE) and applications involving organic solvents; however, they require expensive microfabrication [5]. Those substrates are usually processed using the *photolithography* technique (cf. Table 7.1). It involves sensitization of a substrate surface by coating it with a photoresist, exposure to light (e.g., UV), dissolving the photoresist in appropriate solvent, and in some cases also etching the exposed surface in acids, and removal of the photoresist residues.

The use of polymeric materials in microchips has gained popularity, especially in the fabrication of low-cost, disposable devices [3]. It is particularly common to prototype polydimethylsiloxane (PDMS) chips, which can easily be fabricated using the *soft lithography* technique. This fabrication strategy is considered as a low-expertise route of microscale prototyping. It facilitates creation of micropatterns on a surface or within a microfluidic channel without the need for using photochemical processes [9]. Fabrication

Table 7.1 Examples of methods used to fabricate microfluidic chips for application in analytical chemistry, and perceived characteristics

Method	Examples of materials	Typical features	Typical feature size	Capital cost[a]	Labor[b]	Mass fabrication[c]
Photolitho-graphy	Silicon, glass	Channels, patterns, elec-trodes	Sub-μm to a few tens of μm	+++++	+++++	+++++
Soft lithog-raphy	Polymers (e.g., PDMS, PEG)	Channels, patterns	A few tens of μm to a few hundred μm	+++	++++	+++
Laser ablation	Metals, polymers (e.g., PMMA)	Channels, patterns, elec-trodes	A few tens of μm to a few hundred μm	++++	+++	+
Engraving	Metals, polymers	Channels, patterns	A few hundred μm	++	+++	++
3D printing	Polymers (e.g., ABS)	Channels	A few hundred μm	++	++	++

ABS, acrylonitrile butadiene styrene; PDMS, polydimethylsiloxane; PEG, polyethylene glycol; PMMA, poly(methyl methacrylate).
[a] +, low; +++++, high.
[b] +, little; +++++, much.
[c] +, unsuitable; +++++, suitable.

by *laser ablation* is another convenient method of prototyping. Patterns designed with *computer-aided design* (CAD) software can be transformed into chips without tedious preparation of photomasks and reagents. The wavelength of laser light needs to be selected considering the absorption properties of the substrate material. Infrared lasers are commonly utilized for that purpose. However, laser ablation often involves the use of expensive laser instruments with precise mechanical positioning systems, which may not be available to many mass spectrometrists. It is not suitable for large-scale production because one microchip is normally fabricated at a time. Recently, a number of facile microfabrication techniques have emerged. Some of them are based on the *three-dimensional (3D) printing* concept [10]. The available 3D printing techniques include stereolithography, inkjet printing, selective laser sintering, fused deposition modeling, and laminated object manufacturing [10]. 3D printers are being improved to allow for printing microstructures at higher resolutions, higher speed, and using various materials with distinct mechanical, optical, and electrical properties [11]. One-step fabrication of transparent 3D microfluidic devices using a 3D printer has been demonstrated [12]. 3D printing also facilitates prototyping interfaces for coupling sample delivery systems and microchips with mass spectrometry (MS) [13, 14].

Overall, photolithography is normally chosen to fabricate high-quality chips in large numbers, while laser ablation and 3D printing are currently more suitable for prototyping. Soft lithography is a convenient method for experimenting with chip designs. Due to the physical properties of the polymers used in soft lithography, it enables the use of microscale actuators (micro-valves, micro-pumps), which can be integrated with the chips [15].

7.3 Microreaction Systems

Microchips are commonly used to mix reagents, and prime chemical reactions [1, 16]. The resulting products are detected downstream of the channel by one or several of the available detection systems. In laminar flow, mixing is diffusion-limited [3]. Thus, a major bottleneck of microfluidic reaction systems with narrow channels is the efficiency of mixing. It is often necessary to enhance mixing by patterning channels in specific ways or incorporating microstructures. Microfluidic mixers can be classified into two categories: passive and active [3]. Passive mixers include T- or Y-shaped mixers as well as mixers based on lamination and chaotic advection. Active mixers include electrokinetic, acoustic and magnetohydrodynamic mixers [3]. For example, in the passive mixers, a variety of two-dimensional (2D) and 3D structures may be incorporated into microchannels in order to induce turbulence or "accelerate" diffusion, and thus facilitate mixing in continuous hydrodynamic flow. Micromixers utilizing the principle of flow lamination (made of glass and silicon) could provide mixing times in the millisecond range [17]. With specially designed silicon microchips (two T-mixers connected by a channel), and by implementing *quench-flow* strategy, submillisecond events could be observed [18]. While microfluidic chips are generally small in size, they do require a considerable effort on the interfacing side. The inlet ports of microchips are normally connected to liquid supply lines linked to pumps and valves, adding to the complexity of microfluidic devices.

Microfluidic reactors have been used to obtain important information on chemical processes conducted in microscale volumes [19]. Indeed, the main product of research-grade microreactors is not a chemical compound but the *information* on the optimum conditions of the reaction (e.g., type of catalyst which provides highest reaction yield). Once this information is obtained with such microscale systems, one can readily *scale up* (or *number up*) the synthesis process to produce industry-scale amounts of valuable chemicals. However, to make full use of the available microfluidic technology, microreactors need to be used in conjunction with information-rich on-line detection systems. In some cases, microreactors are coupled with detection systems in the off-line arrangement [20], which limits the amount of the information that can be obtained in a short period of time. Screening reaction conditions can be done this way using "safe" amounts of reagents and minimizing the costs. In fact, the considerations of cost and availability are especially apparent when screening enzyme samples [21, 22]. Enzymes are expensive catalysts. Thus, using minute amounts of enzymes for assays is desirable. Fast determination of the kinetics of surface-immobilized enzymes can be carried out on a chip [23]. Using microreactors, multiple chemical reactions can be carried out sequentially by different enzymes immobilized in series [23]. In one approach, the microfluidic device for drug metabolism studies comprised three components: (1) bioreactors with encapsulated human liver microsomes (vesicles formed from endoplasmatic reticulum while processing cells before experiments); (2) cell culture chambers; and (3) solid-phase extraction columns [24]. The products of the reactions were directly detected by electrospray ionization (ESI)-MS following sample pre-treatment on an integrated micro-solid-phase extraction (SPE) column.

There exist various examples of microfluidic systems coupled to MS with the purpose of monitoring dynamic processes – in particular, chemical reactions. While this technology has not been perfected yet, it is instructive to review the previous applications of microchip-MS systems to learn about the advantages and limitations of the existing technology. Certainly, more effort is needed to enable routine monitoring of reactions in real time with microscopic volumes of reactants on microchips. The following sections briefly outline the theory underlying flow of fluids in microfluidic channels, and they introduce some of the significant achievements in coupling microfluidic chips with MS emphasizing the importance of MS measurements with temporal resolution.

7.4 Hydrodynamic Flow

Many of the fundamental concepts and technical solutions used nowadays in microfluidics have been "borrowed" from the older techniques – continuous flow analysis (CFA) and flow-injection analysis (FIA) [25]. However, the implementation of microfluidic chips has encouraged further miniaturization. Therefore, there have been attempts to scale down the macroscopic fluid handling devices (e.g., peristaltic pumps with footprints of a few square decimeter) to the dimensions of microchips (square millimeters).

Fluids are often pumped hydrodynamically to exert the flow. Various pumps are used, including syringe pumps, peristaltic pumps, piezoelectric pumps, and gas-pressure-driven hydrodynamic pumps. In the case of hydrodynamic pumps, an inert gas is pressurized in the headspace of the vial containing the sample or carrier fluid. The force exerted by the gas on the liquid phase sustains flow of the liquid in the channel. When a liquid moves along the circular cross-section channel, the Poiseuille equation can be used to relate the pressure difference (ΔP) with flow rate (Q):

$$\Delta P = \frac{8\mu LQ}{\pi r^4} \qquad (7.1)$$

where μ is dynamic viscosity of the liquid, L is the length of the channel and r is its radius. If the channel length and pressure difference are known, Equation 7.1 can be used to estimate the flow rate of hydrodynamic flow, or calculate sample injection volume (in CE; see Chapter 6).

Alternatively, electrolyte solutions can be moved along microfluidic channels using electroosmotic flow (EOF; see Chapter 6). In this case, an electric field rather than mechanical force is used to move the liquid. Implementation of EOF allows one to handle the back-pressure problem associated with hydrodynamic pumping. However, EOF is mainly used to transport dilute aqueous electrolyte solution, and is sensitive to pH change.

To characterize the behavior of hydrodynamic flow in microchips, it is helpful to introduce the most important *dimensionless parameters*. The Reynolds number (Re) is defined as the ratio of inertial to viscous forces acting on the fluid (cf. [7]):

$$\mathrm{Re} = \frac{\rho v D}{\mu} \qquad (7.2)$$

where ρ represents density of the fluid, v is velocity, and D is the characteristic dimension of the channel (hydraulic diameter). In the case of channels with circular cross-section, the geometrical diameter can be used as D. In the case of non-circular channel cross-section, the hydraulic diameter can be estimated with the equation [26]:

$$D = \frac{4A}{p} \qquad (7.3)$$

where A is the area of channel cross-section and p is its wetted perimeter. At low Re, viscous forces are dominant while high Re values are associated with a large contribution of inertial forces. Re is used to determine the type of flow. Laminar flow occurs at low Re values (<100) while turbulent flow occurs at high Re values ($>>1000$). Microfluidic systems operate at low Reynolds numbers, which means that the flow is dominated by laminar flow with little turbulence. Thus, diffusion accounts most for mixing solutes in such systems. Times of diffusion of small molecules in aqueous solutions over a distance of 1 mm are in the order of several minutes. The times go down to milliseconds, when the distance is as short as 1 μm. Thus, in the absence of turbulence, using very thin channels, one can – in some cases – mix solutes using diffusion. Differences in diffusion coefficients can also be used to separate solutes (small and large molecules).

Please note that microfluidics is sometimes defined as the science and technology of systems that process or manipulate small ($10^{-9} - 10^{-18}$ l) amounts of fluids, using channels with dimensions of tens to hundreds of micrometers [7]. Nonetheless, practical dimensions of channels used in many prototype microchips range from 10 μm to 1 mm, and beyond. Thus, the laminar character of flow may vary to some extent even within one microchip that incorporates features with different characteristic dimensions.

The Péclet number (Pe) is defined as the ratio of the rate of advection of a physical quantity to the rate of diffusion:

$$\mathrm{Pe} = \frac{Dv}{D_c} \qquad (7.4)$$

where D_c is the diffusion coefficient. Microfluidic systems with high Pe values lead to formation of radial concentration gradients across channels. In systems with low Pe values, diffusion equalizes concentration differences across the channels. The Péclet number is used as one of several scaling factors – in the situations when microscale devices are sized up or macroscale units are sized down.

7.5   Coupling Microfluidics with Mass Spectrometry

Combining numerous steps (reaction or sample preparation) into one miniature device provides immediate benefits for studying chemical reactions. Products of such reactions can promptly be detected by MS while reducing the time lag between the reaction and the detection. The typical flow rates in microfluidic chips are measured in microliters down to nanoliters per minute; thus, the microfluidic chips are compatible with some of the atmospheric pressure ion sources.

Miniaturized ESI interfaces (nanospray electrospray ionization, nanoESI) match the dimensions of microfluidic chips. On-line coupling of microchips with ESI can be accomplished using different interface geometries: blunt end, corner outlet, external capillary, external emitter, or monolithic emitter [27], some of which resemble the nanoESI emitters used in CE-MS (see also Chapter 6). In fact, on-chip capillary channels are often used as CE or LC separation columns, and directly linked with the nanoESI emitters. Atmospheric pressure chemical ionization (APCI) and photoionization (APPI) have also been subject to miniaturization but they have not attracted as much attention when it comes to hyphenation with microchips [28]. This situation may change when the novel nanoAPCI interfaces [29] are perfected, providing the way to transmit and ionize non-polar analytes at low flow rates.

In one approach, the microchip interface was constructed from modified "1/16-inch" high-performance liquid chromatography (HPLC) fittings [30]. It incorporates a free-standing liquid junction formed via continuous delivery of a flow of suitable solvent which carries the separation effluent through a pneumatically assisted electrospray needle located in front of the MS orifice. In some cases, structural features of microchips can be utilized as parts of the ion source (e.g., ESI emitter). Thus, the resulting coupled microchip-MS systems are more compact, and the delay time between the on-chip incubation and MS detection can be decreased. For example, a capillary nanoESI emitter was successfully incorporated into a microchip CE channel for on-line CE-MS analysis [31]. Such microchip-MS systems do not require the use of external pumps because analytes can be driven toward the ion source (ESI or nanoESI) by means of electroosmosis and electrophoresis [32].

Setting up microchip-MS systems also incurs certain technical difficulties which need to be overcome. The ion source emitter needs to be aligned with the MS orifice to prevent substantial losses of analytes/ions. The small size of microchannels and emitters leads to clogging by sample residues. While the clogging issue can be solved when implementing folded polyimide tape emitters [33], such emitters may not be compatible with many types of microchips. Microchips are often fabricated in clean rooms but in MS laboratories they are exposed to a "dusty" environment, which – in some cases – can affect their performance. To minimize this effect, a microchip–MS interface has been developed which integrates a fan and air filter to assure low-dust conditions for operation of the dust-sensitive microchip [14].

When looking at images of microfluidic chips reproduced in the recent scientific literature, the labyrinth-like arrangement of hair-thin channels is one of the first features that attracts the reader's attention. Why so complex? Since the time spent on guiding fluids along channels has to be reduced to a minimum, the microchip channels should – in principle – be as short as possible. Extending the channel's length is often done to achieve a better mixing of merged flow lines in the laminar flow, or to increase the efficiency of electrophoretic or chromatographic separation. Some channels are used to guide the actual samples while others supply ancillary media (e.g., compressed gas to operate on-chip valves or pumps). T-junction, Y-junctions, and crosses are often used to merge flow lines or inject nanoliter samples to separation/reaction channels. Serpentine channels are implemented to control the incubation time following the merger of substrate solutions. In addition, microchips often integrate microelectrodes (e.g., contacts for CE, electrochemical reactions, or to induce electrowetting). All these features make their architecture intrinsically complex. However, beside those complex architectures, there exist simplistic designs of microchips which satisfy a very specific purpose. For instance, micropillar array microchips were fabricated in the course of a multi-step process to facilitate ESI of low-volume samples (Figure 7.1) [34]. This microchip includes a sample reservoir, channel, and outlet for ESI. Deep reactive ion etching (DRIE) was used during fabrication. Analysis with these microchips showed high sensitivity (limit of dectection for verapamil: 30 pM) [34]. Since these are open microchips (no lid), they can readily be used to conduct off-line analysis of samples of reaction mixtures obtained from chemical reactors without the need for refilling syringes.

Another approach takes advantage of multi-nozzle ESI chips which enable introduction of samples in a reproducible manner [35]. Every sample is analyzed using an individual ESI nozzle, which minimizes carry-over effects (cross-contamination of sample aliquots ionized in front of the mass spectrometer). Multitrack ESI chips have also been introduced. They enable analysis of several samples in a short period of time [36]. Moreover, they can facilitate internal mass calibration, reaction evaluation, testing ionization efficiency and quantification of analytes. One could imagine that multiple reactions could be performed in separate on-chip reservoirs and probed by MS. By moving the multi-nozzle chip perpendicular to the ion axis, one might be able to select the reaction mixture for monitoring. This would increase throughput of reaction monitoring in the case of slow reactions. Mao *et al.* [37] reported the silicon-based monolithic multi-nozzle emitter array and demonstrated its applications in high-sensitivity and high-throughput nanoESI-MS. It consists of 96 identical 10-nozzle emitters in a circular array on a silicon chip. The authors suggested that such a design will enable development of fully integrated microfluidic systems for ultra-high-sensitivity and ultra-high-throughput proteomics and metabolomics.

Electrophoretic and chromatographic separations conducted on microchips are often fast; therefore, they do not lower the temporal resolution of the whole analytical process as much as most conventional hyphenated systems (see Chapter 6). Although microscale separations are not always expected to present as good resolution as the conventional ones, they can often contribute to the selectivity of the analytical process. For example, MS cannot directly be used to distinguish ions of optical isomers. Chiral separations can be carried

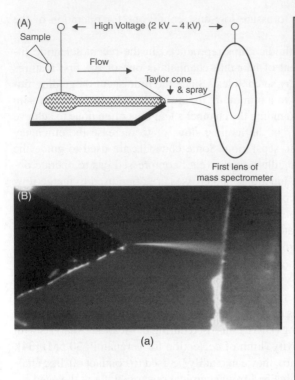

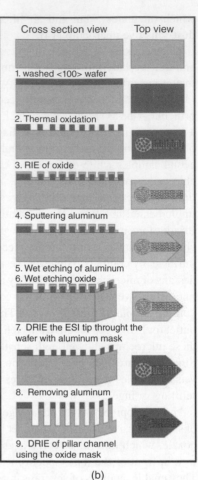

Figure 7.1 *Silicon micropillar array electrospray ionization (μPESI) chip. (a) (A) Setup of the μPESI measurements. (B) Formation of the Taylor cone and spray from the small (4 mm × 9 mm) μPESI chip. The diameter of the pillars was 60 μm and the distance between the pillars was 15 μm. The liquid sample contained 95% acetonitrile, 4.9% water and 0.1% formic acid. The high voltage applied to the chip was 3 kV. (b) Schematic of the cross-sectional and top views of the chip design during fabrication [34]. Reproduced from Nissilä, T., Sainiemi, L., Sikanen, T., Kotiaho, T., Franssila, S., Kostiainen, R., Ketola, R.A. (2007) Silicon Micropillar Array Electrospray Chip for Drug and Biomolecule Analysis. Rapid Commun. Mass Spectrom. 21: 3677–3682 with permission from John Wiley and Sons*

out at a high speed using on-chip CE [38]. In one study, enantiomeric separations of 3,4-dihydroxyphenylalanine, glutamic acid, and serine were accomplished in 130 s [38]. However, in some cases, baseline separations of enantiomers can be accomplished within 1 s [39]. Achieving high separation efficiency and analytical sensitivity of microchip-CE-MS are important goals [40]. Some microchip-CE-MS systems enable particularly fast separations of microscopic samples. In one representative work, a microchip was implemented in on-line analysis of individual cells [41]. Cells were lysed due to rapid buffer exchange and an increase in electric field strength. The system enabled detection of the dissociated heme group and subunits of hemoglobin from erythrocytes at a rate of ~12 cells min^{-1} [41].

A prototype of a microfluidic chip for gradient LC-MS/MS has been fabricated using the photolithography technique (Figure 7.2) [42]. It comprises three electrolysis-based electrochemical pumps (one for loading the sample and two for solvent). It could be operated at pressures >250 psi (~1.7 MPa) [42]. Microchips for nanoflow LC, integrating ESI emitters, have been commercialized [43, 44]. The developers of such microchips have addressed the key obstacles which prevented this technology from getting into common use. These microchips incorporate injection of nanoliter aliquots (plugs) of samples, sample enrichment, and interfacing the nanoLC column with the nanoESI emitter. Overall, the microchip-HPLC-nanoESI-MS systems show good reproducibility, analytical sensitivity, and are a convenient tool for proteomic applications related to

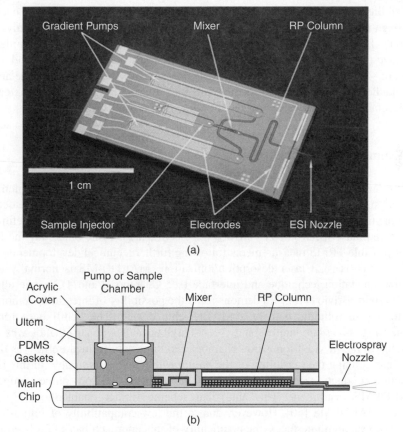

Figure 7.2 *Microfluidic chip for LC-MS analysis of peptides. (a) Photograph of the fabricated LC chip. (b) Diagram of the LC chip showing the placement of the solvent reservoir and cover plate on top of the main chip [42]. Adapted with permission from Xie, J., Miao, Y., Shih, J., Tai, Y.-C., Lee, T.D. (2005) Microfluidic Platform for Liquid Chromatography-Tandem Mass Spectrometry Analyses of Complex Peptide Mixtures. Anal. Chem. 77: 6947–6953. Copyright (2005) American Chemical Society*

biomarker discovery. On-chip sample treatment cells (e.g., electrochemical reactors) can also be coupled with conventional LC-MS systems; for example, to enable studies on electrochemical metabolism pathways [45].

Apart from CE and LC, free-flow electrophoresis is yet another separation technique implemented on microfluidic chips. It enables preparative separations of samples delivered continuously. If the separated analytes are not fluorescent, the separation can be monitored by ESI-MS (Figure 7.3) [46]. Here, on-line MS enables optimization of the separation conditions and identification of the separated zones. Subsequently, the characterized free-flow electrophoretic process may be used for off-line isolation of the required amounts of target compounds (e.g., biomolecules).

Without any doubt, microfluidics offers advantages with respect to miniaturization. While microchips are small in dimensions, the mass spectrometers are often few orders of magnitude larger in volume. Hence, the benefits of miniaturization cannot be seen immediately. Although, at present, one cannot take full advantage of microchip-MS miniaturization, the possibility to miniaturize mass spectrometers has been demonstrated by several scientists (e.g., [47, 48]). Commercial small-scale mass spectrometers have already entered the market. This suggests that the miniaturization trend affects not only the sample preparation stage (microfluidics) but also the separation of gas-phase ions (MS), and – in the near future – small integrated microchip-MS apparatus may be used for routine analyses. Recent studies already suggested the suitability of microscale mass spectrometers fitted with microfabricated ion sources for the monitoring of chemical reactions [49].

7.6 Examples of Applications

We will briefly introduce examples of applications of microfluidic devices which enable rapid analysis of small volumes of dynamic samples with mass spectrometric detection. Please note that some of the discussed approaches are covered by the broad definition of time-resolved mass spectrometry (TRMS) (cf. Section 1.3), while others lay the foundations for possible TRMS measurements following further technical developments.

While matrix-assisted laser desorption/ionization (MALDI)-MS is normally used as an off-line ionization technique and interface (see Chapters 2 and 4), in one attention-grabbing report, Brivio *et al.* [50] demonstrated the possibility of integrating a monitoring port into the microfluidic path of a MALDI-chip device. The Schiff base formation reaction using 4-*tert*-butylaniline and 4-*tert*-butylbenzaldehyde in ethanol was carried out on-chip in the MALDI ionization chamber and the newly formed imine was detected in real time, proving the feasibility of the *monitoring window approach*. In this method, the reaction products can be probed by the laser inside the ion-source compartment of MALDI-MS instrument. This analytical scheme further enabled on-chip kinetic studies by MALDI-MS [50]. However, due to the low compatibility of liquid-handling systems with vacuum interfaces, the practicality of this approach has so far been limited. Compared with MALDI, atmospheric pressure ion sources (such as ESI) have much greater performance when it comes to collecting kinetic data on the processes which take place in microfluidic channels. In another study conducted by Brivio *et al.* [51] the reactions of propyl isocyanate, benzyl isocyanate, and toluene-2,4-diisocyanate with 4-nitro-7-piperazino-2,1,3-benzoxadiazole (NBDPZ) were carried out in a glass microfluidic chip (Figure 7.4; [51]). NanoESI-MS was used to monitor these reactions

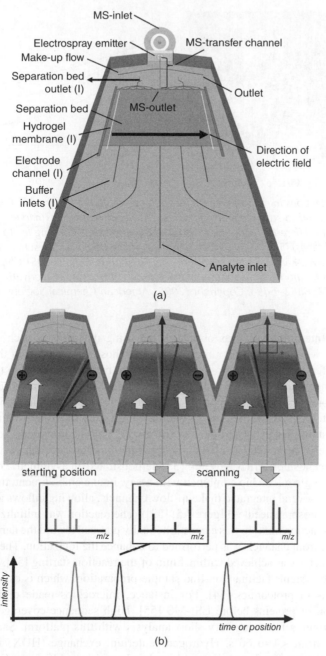

Figure 7.3 *Coupling chip-based free-flow electrophoresis nanoESI-MS. (a) Layout of a microfluidic free-flow electrophoresis-MS chip (I = left). (b) The analysis principle. The separated analytes are directed towards the mass spectrometric outlet by alteration of the buffer's hydrodynamic flow. Arrows indicate relative flow rates and the rectangle (*) labels the area visualized by fluorescence imaging [46]. Reproduced from Benz, C., Boomhoff, M., Appun, J., Schneider, C., Belder, D. (2015) Chip-based Free-Flow Electrophoresis with Integrated Nanospray Mass-Spectrometry. Angew. Chem. Int. Ed. 54: 2766-2770 with permission from John Wiley and Sons*

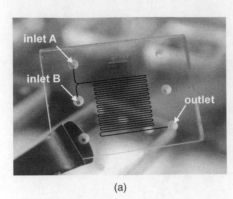
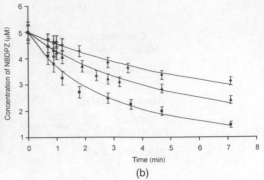

(a)   (b)

Figure 7.4   *Reactions in microfluidic chips coupled with a MS system. (a) Photograph of the two-inlet chip used to study reaction kinetics. (b) Reaction profile (markers) of the on-chip derivatization of different compounds with NBDPZ and corresponding fits to a second-order kinetics model (lines) [51]. Adapted with permission from Brivio, M., Liesener, A., Oosterbroek, R.E., Verboom, W., Karst, U., van den Berg, A., Reinhoudt, D.N. (2005) Chip-based On-line Nanospray MS Method Enabling Study of the Kinetics of Isocyanate Derivatization Reactions. Anal. Chem. 77: 6852–6856. Copyright (2005) American Chemical Society*

in real time. Interestingly, it was observed that using macroscale batch conditions, the rate constants were three to four times lower than those obtained using the microfluidic setup. This effect was attributed to the more efficient molecular diffusion that takes place in the micrometer-sized channel [51]. Abonnenc *et al.* [52] constructed an electrospray micromixer for derivatization and on-line ESI-MS detection. The microchip was made of polyethylene terephthalate (PET) using photoablation with an ArF excimer laser. Electric contact was established using carbon ink. The mixing unit incorporated a series of parallel oblique grooves. The system enabled derivatization of cysteine residues in peptides [52].

In a more recent study, Fritzsche *et al.* [53] demonstrated an asymmetric organocatalytic reaction on a single microchip combined with assay. In one embodiment, they employed a microchip with several integrated makeup-flow channels, allowing inflows at two positions ahead of the nanospray needle (Figure 7.5; [53]). The reaction was initialized in the upper inlets and the reaction mixture was electrokinetically pumped toward the nanospray emitter. Dilution in the front channels was performed to enhance the ionization. The analysis in the mass spectrometer was achieved within 1 min of the reaction starting [53].

Microfluidic systems facilitate on-line sample preparation, which is especially relevant to applications in proteomics [54]. For instance, microchips made of PMMA enable rapid digestion of proteins before ESI-MS [55]. High sequence coverage was achieved for various proteins following very short analysis with this platform. Sample residence times ranged from <4 to 60 s. Hydrogen/deuterium exchange (HDX) is occasionally used to enhance structural analysis of biomolecules prior to MS (see also Chapter 12). The use of microchips in HDX experiments may limit the unfavorable isotope back-exchange effects [55]. Microchips coupled with ESI-MS enabled monitoring protein folding/unfolding kinetics on the millisecond timescale [56, 57].

As mentioned above, microfluidic devices coupled with ESI-MS facilitate *in vitro* metabolic studies [24, 58]. Products of enzymatic reactions can be desalted and

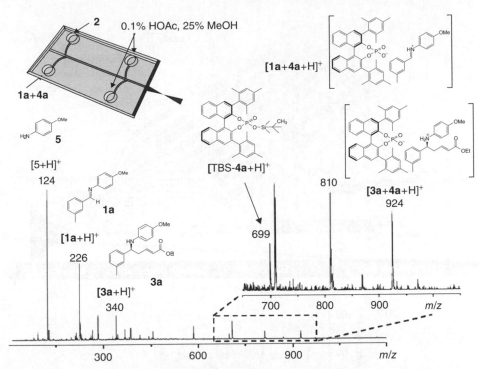

Figure 7.5 *Investigation of a reaction mechanism by microsynthesis-nanoESI-MS [53]. Reproduced from Fritzsche, S., Ohla, S., Glaser, P., Giera, D.S., Sickert, M., Schneider, C., Belder, D. (2011) Asymmetric Organocatalysis and Analysis on a Single Microfluidic Nanospray Chip. Angew. Chem. Int. Ed. 50: 9467–9470 with permission from John Wiley and Sons*

concentrated on a guard column (off-chip or on-chip) before infusion to the mass spectrometer [58]. Michaelis constants (see Chapter 13) and inhibition constants can be determined. Tracking cellular metabolism on chips can be enhanced by stable isotope labeling and ESI-MS detection (Figure 7.6) [59]. This method was used to follow the response of cells upon stimulation with drugs. Thus, it has the potential for implementation in metabolomic studies and drug screening. In fact, the dynamic range of quantitative analysis of genistein extended over almost two orders of magnitude which can be regarded as satisfactory in the case of ESI-MS detection [59].

Miniaturization provides a significant increase of analysis throughput of low-volume samples [54]. Microfluidic systems enable manipulation of small droplets, which is a convenient way of conducting a large number of discrete chemical assays [3]. Samples are made discrete by introducing two immiscible phases (liquid/gas or two immiscible liquids). Such droplet-based microfluidic systems (using optical detection) enable studies of fast reaction kinetics [60]. By means of the droplet microfluidics approach, high-throughput analyses can be conducted. In this case, the high temporal resolution of MS is required to obtain quality spectra for every sample droplet delivered to the ion source. While segmented flow prevents dispersion of individual sample aliquots, it also incurs some technical problems, especially when the successive operations carried out on the chip require a homogeneous one-phase flow [61]. If the segmentation is conducted

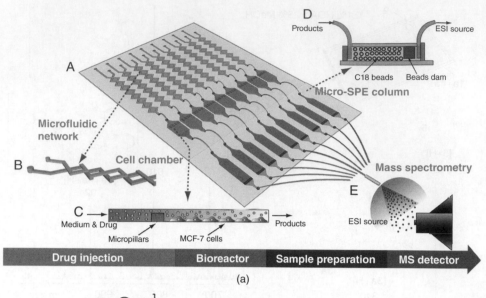

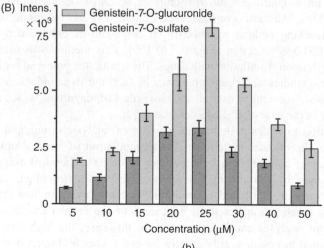

(a)

(b)

Figure 7.6 *(a) Schematic diagram of the chip-ESI-MS system. (A) The system consists of a microfluidic network for culture medium and drug injections, cell culture chambers, on-chip micro-SPE columns for sample desalting and purifying, and an ESI-quadrupole–time-of-flight-MS device. (B) Microfluidic network design for the concentration gradient generator during cell culture and drug screening. (C) Magnified view of the MCF-7 cells culture chamber for cell metabolism on the microfluidic chip. (D) Magnified view of an integrated micro-SPE column on the microdevice for sample pretreatment prior to ESI-MS detection. (E) The ESI source was coupled with the micro-SPE column together by capillaries. (b) (A) Time-dependent accumulation of genistein absorption in cultured MCF-7 cells on a microfluidic chip. (B) The accumulation of genistein-7-O-glucuronide and genistein-7-O-sulfate in the culture medium of MCF-7 cells incubated with varied concentrations of genistein from 0 to 50 μM for 30 h on-chip [59]. Adapted with permission from Chen, Q., Wu, J., Zhang, Y., Lin, J.-M. (2012) Qualitative and Quantitative Analysis of Tumor Cell Metabolism via Stable Isotope Labeling Assisted Microfluidic Chip Electrospray Ionization Mass Spectrometry. Anal. Chem. 84: 1695–1701. Copyright (2012) American Chemical Society*

by introducing air bubbles, there exist ways to remove them on the chip. Since the oil phase is not compatible with most ion sources, aqueous droplets in the oil medium can be analyzed by ESI following an on-chip transfer. In one implementation, the transfer of aqueous droplets in oil to the stream of aqueous ESI electrolyte is evoked by the electric field present between on-chip microelectrodes [62]. Zhu and Fang [63] described an integrated microchip-based system for droplet analysis by ESI-MS. It combined multiple modules including a droplet generator, a droplet extraction interface, and a monolithic ESI emitter. The device was applied in the on-line monitoring of a droplet-based microreaction (alkylation of a peptide). Volný *et al.* [64] reported a microfluidic device with segmented flow, which delivered aqueous droplets into a MS system. Sample ionization was assisted by multiple passes of an infrared laser beam in the interface. Interestingly, it was insensitive to the presence of buffer salts and other matrices. Gasilova *et al.* [65] applied so-called *electrostatic spray ionization* MS, which can be considered as a modified nanoESI interface. The system enabled efficient coupling between droplet-based microfluidics with MS via a "spyhole" drilled on the top of a microchip. ESI occurred at the spyhole, and the droplet content was analyzed by MS without dilution or an oil removal step. Using this system, they monitored a droplet-based tryptic digestion, and a biphasic reaction between β-lactoglobulin in water and α-tocopheryl acetate in 1,2-dichloroethane [65]. These studies exemplify various ways of coupling microchips with MS to enable analysis of large numbers of aliquoted low-volume samples in a short period of time.

7.7 Digital Microfluidics

Disadvantages of conventional microchips include resistance to hydrodynamic flow (backpressure), adsorption of some biomolecules on walls, and limited robustness (e.g., clogging narrow channels). *Digital microfluidic* systems overcome the problems with mixing reagents, which are normally associated with conventional microfluidic devices that use laminar hydrodynamic flow. Such microscale platforms are normally based on the electrowetting-on-dielectric (EWOD) principle [66]. In these digital microchips, samples and reagents are digitized into confined volumes – microliter and nanoliter droplets – which can readily be manipulated by application of electric potentials. These droplets are moved over the hydrophobic surface of the microchip by applying electric potentials to microelectrodes "printed" under a dielectric layer. Mixing droplets on open digital microchips is very fast. It can be accelerated by switching potentials between the electrodes near a sample droplet. The resulting mixtures can readily be transported to the ion source compartment of a mass spectrometer.

Fabrication of EWOD systems usually involves a photolithographic process although laser ablation can also be used in fast prototyping of such devices. A solid support is coated with a thin layer of metal, which is subsequently structured into electrodes by using one of the above fabrication approaches. The electrodes are then covered with a layer of photoresist which fulfills the role of dielectric in the EWOD method. If the photoresist is not hydrophobic enough, an additional superhydrophobic layer can subsequently be coated on top of the photoresist layer. Hydrophobicity of the surface is evaluated by measuring the contact angle between the surface of a solvent droplet and the chip surface. High contact angles (≫90°) indicate a high level of hydrophobicity.

Digital microchips can readily be interfaced to other auxiliary systems, including MS (e.g., [14, 67, 68]). Coupling digital microchips with MS benefits proteomics, chemical synthesis, and clinical diagnostics [69]. For instance, such systems enable off-line and on-line MS analysis of dried blood spots [70]. Analytes extracted from dry blood spots can be processed on the digital microchips and analyzed by nanoESI-MS [71]. Importantly, using a digital microchip-MS system, one can carry out multiple analysis steps, including extraction, mixing with internal standards, derivatization, and reconstitution of sample in a suitable MS-friendly solvent.

Applicability of digital microfluidics to TRMS is apparent in several studies focused on kinetic profiling of enzyme-catalyzed reactions. For example, Nichols and Gardeniers [72] presented a digital microfluidic system based on the EWOD concept to facilitate the investigation of pre-steady-state reaction kinetics using rapid quenching and off-line MALDI-MS detection. In other work, an enzyme and substrate droplets were merged in a porous silicon microfluidic channel. Residues of the reactants were deposited on the channel walls, and analyzed off-line using the desorption/ionization on silicon (DIOS)-MS approach [73]. Digital microchips also enabled real-time monitoring of the Morita–Baylis–Hillman reaction [74]: the prototype device contained a two-plate-to-one-plate digital microfluidic interface, which allowed for straightforward coupling of micro-reaction and product delivery to the "folded" nanoESI emitter. Luk *et al.* [75] used enzymes embedded in hydrogels to process protein samples in digital microfluidic systems prior to MS. Using this approach, a higher sequence coverage was achieved compared with conventional homogenous processing. Interestingly, the digital microfluidic platforms can be used for culturing and analysis of cells [76]. It would be appealing to combine these devices with on-line MS to enable molecular characterization of the molecules excreted by cells in real time. However, to enable such TRMS studies, various technical problems need to be solved which relate to robustness of digital microchips (while handling biological samples) as well as the ion suppression caused by cell growth media.

Universal electronic modules have recently gained huge popularity among analytical chemists because their implementation does not require expert knowledge and investment of funds [77]. Single-board microcontrollers and micro-computers such as Arduino,

Teensy, Raspberry Pi, BeagleBone, or Edison enable experimental data to be collected with high precision as well as efficient control of electric potentials, and actuation of mechanical systems. They are readily programmed using high-level languages, such as C, C++, JavaScript, or Python. They can also be coupled with mobile consumer electronics, including smartphones as well as teleinformatic networks. More demanding analytical tasks require fast signal processing. For instance, field-programmable gate arrays enable efficient and inexpensive prototyping of high-performance analytical platforms, which are thus becoming increasingly popular among analytical chemists. Open-source electronics enhances the operation of EWOD systems because it enables application of voltages to the EWOD chip electrodes without the need to use expensive instruments [14, 71].

7.8 Concluding Remarks

To reach the full potential of microfluidics, co-ordination between different fields is necessary [3], including engineering, physics, computer science, chemistry, and biology. Microfluidic chips are occasionally utilized by chemists to carry out mass spectrometric analysis of dynamic samples. They bring several advantages to MS since they enable automated handling of microscale samples (nano- and picoliter volumes). They facilitate sample processing at high speeds, and thus ascertain high-throughput operation. The diffusion times in the microscale are in the order of seconds down to milliseconds. Therefore, reactions can proceed fast, and the reaction conditions can be reproduced precisely in the absence of turbulence. For these reasons, microfluidic systems are a convenient tool for studying chemical processes in short time scales. Microchips allow us to limit the time between the studied processes (e.g., reactions) and the sample delivery to the ion source. Therefore, their performance in microscale and high-throughput assays is superior, and suitable for various R&D applications. However, microchip-MS systems are less suitable for industrial and environmental applications – for example where liquid samples with high concentration or large-volume gas samples need to be processed.

The main obstacle to the widespread use of microfluidics in conjunction with MS is related to the fabrication of such systems. It still requires a considerable effort and costly equipment. Implementation of microchips often necessitates expert knowledge about the device assembly, maintenance, and trouble-shooting. However, the rapidly expanding 3D printing tools, and open-source electronic modules, can cut the costs of fabrication and promote the use of microfluidic devices in kinetic studies conducted with MS detectors. We anticipate that, in the near future, a microscale total analysis system (µTAS) combining microfluidics and miniature mass spectrometers will become an omnipresent piece of the laboratory toolkit.

For a mass spectrometrist willing to utilize microfluidic technology in MS measurements with temporal resolution, it may be difficult to figure out which approach (e.g., mode of microfluidics, type of flow, material, fabrication technique) is most suitable for a given application. Thus, when starting microfluidics-related projects, it may be helpful to answer the following questions which are related to the intended purpose:

- Is the small scale essential to my analysis? Considering the large size of commercial mass spectrometers miniaturization of the sample preparation/delivery stage is not always justified. Microscale devices should be combined with MS if there is a clear benefit for the measurement (e.g., analysis of volume-limited samples, taking advantage of short diffusion times).

- Is the use of a microfluidic chip essential to my analysis? In many cases microscale capillaries can be used as nanoliter-volume reaction vessels in place of microchips [78].
- Is the investigated process continuous or discrete? Answering this question may help us to decide whether continuous flow of a homogeneous solution should be used or whether droplet-based microfluidics would be more suitable.
- What solvent/medium will be used? The answer to this question will help us to select the right material for fabrication of the microchip, and decide what pumping force can be used. Aqueous solutions are compatible with many materials (inorganic and organic) while organic solvents are less compatible with polymers which are often utilized in soft lithographic techniques. Moreover, using polymer microchips may invite contamination of the mass spectrometer. Thus, it is preferable to use glass or silicon substrates. If the medium is a dilute electrolyte, then EOF may be used as the driving force. EOF can also be used in the on-chip CE separations. If the medium contains organic solvents, it is suggested to use hydrodynamic flow (e.g., exerted by a syringe pump).

References

1. deMello, A.J. (2006) Control and Detection of Chemical Reactions in Microfluidic Systems. Nature 442: 394–402.
2. Stanley, C.E., Wootton, R.C.R., deMello, A.J. (2012) Continuous and Segmented Flow Microfluidics: Applications in Highthroughput Chemistry and Biology. Chimia 66: 88–98.
3. Nge, P.N., Rogers, C.I., Woolley, A.T. (2013) Advances in Microfluidic Materials, Functions, Integration, and Applications. Chem. Rev. 113: 2550–2583.
4. Sivagnanam, V., Gijs, M.A.M. (2013) Exploring Living Multicellular Organisms, Organs, and Tissues Using Microfluidic Systems. Chem. Rev. 113: 3214–3247.
5. Ren, K., Zhou, J., Wu, H. (2013) Materials for Microfluidic Chip Fabrication. Acc. Chem. Res. 46: 2396–2406.
6. Belder, D. (2009) Towards an Integrated Chemical Circuit. Angew. Chem. Int. Ed. 48: 3736–3737.
7. Whitesides, G.M. (2006) The Origins and the Future of Microfluidics. Nature 442: 368–373.
8. Sackmann, E.K., Fulton, A.L., Beebe, D.J. (2014) The Present and Future Role of Microfluidics in Biomedical Research. Nature 507: 181–189.
9. Kim, P., Kwon, K.W., Park, M.C., Lee, S.H., Kim, S.M., Suh, K.Y. (2008) Soft Lithography for Microfluidics: a Review. Biochip J. 2:1–11.
10. Gross, B.C., Erkal, J.L., Lockwood, S.Y., Chen, C., Spence, D.M. (2014) Evaluation of 3D Printing and Its Potential Impact on Biotechnology and the Chemical Sciences. Anal. Chem. 86: 3240–3253.
11. Tseng, P., Murray, C., Kim, D., Di Carlo, D. (2014) Research Highlights: Printing the Future of Microfabrication. Lab Chip 14: 1491–1495.
12. Shallan, A.I., Smejkal, P., Corban, M., Guijt, R.M., Breadmore, M.C. (2014) Cost Effective 3D-Printing of Visibly Transparent Microchips within Minutes. Anal. Chem. 86: 3124–3130.

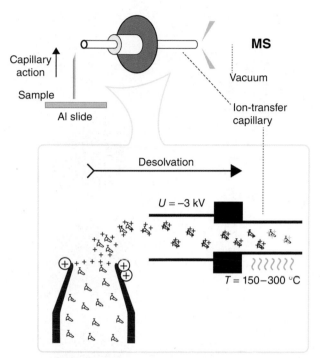

Figure 2.15 *Schematic representation of C-API MS, with sample delivery enabled by capillary action. A short tapered silica capillary [length, 1 cm; base o.d., 363 μm (or 323 μm without polyimide); tip o.d., 10 μm] was positioned vertically above an electrically isolated aluminum slide, with the outlet end placed orthogonal to the inlet of a metal capillary attached to the orifice of an ion trap mass spectrometer. The distance between the outlet of the silica capillary and the inlet of the metal capillary, attached to the MS orifice, was ~ 1 mm. Before the measurements, the silica capillary was filled with a makeup solution [deionized water/acetonitrile (1 : 1, v/v)] by means of capillary action. The inlet end of the silica capillary was then dipped into a droplet of a sample (10 μl) put onto the surface of the aluminum slide. The inset provides an illustration of the hypothetical mechanism of C-API [109]. Reproduced with permission from Hsieh, C.-H., Chang, C.-H., Urban, P.L., Chen, Y.-C. (2011) Capillary Action-supported Contactless Atmospheric Pressure Ionization for the Combined Sampling and Mass Spectrometric Analysis of Biomolecules. Anal. Chem. 83: 2866–2869. Copyright (2011) American Chemical Society*

Time-Resolved Mass Spectrometry: From Concept to Applications, First Edition.
Pawel Lukasz Urban, Yu-Chie Chen and Yi-Sheng Wang.
© 2016 John Wiley & Sons, Ltd. Published 2016 by John Wiley & Sons, Ltd.

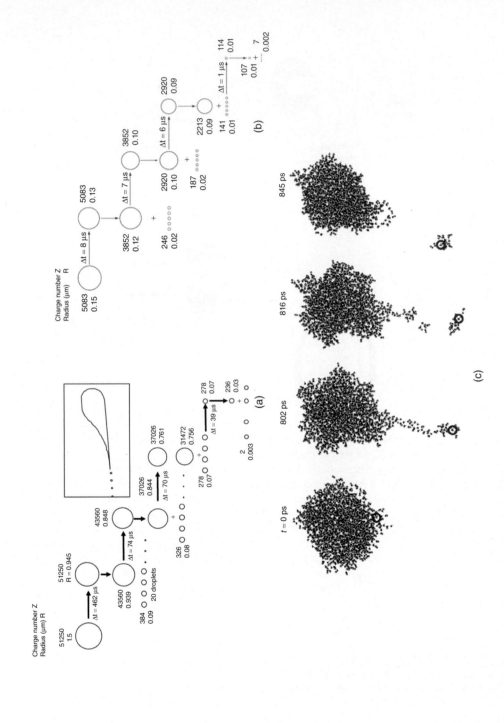

Charge number Z
Radius (µm) R

51250
1.5

51250
R = 0.945

Δt = 462 µs

43560
0.939

Δt = 74 µs

43560
0.848

37026
0.844

Δt = 70 µs

37026
0.761

384
0.09

20 droplets

326
0.08

31472
0.756

278
0.07

Δt = 39 µs

278
0.07

236
0.03

2
0.003

(a)

Charge number Z
Radius (µm) R

5083
0.15

Δt = 8 µs

5083
0.13

Δt = 7 µs

3852
0.10

3852
0.12

2920
0.10

Δt = 6 µs

2920
0.09

246
0.02

187
0.02

2213
0.09

Δt = 1 µs

114
0.01

141
0.01

107
0.01

7
0.002

(b)

t = 0 ps

802 ps

816 ps

845 ps

(c)

Figure 2.8 *Time spent on ion formation in ESI. (a) Schematic representation of time history of parent and offspring droplets. Droplet at the top left is a typical parent droplet created near the ESI capillary tip at low flow rates. Evaporation of solvent at constant charge leads to uneven fission. The number beside the droplets gives radius $R(\mu m)$ and number of elementary charge N on droplet; Δt corresponds to the time required for evaporative droplet shrinkage to the size where fission occurs. Only the first three successive fissions of a parent droplet are shown. At the bottom right, the uneven fission of an offspring droplet to produce offspring droplet is shown. The timescale is based on $R = R_0 - 1.2 \times 10^{-3}t$ (R and R_0, radius of droplet; t, time), which produces only a rough estimate. Inset: Tracing of photograph by Gomez and Tang [36] of droplet undergoing "uneven" fission. Typical droplet loses 2% of its mass, producing some 20 smaller droplets that carry 15% of the parent charge [32]. Reprinted with permission from Kebarle, P., Tang, L. (1993) From Ions in Solution to Ions in the Gas Phase. The Mechanism of Electrospray Mass Spectrometry. Anal. Chem. 64: 972A–986A. Copyright (1993) American Chemical Society. (b) Droplet histories for charged water droplets produced by nanoESI. The first droplet is one of the droplets produced at the spray tip. This parent droplet is followed for three evaporation and fission events. The first generation droplets are shown as well as the fission of one of these to lead to second generation offspring droplets [40]. Reproduced from Peschke, M. et al. (2004) [40] with permission of Springer. Data shown based on experimental results [11] and calculations [40]. (c) In this molecular dynamic simulation, an unfolded protein chain that was initially placed within a Rayleigh-charged water droplet gets ejected via the chain ejection model. Side chains and backbone moieties are represented as beads [42]. Reproduced with permission from Konermann, L., Ahadi, E., Rodriguez, A.D., Vahidi, S. (2013) Unraveling the Mechanism of Electrospray Ionization. Anal. Chem. 85: 2–9. Copyright (2013) American Chemical Society*

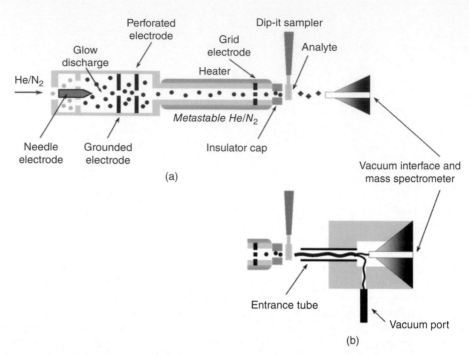

Figure 2.18 (a) Scheme of DART-ion source; and (b) scheme of a gas-ion separator (Vapur interface) equipped with a vacuum pump [124]. Reproduced with permission from Cody, R.B., Laramée, J.A., Durst, H.D. (2005) Versatile New Ion Source for the Analysis of Materials in Open Air under Ambient Conditions. Anal. Chem. 77: 2297–2302. Copyright (2005) American Chemical Society

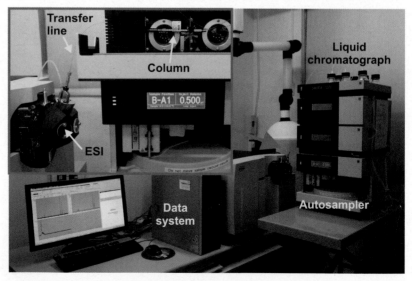

Figure 6.4 A modern liquid chromatograph hyphenated with an ion-trap mass spectrometer using an ESI interface. Courtesy of E.P. Dutkiewicz

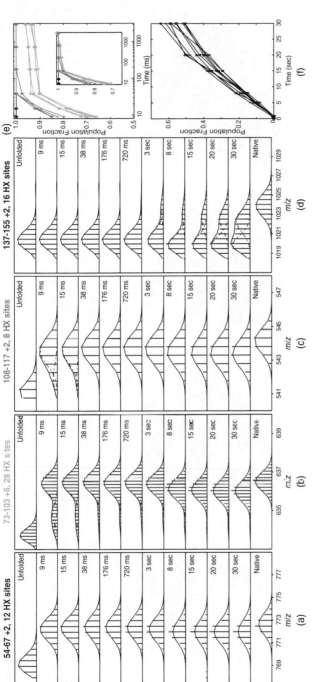

Figure 12.3 Monitoring the folding process by hydrogen/deuterium pulse labeling. (a–d) Illustrative MS spectra versus folding time. Peptides shown cover each helical segment plus some neighboring sequence in the native protein. The top and bottom frames show control experiments in which the unfolded and native proteins were subjected to the same labeling pulse and analysis. Fitted envelopes separate the fractional populations of the unfolded, intermediate, and native state present at the time of the labeling pulse. Deuterons on side chains and the first two residues of each peptide are lost during sample preparation. The subpeaks within each isotopic envelope are caused by the natural abundance of ^{13}C (∼1%) convolved with the carried number of deuterons. A leftward drift in folded peptide mass at long folding times (d) occurs because not-yet-protected sites are exposed to D-to-H exchange during the prepulse folding period (pH 5, 10°C). (e and f) The time dependence for the formation of the protected state of different protein regions. (Inset) The unblocked folding phase of the lower set of curves is renormalized to 100% to allow direct comparison with the folding time of the upper set of curves. For this comparison, the experiment was replicated in triplicate, and only the highest-precision peptides were used. The segments highlighted in (e) fold in detectably different phases [76]. Reproduced from Hu, W., Walters, B.T., Kan, Z.-Y., Mayne, L., Rosen, L.E., Marqusee, S., Englander, S.W. (2013) Stepwise Protein Folding at Near Amino Acid Resolution by Hydrogen Exchange and Mass Spectrometry. Proc. Natl. Acad. Sci. USA 110: 7684–7689 with permission from PNAS

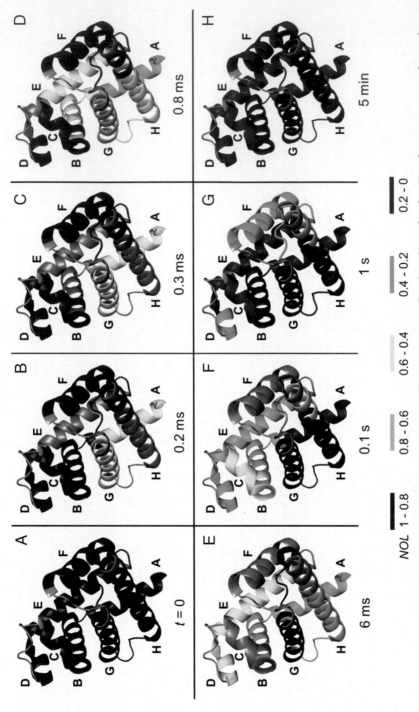

Figure 12.5 Structural changes during aMb folding as measured by FPOP. Normalized oxidation level (NOL) data were mapped onto the crystal structure of native hMb (PDB code 1WLA) [100] using a five-level color code. Regions for which no structural data are available appear in gray. (A–H) Regions within the aMb polypeptide [18]. Adapted with permission from Vahidi, S., Stocks, B.B., Liaghati-Mobarhan, Y., Konermann, L. (2013) Submillisecond Protein Folding Events Monitored by Rapid Mixing and Mass Spectrometry-Based Oxidative Labeling. *Anal. Chem.* 85: 8618–8625. Copyright (2013) American Chemical Society

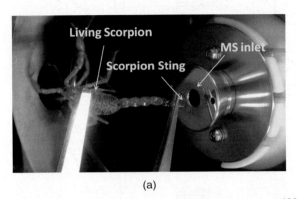

(a)

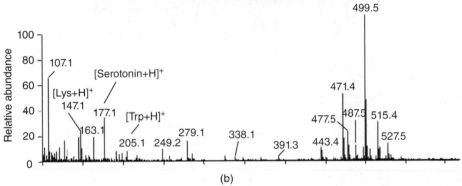

(b)

Figure 13.8 Real-time monitoring of animal metabolites by mass spectrometry. (a) Photograph of in-vivo analysis of a living scorpion by field-induced direct ionization mass spectrometry. (b) Mass spectrum obtained by field-induced direct ionization mass spectrometry analysis of the secretion released from a living scorpion upon stimulation [83]. Reprinted by permission from Macmillan Publishers Ltd: Scientific Reports [Hu, B., Wang, L., Ye, W.-C., Yao, Z.-P. (2013) In Vivo and Real-time Monitoring of Secondary Metabolites of Living Organisms by Mass Spectrometry. Sci. Rep. 3: 2104]. Copyright (2013)

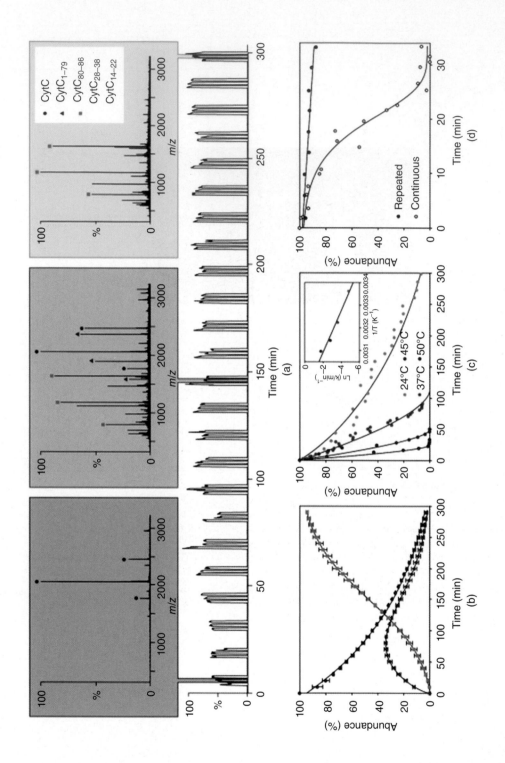

Figure 13.1 *Automated nanoESI monitoring of the tryptic digestion of cytochrome c (CytC). (a) Total-ion current chromatogram of CytC digestion monitored in triplicate over the course of 300 min. Data in the inset are spectra obtained at the beginning, middle, and end of the time course. At the beginning, the predominant species corresponds to full-length CytC (circles), and at the end numerous peptides are observed, the most prominent being $CytC_{80-86}$, $CytC_{28-38}$, and $CytC_{14-22}$ (squares). Halfway through the reaction, an intermediate fragment, $CytC_{1-79}$ (triangles), can also be detected. (b) Plotting the relative abundances of these peptides allows the quantitative monitoring of the digestion reaction. Error bars correspond to three standard deviations from the mean. The amount of CytC decreases exponentially, enabling the determination of first-order rate constants. (c) Monitoring this reaction, specifically the disappearance of intact CytC, at different temperatures (darker markers correspond to higher temperatures while lighter markers correspond to lower temperatures) demonstrates how the reaction velocity increases at higher temperatures. From the Arrhenius plot (inset), the activation energy and pre-exponential factor can be determined. (d) In protein-destabilizing solution conditions, a different reaction profile for the disappearance of intact CytC is determined when the solution is sampled continuously versus repeatedly. This result likely is due to pH effects in the emitter associated with prolonged electrospraying and highlights the benefits of the repeated sampling method advanced here [18]. Reprinted from Chemistry & Biology, 15, Painter, A.J., Jaya, N., Basha, F., Vierling, E., Robinson, C.V., Benesch, L.P., Real-time Monitoring of Protein Complexes Reveals Their Quaternary Organization and Dynamics, 246–253. Copyright (2008), with permission from Elsevier*

13. Salentijn, G.I., Permentier, H.P., Verpoorte, E. (2014) 3D-printed Paper Spray Ionization Cartridge with Fast Wetting and Continuous Solvent Supply Features. Anal. Chem. 86: 11657–11665.
14. Hu, J.-B., Chen, T.-R., Chang, C.-H., Cheng, J.-Y., Chen, Y.-C., Urban, P.L. (2015) A Compact 3D-Printed Interface for Coupling Open Digital Microchips with Venturi Easy Ambient Sonic-Spray Ionization Mass Spectrometry. Analyst 140: 1495–1501.
15. Unger, M.A., Chou, H.-P., Thorsen, T., Scherer, A., Quake, S.R. (2000) Valves and Pumps by Multilayer Soft Lithography. Science 288: 113–116.
16. Brivio, M., Verboom, W., Reinhoudt, D.N. (2006) Miniaturized Continuous Flow Reaction Vessels: Influence on Chemical Reactions. Lab Chip 6: 329–344.
17. Bessoth, F.G., deMello, A.J., Manz, A. (1999) Microstructure for Efficient Continuous Flow Mixing. Anal. Commun. 36: 213–215.
18. Bökenkamp, D., Desai, A., Yang, X., Tai, Y.-C., Marzluff, E.M., Mayo, S.L. (1998) Microfabricated Silicon Mixers for Submillisecond Quench-Flow Analysis. Anal. Chem. 70: 232–236.
19. Elvira, K.S., Casadevall i Solvas, X., Wootton, R.C.R., deMello, A.J. (2013) The Past, Present and Potential for Microfluidic Reactor Technology in Chemical Synthesis. Nature Chem. 5: 905–915.
20. Ratner, D.M., Murphy, E.R., Jhunjhunwala, M., Snyder, D.A., Jensen, K.F., Seeberger. P.H. (2005) Microreactor-Based Reaction Optimization in Organic Chemistry – Glycosylation as a Challenge. Chem. Commun. 578–580.
21. Urban, P.L., Goodall, D.M., Bruce, N.C. (2006) Enzymatic Microreactors in Chemical Analysis and Kinetic Studies. Biotechnol. Adv. 24: 42–57.
22. Urban, P.L., Goodall, D.M., Bergström, E.T., Bruce, N.C. (2006) On-line Low-volume Transesterification-based Assay for Immobilized Lipases. J. Biotechnol. 126: 508–518.
23. Mao, H., Yang, T., Cremer, P.S. (2002) Design and Characterization of Immobilized Enzymes in Microfluidic Systems. Anal. Chem. 74: 379–385.
24. Mao, S., Gao, D., Liu, W., Wei, H., Lin, J.-M. (2012) Imitation of Drug Metabolism in Human Liver and Cytotoxicity Assay Using a Microfluidic Device Coupled to Mass Spectrometric Detection. Lab Chip 12: 219–226.
25. Růžička, J., Hansen, E.H. (1988) Flow Injection Analysis, 2nd Edition. John Wiley & Sons, Inc., New York.
26. The Engineering ToolBox. Hydraulic Diameter, http://www.engineeringtoolbox.com/hydraulic-equivalent-diameter-d_458.html (accessed April 10, 2015).
27. Ohla, S., Belder, D. (2012) Chip-based Separation Devices Coupled to Mass Spectrometry. Curr. Opin. Chem. Biol. 16: 453–459.
28. Sikanen, T., Franssila, S., Kauppila, T.J., Kostiainen, R., Kotiaho, T., Ketola, R.A. (2010) Microchip Technology in Mass Spectrometry. Mass Spectrom. Rev. 29: 351–391.
29. Raterink, R.J., de Korte, M., van der Linden, H., Hankemeier, T. (2010) Chip-based Heaterless Nano-APCI-MS. 14th International Conference on Miniaturized Systems for Chemistry and Life Sciences, October 3–7, 2010, Groningen.
30. Wachs, T., Henion, J. (2001) Electrospray Device for Coupling Microscale Separations and Other Miniaturized Devices with Electrospray Mass Spectrometry. Anal. Chem. 73: 632–638.

31. Lazar, I.M., Ramsey, R.S., Jacobson, S.C., Foote, R.S., Ramsey, J.M. (2000) Novel Microfabricated Device for Electrokinetically Induced Pressure Flow and Electrospray Ionization Mass Spectrometry. J. Chromatogr. A 892: 195–201.
32. Hoffmann, P., Häusig, U., Schulze, P., Belder, D. (2007) Microfluidic Glass Chips with an Integrated Nanospray Emitter for Coupling to a Mass Spectrometer. Angew. Chem. Int. Ed. 46: 4913–4916.
33. Kirby, A.E., Jebrail, M.J., Yang, H., Wheeler, A.R. (2010) Folded Emitters for Nanoelectrospray Ionization Mass Spectrometry. Rapid Commun. Mass Spectrom. 24: 3425–3431.
34. Nissilä, T., Sainiemi, L., Sikanen, T., Kotiaho, T., Franssila, S., Kostiainen, R., Ketola, R.A. (2007) Silicon Micropillar Array Electrospray Chip for Drug and Biomolecule Analysis. Rapid Commun. Mass Spectrom. 21: 3677–3682.
35. Dethy, J.-M., Ackermann, B.L., Delatour, C., Henion, J.D., Schultz, G.A. (2003) Demonstration of Direct Bioanalysis of Drugs in Plasma Using Nanoelectrospray Infusion from a Silicon Chip Coupled with Tandem Mass Spectrometry. Anal. Chem. 75: 805–811.
36. Dayon, L., Abonnenc, M., Prudent, M., Lion, N., Girault, H.H. (2006) Multitrack Electrospray Chips. J. Mass Spectrom. 41: 1484–1490.
37. Mao, P., Wang, H.-T., Yang, P., Wang, D. (2011) Multinozzle Emitter Arrays for Nanoelectrospray Mass Spectrometry. Anal. Chem. 83: 6082–6089.
38. Li, X., Xiao, D., Ou, X.-M., McCullumn, C., Liu, Y.-M. (2013) A Microchip Electrophoresis-Mass Spectrometric Platform for Fast Separation and Identification of Enantiomers Employing the Partial Filling Technique. J. Chromatogr. A 1318: 251–256.
39. Piehl, N., Ludwig, M., Belder, D. (2004) Subsecond Chiral Separations on a Microchip. Electrophoresis 25: 3848–3852.
40. Schwarzkopf, F., Scholl, T., Ohla, S., Belder, D. (2014) Improving Sensitivity in Microchip Electrophoresis Coupled to ESI-MS/MS on the Example of a Cardiac Drug Mixture. Electrophoresis 35: 1880–1886.
41. Mellors, J.S., Jorabchi, K., Smith, L.M., Ramsey, J.M. (2010) Integrated Microfluidic Device for Automated Single Cell Analysis Using Electrophoretic Separation and Electrospray Ionization Mass Spectrometry. Anal. Chem. 82: 967–973.
42. Xie, J., Miao, Y., Shih, J., Tai, Y.-C., Lee, T.D. (2005) Microfluidic Platform for Liquid Chromatography-Tandem Mass Spectrometry Analyses of Complex Peptide Mixtures. Anal. Chem. 77: 6947–6953.
43. Yin, H., Killeen, K., Brennen, R., Sobek, D., Werlich, M., van de Goor, T. (2005) Microfluidic Chip for Peptide Analysis with an Integrated HPLC Column, Sample Enrichment Column, and Nanoelectrospray Tip. Anal. Chem. 77: 527–533.
44. Yin, H., Killeen, K. (2007) The Fundamental Aspects and Applications of Agilent HPLC-Chip. J. Sep. Sci. 30: 1427–1434.
45. Odijk, M., Baumann, A., Lohmann, W., van den Brink, F.T.G., Olthuis, W., Karst, U., van den Berg, A. (2009) A Microfluidic Chip for Electrochemical Conversions in Drug Metabolism Studies. Lab Chip 9: 1687–1693.
46. Benz, C., Boomhoff, M., Appun, J., Schneider, C., Belder, D. (2015) Chip-based Freeflow Electrophoresis with Integrated Nanospray Mass-Spectrometry. Angew. Chem. Int. Ed. 54: 2766–2770.

47. Hendricks, P.I., Dalgleish, J.K., Shelley, J.T., Kirleis, M.A., McNicholas, M.T., Li, L., Chen, T.C., Chen, C.H., Duncan, J.S., Boudreau, F., Noll, R.J., Denton, J.P., Roach, T.A., Ouyang, Z., Cooks, R.G. (2014) Autonomous *In Situ* Analysis and Real-time Chemical Detection Using a Backpack Miniature Mass Spectrometer: Concept, Instrumentation Development, and Performance. Anal. Chem. 86: 2900–2908.

48. Li, L., Chen, T.-C., Ren, Y., Hendricks, P.I., Cooks, R.G., Ouyang, Z. (2014) Mini 12, Miniature Mass Spectrometer for Clinical and Other Applications – Introduction and Characterization. Anal. Chem. 86: 2909–2916.

49. Browne, D.L., Wright, S., Deadman, B.J., Dunnage, S., Baxendale, I.R., Turner, R.M., Ley, S.V. (2012) Continuous Flow Reaction Monitoring Using an On-line Miniature Mass Spectrometer. Rapid Commun. Mass Spectrom. 26: 1999–2010.

50. Brivio, M., Tas, N.R., Goedbloed, M.H., Gardeniers, H.J., Verboom, W., van den Berg, A., Reinhoudt, D.N. (2005) A MALDI-chip Integrated System with a Monitoring Window. Lab Chip 5: 378–381.

51. Brivio, M., Liesener, A., Oosterbroek, R.E., Verboom, W., Karst, U., van den Berg, A., Reinhoudt, D.N. (2005) Chip-based On-line Nanospray MS Method Enabling Study of the Kinetics of Isocyanate Derivatization Reactions. Anal. Chem. 77: 6852–6856.

52. Abonnenc, M., Dayon, L., Perruche, B., Lion, N., Girault, H.H. (2008) Electrospray Micromixer Chip for On-line Derivatization and Kinetic Studies. Anal. Chem. 80: 3372–3378.

53. Fritzsche, S., Ohla, S., Glaser, P., Giera, D.S., Sickert, M., Schneider, C., Belder, D. (2011) Asymmetric Organocatalysis and Analysis on a Single Microfluidic Nanospray Chip. Angew. Chem. Int. Ed. 50: 9467–9470.

54. Lion, N., Rohner, T.C., Dayon, L., Arnaud, I.L., Damoc, E., Youhnovski, N., Wu, Z.-Y., Roussel, C., Josserand, J., Jensen, H., Rossier, J.S., Przybylski, M., Girault, H.H. (2003) Microfluidic Systems in Proteomics. Electrophoresis 24: 3533–3562.

55. Liuni, P., Rob, T., Wilson, D.J. (2010) A Microfluidic Reactor for Rapid, Low-pressure Proteolysis with On-chip Electrospray Ionization. Rapid Commun. Mass Spectrom. 24: 315–320.

56. Rob, T., Wilson, D.J. (2009) A Versatile Microfluidic Chip for Millisecond Time-scale Kinetic Studies by Electrospray Mass Spectrometry. J. Am. Soc. Mass Spectrom. 20: 124–130.

57. Rob, T., Liuni, P., Gill, P.K., Zhu, S., Balachandran, N., Berti, P.J., Wilson, D.J. (2012) Measuring Dynamics in Weakly Structured Regions of Proteins Using Microfluidics-enabled Subsecond H/D Exchange Mass Spectrometry. Anal. Chem. 84: 3771–3779.

58. Benetton, S., Kameoka, J., Tan, A., Wachs, T., Craighead, H., Henion, J.D. (2003) Chip-based P450 Drug Metabolism Coupled to Electrospray Ionization-Mass Spectrometry Detection. Anal. Chem. 75: 6430–6436.

59. Chen, Q., Wu, J., Zhang, Y., Lin, J.-M. (2012) Qualitative and Quantitative Analysis of Tumor Cell Metabolism via Stable Isotope Labeling Assisted Microfluidic Chip Electrospray Ionization Mass Spectrometry. Anal. Chem. 84: 1695–1701.

60. Song, H., Ismagilov, R.F. (2003) Millisecond Kinetics on a Microfluidic Chip Using Nanoliters of Reagents. J. Am. Chem. Soc. 125: 14613–14619.

61. Belder, D. (2005) Microfluidics with Droplets. Angew. Chem. Int. Ed. 44: 3521–3522.

62. Fidalgo, L.M., Whyte, G., Ruotolo, B.T., Benesch, J.L.P., Stengel, F., Abell, C., Robinson, C.V., Huck, W.T.S. (2009) Coupling Microdroplet Microreactors with

Mass Spectrometry: Reading the Contents of Single Droplets Online. Angew. Chem. Int. Ed. 48: 3665–3668.

63. Zhu, Y., Fang, Q. (2010) Integrated Droplet Analysis System with Electrospray Ionization-Mass Spectrometry Using a Hydrophilic Tongue-based Droplet Extraction Interface. Anal. Chem. 82: 8361–8366.

64. Volný, M., Rolfs, J., Hakimi, B., Fryčák, P., Schneider, T., Liu, D., Yen, G., Chiu, D. T., Tureček, F. (2014) Nanoliter Segmented-flow Sampling Mass Spectrometry with Online Compartmentalization. Anal. Chem. 86: 3647–3652.

65. Gasilova, N., Yu, Q., Qiao, L., Girault, H.H. (2014) On-chip Spyhole Mass Spectrometry for Droplet-based Microfluidics. Angew. Chem. Int. Ed. 53: 4408–4412.

66. Wheeler, A. (2008) Putting Electrowetting to Work. Science 322: 539–540.

67. Baker, C.A., Roper, M.G. (2012) Online Coupling of Digital Microfluidic Devices with Mass Spectrometry Detection Using an Eductor with Electrospray Ionization. Anal. Chem. 84: 2955–2960.

68. Jebrail, M.J., Sinha, A., Vellucci, S., Renzi, R.F., Ambriz, C., Gondhalekar, C., Schoeniger, J.S., Patel, K.D., Branda, S.S. (2014) World-to-Digital-Microfluidic Interface Enabling Extraction and Purification of RNA from Human Whole Blood. Anal. Chem. 86: 3856–3862.

69. Kirby, A.E., Wheeler, A.R. (2013) Digital Microfluidics: An Emerging Sample Preparation Platform for Mass Spectrometry. Anal. Chem. 85: 6178–6184.

70. Jebrail, M.J., Yang, H., Mudrik, J.M., Lafrenière, N.M., McRoberts, C., Al-Dirbashi, O.Y., Fisher, L., Chakraborty, P., Wheeler, A.R. (2011) A Digital Microfluidic Method for Dried Blood Spot Analysis. Lab Chip 11: 3218–3224.

71. Shih, S.C.C., Yang, H., Jebrail, M.J., Fobel, R., McIntosh, N., Al-Dirbashi, O.Y., Chakraborty, P., Wheeler, A.R. (2012) Dried Blood Spot Analysis by Digital Microfluidics Coupled to Nanoelectrospray Ionization Mass Spectrometry. Anal. Chem. 84: 3731–3738.

72. Nichols, K.P., Gardeniers, H.J.G.E. (2007) A Digital Microfluidic System for the Investigation of Pre-steady-state Enzyme Kinetics Using Rapid Quenching with MALDI-TOF Mass Spectrometry. Anal. Chem. 79: 8699–8704.

73. Nichols, K.P., Azoz, S., Gardeniers, H.J.G.E. (2008) Enzyme Kinetics by Directly Imaging a Porous Silicon Microfluidic Reactor Using Desorption/Ionization on Silicon Mass Spectrometry. Anal. Chem. 80: 8314–8319.

74. Kirby, A.E., Wheeler, A.R. (2013) Microfluidic Origami: a New Device Format for In-line Reaction Monitoring by Nanoelectrospray Ionization Mass Spectrometry. Lab Chip 13: 2533–2540.

75. Luk, V.N., Fiddes, L.K., Luk, V.M., Kumacheva, E., Wheeler, A.R. (2012) Digital Microfluidic Hydrogel Microreactors for Proteomics. Proteomics 12: 1310–1318.

76. Srigunapalan, S., Eydelnant, I.A., Simmons, C.A., Wheeler, A.R. (2012) A Digital Microfluidic Platform for Primary Cell Culture and Analysis. Lab Chip 12: 369–375.

77. Urban P.L. (2015) Universal Electronics for Miniature and Automated Chemical Assays. Analyst 140: 963–975.

78. Goodall, D.M., Urban, P.L. (2007) Lab-on-capillary: A Versatile Format for Nanolitre Scale Chemistry and Biochemistry. International Labmate 02.

8

Quantitative Measurements by Mass Spectrometry

8.1 The Challenge of Quantitative Mass Spectrometry Measurements

In a large portion of routine and discovery-oriented analyses, mass spectrometry (MS) is used as a qualitative technique. The obtained qualitative data enable detection and structural elucidation of molecules present in the analyzed samples. However, modern chemistry and biochemistry heavily rely on quantitative information. In biochemistry it is often sufficient to conduct quantification of analytes in biofluids every few hours, days, or even weeks. In the real-time monitoring of highly dynamic samples, it is necessary to collect data points at higher frequencies. When it comes to selection of techniques for quantitative analyses, especially in the monitoring of dynamic samples, MS has not generally been favored. In fact, the performance of MS in quantitative analysis is worse than that of optical spectroscopies – especially, ultraviolet–visible (UV-Vis) absorption and fluorescence spectroscopy.

When conducting real-time MS monitoring of chemical or physical processes, obtaining quantitative data is particularly challenging. In principle, abundances of recorded ions should change as the time goes by. It is a cumbersome task to verify if the changes in ion abundances are due to the outcome of the chemical reaction to be monitored, not the instability of instrument/sample. In some cases, obtaining relative quantity data (values in percent or arbitrary units) rather than absolute quantity data (e.g., in moles) may be possible and sufficient to verify hypotheses.

The limitations of MS in the acquisition of quantitative data (absolute or relative) are due to several factors which can be categorized into two groups encompassing those related to the instrument (I) and the sample (II). The contributions of different factors are outlined below.

Time-Resolved Mass Spectrometry: From Concept to Applications, First Edition.
Pawel Lukasz Urban, Yu-Chie Chen and Yi-Sheng Wang.
© 2016 John Wiley & Sons, Ltd. Published 2016 by John Wiley & Sons, Ltd.

8.1.1   (I) Instrument

8.1.1.1   Setting Ion Source Parameters

Most ion sources offer the possibility to adjust important operation parameters. In the case of electrospray ionization (ESI), the parameters include: sample flow rate, pressure of nebulizing gas, geometrical features of the sprayer (e.g., protrusion of the sample needle with respect to the nebulizing gas tubing, angle of the sample needle relative to the ion guide axis), and the applied voltage. Altering these parameters can affect ionization efficiency, and stability of electrospray. Important parameters of matrix-assisted laser desorption/ionization (MALDI) encompass laser wavelength, relative laser fluence, laser focus, laser frequency, and number of laser shots. It is necessary to optimize these settings in the course of method development, especially when quantitative analysis is conducted.

8.1.1.2   Transmission Efficiency of Ions

Analyte ions need to be transmitted with satisfactory efficiency so that they can arrive in the mass analyzer, and later in the detector. Small ionic species may likely diffuse right after ionization, while very large ions (formed from macromolecules) may not readily respond to electric field of ion guides and lenses. Geometry of ion guides next to the ion source may affect the efficiency of ion capture.

8.1.1.3   Detection Efficiency

Ion detectors used in most mass spectrometers exhibit unequal response toward ions with different mass-to-charge (m/z) ratio. For example, time-of-flight (TOF) instruments can handle separation of large biomolecules but their built-in microchannel plate (MCP) detectors are not adequate for detection of particularly large ions formed by macromolecules (a few hundred kilodaltons). To solve this problem, special *high-mass detectors* are occasionally fitted into TOF tubes to enable reliable detection of high m/z ions. When the abundances of ions are very large, the MS detector may be saturated leading to a non-linear response to the ion quantity. Interestingly, less abundant natural isotopologue signals can be used in liquid chromatography (LC)-MS to extend the linear dynamic range, decreasing the probability of detector saturation [1].

8.1.1.4   Contamination of Mass Spectrometry Parts

Mass spectrometers are sensitive instruments (see also Section 5.3.3), what inherently makes them vulnerable to even minor contamination caused by samples and the environment of operation. The ion source compartment of every mass spectrometer is vulnerable to contamination. In the ion source, samples can be vaporized, and the analytes ionized and transferred toward the mass analyzer. Efficiency of these processes is limited. Thus, many non-ionic and ionic species are deposited on surfaces in the ion source compartment. Contamination of the ion transfer line adjacent to the ion source can affect quantitative capabilities of the instrument in various ways. Contaminant deposits on electrodes and insulator parts of the ion duct influence the intensity and distribution of the electric field, thus affecting analytical sensitivity and resolution of the mass spectrometer. They contribute to the chemical background (noise) in the recorded mass spectra. Thus, they elevate the lower limit of the dynamic range in MS analyses. High chemical background prevents detection and quantification of low-abundance analytes.

Whenever quantitative analyses are conducted, it is of paramount importance to assure adequate maintenance of the instrument and conduct quality control tests by analyzing standard samples periodically to ensure high reproducibility of quantitative analyses. The inner parts of a mass spectrometer (ion guides, mass analyzer, detector) are more difficult to clean. Fortunately, they get contaminated slower because the front parts of the sample/ion path conduct a rudimentary pre-screening of contaminants (e.g., neutral species, solid or liquid particles).

8.1.2 (II) Sample

8.1.2.1 Ionization Efficiencies of Analytes

Ionization efficiencies of various analytes may be different. When using internal standards, it is necessary to assure that the ionization efficiency of the internal standard is similar to that of the analyte. This can be achieved by implementing isotopically labeled standards. However, such chemicals are expensive and not always available.

8.1.2.2 Concentrations of Analytes

Reliable quantitative analysis can be conducted for analytes with concentrations falling within a certain range. Low abundance analytes may not be detected, or their signals would be too low to provide repeatable data for quantitative analysis (if they are below the limit-of-quantification threshold). On the other hand, high abundance analytes may not be quantified reliably because of saturation of the detection system. Therefore, sample dilution is often required. This kind of basic sample preparation (by dilution) can readily be implemented using valves and pumps. However, it may also decrease the temporal resolutions of the real-time monitoring routines.

8.1.2.3 Degradation of Analytes

Some analytes are labile molecules, and not all of them can "survive the journey" from the sample vial to the detector. Decomposition of analyte species can already occur in the sample vial, on the way to the ion source, within the ion source (*in-source decay*), and in the ion guides or mass analyzer (*post-source decay*). All these processes can affect quantification of chemical species involved in the studied dynamic processes. For example, the concentration of adenosine triphosphate (ATP) in aqueous solution decreases due to hydrolysis of the phosphate group. If the sample is contaminated by ATPase enzyme, the decomposition rate can be high. Moreover, ATP gets fragmented in the course of ionization (e.g., MALDI, ESI) forming ions with the same structures as those generated by adenosine diphosphate (ADP) or adenosine monophosphate (AMP). Thus, quantification of ATP, ADP, and AMP, by MS, is particularly challenging. The extent of the in-source decay may vary in the course of analyses. Therefore, selection of one of the peaks related to the analyte may not warrant repeatable quantification.

8.1.2.4   Sample Preparation

Sample preparation can affect analytical sensitivity/detectivity but also quantitative aspects of analysis. In the case of ESI, one needs to select the sample solvent, and make sure that the analytes are within the operating concentration range of the method. While MALDI-MS is generally considered to be a qualitative analysis technique, some studies suggest the possibility of using MALDI in quantitative analyses (e.g., [2]). To this end, it is critical to assure that the matrix/sample deposits are uniform and populated with microcrystals of similar size. Heterogeneous crystallization of matrix favors formation of the so-called *sweet spots* (regions within matrix/sample which produce high analyte signals). In the case of complex samples, simple dilution may not be sufficient sample preparation before quantitative analysis. Redundant sample components, and possible interferents, may need to be removed in the course of off-line or on-line sample treatment steps.

8.1.2.5   Limited Signal Stability and Repeatability

The unsatisfactory stability of ion currents observed in many MS methods can be related to the operation of ion sources and analyzers, which can be affected by sample and environment-related factors. In the off-line techniques using laser beams such as MALDI-MS, the population of ions produced in an instant is strongly related to the quality of the sample/matrix deposit. In fact, heterogeneous sample and matrix distribution, and the so-called *sweet-spot effect*, are the major impediments of the MALDI-MS methodology. The stability of signals in on-line techniques such as ESI-MS is related to the actual design of the spray emitter and the ion conduits. However, possible long-term signal variations can still affect analytical repeatability and reproducibility. They make it difficult to build the calibration plots that could stand for a long time, and be used in quantitative analyses of multiple samples. Normalization of analyte signals can help to compensate for the signal variability, and make the off-line and on-line MS data records more useful in quantitative analysis.

8.1.2.6   Ionization Interferences

Some MS methods possess a linear dynamic range of just one order of magnitude, which puts them far behind many spectroscopic techniques for which linear calibration plots extending over >3 orders of magnitude can be constructed. The non-linear relations between the recorded ion currents and concentrations of analytes can be due to imperfections of sample preparation protocols [laser desorption/ionization (LDI) techniques] as well as non-linear characteristics of ionization, ion analysis, and detection. Ionization of different compounds supplied to the ion source is affected by other components of the sample matrix. Interactions of ions also exist in the mass analyzers. Moreover, detectors in MS can get saturated when a large number of ions arrive at the same time. However, interferences are especially significant during ionization. If the sample contains a large amount of inorganic salt, the salt's liquid-phase ions may "compete" for ionization in the gas phase, affecting ionization of the larger, and less readily ionized, species of interest. When complex samples are analyzed, the effect of interfering species may be significant. In some cases, the recorded ion currents can be more dependent on the sample matrix components than on the concentration of the target analyte. Such a situation can occur when one intends to look at minor changes of the target analyte while the sample matrix is highly variable. Examples include analysis of biological (e.g., urine) and environmental samples (e.g., waste water). Special precautions need to be taken in these cases in order to avoid misinterpretation of the analytical data. Ionization interferences are addressed by separating complex samples prior to ionization using chromatographic or electrophoretic techniques (see Chapter 6).

8.1.2.7   Detection Interferences

Interferences between the analyzed ions can also occur at the detector. In the case of TOF analyzers, the low m/z ions arrive at the detector before the higher m/z ions. A large number of low m/z ions can saturate the detector rendering it less sensitive for the ions with higher m/z, which arrive afterwards. In some commercial instruments, there exist hardware and software patches which allow the influence of the low m/z ions on detection of higher m/z ions to be limited. In addition to the saturation of the detector with low mass ions, detection efficiency is mass-dependent even when the detector is not saturated.

8.1.2.8   Spectral Interferences

When using mass spectrometers with lower resolutions [e.g., quadrupole (Q), ion trap (IT)] it is particularly challenging to quantify species generating ions with very similar m/z values (< 5). Normally quantification is conducted based on the highest peak within the isotopic distribution feature in the mass spectrum. However, lower intensity isotopologues can affect the intensity and area of nearby signals produced by different analytes. In this case, post-analysis deconvolution of signals may be conducted. A better solution would be to carry out pre-MS separation of different analyte species. This can be done using chromatography, electrophoresis, or ion mobility separation. Considering the typical separation times in chromatography (minutes), this would decrease the temporal resolution of the MS monitoring. However, ion mobility separations are fast enough to accommodate uninterrupted analysis of dynamic samples in millisecond-range time intervals (see Chapter 6).

8.2   Selection of Instrument

The available MS instruments and analysis approaches have different capabilities in terms of quantitative analysis. Table 8.1 grades the quantitative capabilities of various ion sources and analyzers taking into account the specific requirements of the real-time MS monitoring. In general, continuous operation mass analyzers (S, Q) are expected to possess superior quantitative capabilities when compared with batch mass analyzers (IT, ICR, TOF). Coupling separation techniques to MS enhances quantitative capabilities of the technique by reducing ion suppression, contamination effects, and minimizing the effect of sample matrix variability.

8.3   Solutions to Quantitative Mass Spectrometry

8.3.1   Quantification with Separation

The most widely recognized approach to quantitative analysis by MS is performing separation of analytes and sample matrix components prior to their introduction to the ion

Table 8.1 *Perceived suitability of selected MS components for quantitative and real-time MS measurements*

Instrument	Suitability for quantitative MS		Suitability for real-time MS
	Without separation	With separation	
Ion source			
EI	++	+++++	++++
CI	++	+++++	++++
ESI	++	+++++	++++
V-EASI	+++	n.d.	+++++
EESI	+++	n.d.	+++++
MALDI	++	+++	+
SALDI	++	n.d.	+
Analyzer			
S	+++++		+++++
Q	+++++		+++++
QqQ	+++++		++++
IT	++		++
TOF	+++		+++
Q-TOF	+++		+++
ICR	++		++
OIT	++		++

CI, chemical ionization; EESI, electrosonic spray ionization; EI, electron ionization; ICR, ion cyclotron resonance; OIT, orbital ion trap; QqQ, triple quadrupole; S, sector; SALDI, surface-assisted laser desorption/ionization; V-EASI, Venturi easy ambient sonic-spray ionization. n.d., No data.

source (see Chapter 6). Complex samples are normally separated by gas chromatography (GC), LC, or capillary electrophoresis (CE), which can readily be coupled on-line with MS. Separation of analytes from other sample components decreases or even eliminates ionization interference. The increased analytical sensitivity allows small amounts of relatively dilute samples to be analyzed, which furthermore mitigates contamination to MS systems. In the case of separation techniques coupled on-line with MS, quantitative results are normally obtained by measuring the area under the temporal peaks within total ion currents or extracted ion currents – representing integrated spectral intensities over a wide or a narrow m/z range, respectively (see also Chapter 9).

It is practical to carry out analysis on on-line systems with two detectors. Since quantitative capabilities of UV-Vis absorption spectroscopy are typically greater than those of MS, in some protocols, UV absorption detectors are used to carry out quantitative analysis (after separation on a chromatographic column) while MS data are used for qualitative analysis (peak assignment) [3]. In one study, the effluent from a high-performance liquid chromatography (HPLC) instrument was transferred to the UV detector. After UV detection, the flow was split, and part of it was transferred to the ESI-MS system [4]. Using LC separations coupled with UV and MS is also useful in the characterization of products of electrochemical reactions, and considered complementary to on-line monitoring [5, 6].

One inconvenience in coupling UV-Vis absorption detectors on-line with MS is that these devices are quite bulky – it is difficult to mount them in front of the ion sources (e.g., ESI), and at the same time keep the transfer line short <20 cm. A solution to this problem would be the use of miniature fiber optic detectors which are nowadays available commercially. It is straightforward to install sensor heads of such detectors on the flow line between the separation column and the ion source while maintaining the length of transfer line between the two parts as short as possible. When using microchips for separations, fluorescence detection is the preferred mode because it exhibits superior analytical sensitivity and quantitative capabilities. Identification of unknown molecules present in real samples can readily be done by coupling the microchip with MS [7]. Electrochemical detection is also orthogonal to MS, and it can support metabolomic experiments providing quantitative results [8].

Obtaining absolute quantity or concentration of an analyte typically requires construction of a calibration plot based on analyses of standard samples with different concentrations. When the number of analyzed samples is small, or when the separation is not perfect, and it does not completely eliminate interference, it is advised to carry out quantification using multiple-standard addition. In the case of the standard addition method, the sample is analyzed first, followed by analyses of the same sample supplemented with the chemical standard at different concentrations. If the amounts of added standards are chosen correctly, one can estimate the amount of target analyte initially present in the sample. When carrying out quantification by multiple standard addition, one still has to make sure that the peaks of the supplemented samples are not overloaded. Verification of this point may require conducting additional preliminary tests.

Protein quantification is an important but difficult task. It can be done using several methods, including two-dimensional gel electrophoresis [9]. Simple LC-MS methods can already allow relative quantification of proteins in complex mixtures [9]. Using LC-MS one can obtain precision that is acceptable for various applications variations <20%. A few instrumentation-related MS methods for the collection of quantitative data have been popularized recently, including product ion monitoring, neutral loss scanning, inclusion list scanning, immonium ion detection and multiple reaction monitoring (MRM) (cf. [10]). Typically, the MRM approach takes advantage of fragmentations occurring in the collision-induced disassociation cell of a tandem mass spectrometer [10] (see also Section 3.3.1.1). It provides an increased productivity when compared with its predecessor – selected reaction monitoring (SRM). Robust quantification of peptides can be accomplished, for example, by implementing the isotope-dilution method [11]. (See the review by Yocum and Chinnaiyan [10] and the references cited therein.)

There exist several general strategies for quantification of proteins by MS that cover not just the analysis step but the whole experimental procedures, including treatment of live biological material and sample preparation. They include metabolic stable-isotope labeling, isotope tagging by chemical reaction, and stable-isotope incorporation via enzyme reaction (Figure 8.1; [12]). The so-called *stable isotope labeling by amino acids in cell culture* (SILAC) is based on the supplementation of cell culture media with amino acids containing stable isotopes which are heavier than the naturally occurring ones [13, 14]. In typical experiments, two cell populations are grown on similar culture media that only differ by the presence of heavy isotope forms in some of the components. SILAC is an example of metabolic labeling; so it requires special treatment of live specimens (supply

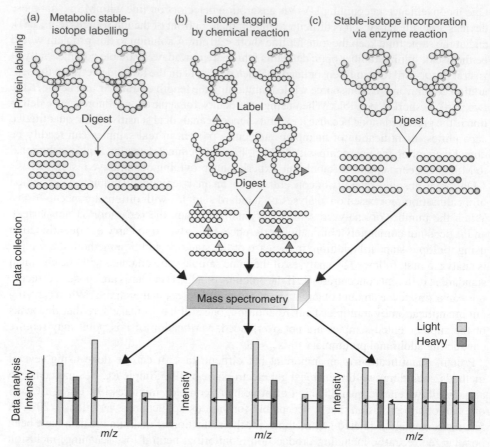

Figure 8.1 *Schematic representation of methods for stable-isotope protein labeling for quantitative proteomics. (a) Proteins are labeled metabolically by culturing cells in media that are isotopically enriched (e.g., containing ^{15}N salts, or ^{13}C-labelled amino acids) or isotopically depleted. (b) Proteins are labeled at specific sites with isotopically encoded reagents. The reagents can also contain affinity tags, allowing for the selective isolation of the labeled peptides after protein digestion. The use of chemistries of different specificity enables selective tagging of classes of proteins containing specific functional groups. (c) Proteins are isotopically tagged by means of enzyme-catalyzed incorporation of ^{18}O from ^{18}O water during proteolysis. Each peptide generated by the enzymatic reaction carried out in heavy water is labeled at the carboxy terminal. In each case, labeled proteins or peptides are combined, separated and analyzed by MS and/or tandem MS for the purpose of identifying the proteins contained in the sample and determining their relative abundance. The patterns of isotopic mass differences generated by each method are indicated schematically. The mass difference of peptide pairs generated by metabolic labeling is dependent on the amino acid composition of the peptide and is therefore variable. The mass difference generated by enzymatic ^{18}O incorporation is either 4 or 2 Da, making quantitation difficult. The mass difference generated by chemical tagging is one or multiple times the mass difference encoded in the reagent used [12]. Reprinted by permission from Macmillan Publishers Ltd: Nature [Aebersold, R., Mann, M. (2003) Mass Spectrometry-based Proteomics. Nature 422: 198–207], copyright (2003)*

of heavy isotopes to the medium). Other methods focus on the sample treatment *ex vivo*. It is known that quantitative features, including intensity, timing, and duration of phosphorylation might determine cellular response [15]. Thus, in one work, it was necessary to quantify tyrosine phosphorylation of specific residues in a time-resolved manner [15]. To achieve that goal, tryptic peptides were labeled with four isoforms of the so-called iTRAQ reagent [16]. Further sample preparation involved immunoprecipitation and enrichment on immobilized metal ion affinity chromatography (IMAC) column for LC-MS analysis [15]. The data generated in this way could facilitate modeling of receptor tyrosine kinase-initiated signal transduction, trafficking, and regulation. In another representative study, Verano-Braga *et al.* [17] implemented proteomic methodology to reveal the role of angiotensin-(1-7) signaling in human endothelial cells. The protocol involved protein digestion, iTRAQ labeling, phosphopeptide enrichment, and LC-MS/MS analysis. The temporal resolution in this study was in the order of a few minutes. Also metabolomics research heavily relies on LC-MS and isotopic labeling for kinetic flux profiling [18] and absolute quantification of intracellular metabolites [19] among others.

If separation and MS conditions are optimized properly, using LC-MS one can obtain quantitative data with satisfactory quality. Such hyphenated systems take advantage of the temporal characteristics of MS, which records data points at a high speed plotting baselines and chromatographic/electrophoretic peaks. However, this kind of analyses do not enable monitoring dynamic chemical processes with high temporal resolution. Aliquots are normally obtained from dynamic matrices, and subsequently injected to the separation columns. Depending on the sampling/separation scheme, the temporal resolution of such analyses can be limited by the duration of sampling or the duration of the separation routine. For example, in one approach, an automated sampler device incorporating two pinch valves was used to transfer aliquots of a dynamic sample to the injection port of a GC-MS system (Figure 8.2; [20]). The system enabled quantitative monitoring of the progress of transesterification catalyzed by immobilized lipase as well as extraction of metabolites from plant tissue samples. The monitoring extended over 2 h while the sampling interval was about 20 min, which resulted from the combined durations of sampling and chromatographic separation. An internal standard was used to compensate for technical variability. The measured values were fed to the calibration functions to obtain absolute quantities of products at different time points, and calculate reaction rates. Similar accessories – incorporating injection valves – for recurrent sample injections are available commercially. Chromatographic separation is the limiting factor for the sampling rate. A possible improvement would be coupling such interfaces with fast gas chromatographic separations.

The above example shows an application of GC-MS in the monitoring of a dynamic process with a relatively low temporal resolution. Other processes studied by MS are much faster than the transesterification reaction, and it would be inadequate to implement lengthy chromatographic separation to obtain every data point in the time plot. In order to minimize the dead time between sampling and analysis, it is appealing to integrate the studied dynamic processes with separations by means of automated and miniature systems. In one study, products of enantioselective Mannich reaction catalyzed by phosphoric acid were separated by on-chip electrophoresis with MS detection [21]. While fast chromatographic and electrophoretic separations can be carried out using short columns and microchips, they are not in common use now.

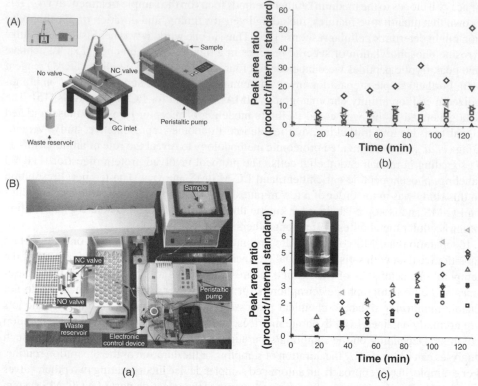

Figure 8.2 *Pinch-valve interface for automated sampling and monitoring of dynamic processes by GC-MS. (a) System for online sampling and sample introduction to GC-MS using pinch valves: (A) device layout; (B) view of the assembled device. NO, normally open valve; NC, normally closed valve. Note, the commercial autosampler (left-hand side) was not used in this study. The lid of the mini-thermoshaker (right-hand side) was closed during the experiment. (b) Synthesis of 1-butyl acetate catalyzed by a small number of macroporous resin microbeads with immobilized lipase (from Candida antarctica; expressed in Aspergillus oryzae) monitored by this setup. Size range of microbeads: ~400–600 μm. Markers: (△) 0 microbeads; (○) 1 microbead; (□) 2 microbeads; (▷) 3 microbeads; (▽) 4 microbeads; (◁) 5 microbeads; (◇) 10 microbeads. (c) Single-microbead biocatalysis. Each series corresponds to one lipase microbead [20]. Adapted from Ting, H., Hu, J.-B., Hsieh, K.-T., Urban, P.L. (2014) A Pinch-Valve Interface for Automated Sampling and Monitoring of Dynamic Processes by Gas Chromatography–Mass Spectrometry. Anal. Meth. 6: 4652–4660 with permission of The Royal Society of Chemistry*

8.3.2   Quantification without Separation

To attain high sampling rates, a compromise may need to be made leading to poorer quantitative capability of the MS method. For instance, one can carry out monitoring of a dynamic sample directly, without separation – by sampling molecules from the reaction chamber, and delivering them to the ionization area via an on-line interface. In fact, many such approaches have been listed in Table 4.1. If the sample matrix is constant, and the analytes change within the dynamic range of the MS method, it may be justified to use the data from such direct monitoring as quantitative. However, the above pre-conditions may not always be fulfilled: for example, when monitoring the extraction or dissolution process, where sample matrix changes significantly and may influence ionization of the analyte molecules. In the case of on-line monitoring of organic reactions, the stability of the matrix can be taken for granted in many cases but not always. Therefore, preliminary tests need to be carried out whenever quantitative information is to be extracted from the MS monitoring data. It is also a good practice to carry out comparative measurements using complementary detection techniques with quantitative capabilities (e.g., UV-Vis absorption, nuclear magnetic resonance [22]).

Internal standards can be used to mitigate the instabilities of ionization during long-term process monitoring. However, different molecules have different ionization efficiencies. Matrix effects may be manifested in different ways for different analytes. Therefore, it is recommended to use isotopically labeled standards. Isotopic effects during ionization are negligible; therefore, isotopologues are good standards for correcting ionization instabilities and removing the effects of sample matrix interference. One needs to ensure that the abundance of the isotopologue standard and the analyte are comparable, and none of them overloads the detector. A major challenge is that isotopologues of the studied analytes are not always available. Chemical vendors offer a small number of isotopically labeled standards. In other cases, analysts may need to synthesize the compounds for themselves, which is far from ideal – considering the complexity of the synthesis and the purity requirements. Internal standards may not always be compatible with the studied processes. For example, they may inhibit enzymes or react with other components of the reaction mixture. In some cases, it may be possible to introduce an internal standard to the interface (not directly to the sample, e.g., reaction mixture) which can mitigate that risk. One valuable approach to quantitative MS analysis (with or without separation) is the use of a dual ESI emitter to introduce the real sample and standard sample simultaneously [23]. This approach can enable mass and abundance calibration continuously in real time. It may – to some extent – compensate for system instabilities in long-term experiments.

While LDI-MS cannot be considered to be the prime technique for time-resolved and quantitative measurements, careful sample preparation and choice of appropriate matrix which can provide homogeneous sample deposits enables calibration plots to be obtained with sufficient quality to record kinetic data with absolute quantities/concentrations of reactants (see, e.g., [24]).

8.4   Data Treatment

Mass spectrometers record spectra which relate relative ion abundances to m/z ratios. Data obtained after long-term monitoring of dynamic samples often require post-run processing to extract quantitative information. This kind of processing can be conducted using the software provided by the instrument manufacturer, customized third-party software, or generic third-party software packages. Customized software programs are particularly useful when handling data obtained by hyphenation of a mass spectrometer to a separation device (chromatograph, electropherograph, ion mobility spectrometer). However, they do not normally include many features required for processing kinetic data obtained in real-time monitoring. Therefore, generic software packages are often used. Popular programs include Excel

(Microsoft), Origin (OriginLab), and SPSS (IBM). For example, non-linear fitting of experimental data points with custom equations can be conducted using SPSS. Chromatographic and electrophoretic peaks can readily be fitted with symmetrical and asymmetrical distribution functions using PeakFit (Systat Software). The fitted functions can be integrated seamlessly to obtain peak areas, which are further used in absolute quantification. This kind of peak area estimation is particularly applicable when handling a few pieces of data of lower quality (e.g., peaks with irregular edges, noisy baseline). A more common method of peak area determination (implemented in many commercial programs) involves computing the absolute area under the peak outline. This method is easy to use but attention needs to be paid to the proper definition of the baseline: for example, a conscious decision has to be made as to whether or not the points between the baseline and horizontal axis should be included in the peak area. The Matlab software (MathWorks) is particularly suitable for processing large multi-dimensional tabulated datasets. It uses an easy-to-learn high level programming language. Scripts can be constructed and used for automated unsupervised processing of large numbers of datasets. Analysts with a greater inclination for computer science can consider using other high-level languages such as Python, C++, or C#, which provide high flexibility. Compilers are available for the most popular operating systems (Windows, Linux, and Mac OS). Less experienced but eager to learn mass spectrometrists will find numerous tutorials on programming in these languages on the internet. No matter how sophisticated the data processing procedure is, inspection of intermediate data product at every step is crucial in order to avoid systematic, random, or even gross errors.

8.5   Concluding Remarks

For a long time MS has been perceived as a qualitative analytical technique. Its selectivity made it a first-choice platform for identification of unknown molecules. Hyphenation of MS with separation techniques encouraged wide-spread use of MS in quantitative analyses. LC-MS and GC-MS are the leading hyphenated systems for quantitative measurements. However, low throughput of these combined systems, and poor compatibility of chromatography with some ionization techniques (e.g., LDI) triggered research that led to the invention of simple mass spectrometric approaches providing analytical results at higher speeds. In some cases it is possible to circumvent separation while performing quantitative MS analyses. Such methods need to be characterized thoroughly before use in order to ascertain that they are free of artifacts. The benefit of such additional effort is the possibility to perform quantitative real-time measurements circumventing the delay caused by time-consuming sample pre-treatment/separation. While some of the biochemical measurements (e.g., those related to enzyme kinetics) may require absolute quantities of analytes to be obtained, in many cases, relative quantities are sufficient to gain valuable insights on the studied dynamic processes.

References

1. Liu, H., Lam, L., Yan, L., Chi, B., Dasgupta, P.K. (2014) Expanding the Linear Dynamic Range for Quantitative Liquid Chromatography–High Resolution Mass Spectrometry Utilizing Natural Isotopologue Signals. Anal. Chim. Acta 850: 65–70.

2. Vaidyanathan, S., Goodacre, R. (2007) Quantitative Detection of Metabolites Using Matrix-assisted Laser Desorption/Ionization Mass Spectrometry with 9-Aminoacridine as the Matrix. Rapid Commun. Mass Spectrom. 21: 2072–2078.

3. Avula, B., Wang, Y.-H., Wang, M., Avonto, C., Zhao, J., Smillie, T.J., Rua, D., Khan, I.A. (2014) Quantitative Determination of Phenolic Compounds by UHPLC-UV-MS and Use of Partial Least-square Discriminant Analysis to Differentiate Chemo-types of Chamomile/*Chrysanthemum* Flower Heads. J. Pharm. Biomed. Anal. 88: 278–288.

4. Zhao, X., Wang, Y., Sun, Y. (2007) Quantitative and Qualitative Determination of Liuwei Dihuang Tablets by HPLC-UV-MS-MS. J. Chromatogr. Sci. 45: 549–552.

5. Karst, U. (2004) Electrochemistry/Mass Spectrometry (EC/MS) – A New Tool to Study Drug Metabolism and Reaction Mechanisms. Angew. Chem. Int. Ed. 43: 2476–2478.

6. Jahn, S., Lohmann, W., Bomke, S., Baumann, A., Karst, U. (2012) A Ferrocene-based Reagent for the Conjugation and Quantification of Reactive Metabolites. Anal. Bioanal. Chem. 402: 461–471.

7. Ohla, S., Schulze, P., Fritzsche, S., Belder, D. (2011) Chip Electrophoresis of Active Banana Ingredients with Label-free Detection Utilizing Deep UV Native Fluorescence and Mass Spectrometry. Anal. Bioanal. Chem. 399: 1853–1857.

8. Gamache, P.H., Meyer, D.F., Granger, M.C., Acworth, I.N. (2004) Metabolomic Applications of Electrochemistry/Mass Spectrometry. J. Am. Soc. Mass Spectrom. 15: 1717–1726.

9. Silva, J.C., Denny, R., Dorschel, C.A., Gorenstein, M., Kass, I.J., Li, G.-Z., McKenna, T., Nold, M.J., Richardson, K., Young, P., Geromanos, S. (2005) Quantitative Proteomic Analysis by Accurate Mass Retention Time Pairs. Anal. Chem. 77: 2187–2200.

10. Yocum, A.K., Chinnaiyan, A.M. (2009) Current Affairs in Quantitative Targeted Proteomics: Multiple Reaction Monitoring–Mass Spectrometry. Brief. Funct. Genom. Proteom. 8: 145–157.

11. Gerber, S.A., Rush, J., Stemman, O., Kirschner, M.W., Gygi, S.P. (2003) Absolute Quantification of Proteins and Phosphoproteins from Cell Lysates by Tandem MS. Proc. Natl. Acad. Sci. USA 100: 6940–6945.

12. Aebersold, R., Mann, M. (2003) Mass Spectrometry-based Proteomics. Nature 422: 198–207.

13. Warscheid, B. (ed.) (2014) Stable Isotope Labeling by Amino Acids in Cell Culture (SILAC): Methods and Protocols (Methods in Molecular Biology). Humana Press, New York.

14. SILAC. Stable Isotope Labeling by Amino Acids in Cell Culture, http://www.silac.org/ (accessed May 17, 2015).

15. Zhang, Y., Wolf-Yadlin, A., Ross, P.L., Pappin, D.J., Rush, J., Lauffenburger, D.A., White, F.M. (2005) Time-resolved Mass Spectrometry of Tyrosine Phosphorylation Sites in the Epidermal Growth Factor Receptor Signaling Network Reveals Dynamic Modules. Mol. Cell. Proteom. 4: 1240–1250.

16. SCIEX. iTRAQ Reagents, http://sciex.com/products/standards-and-reagents/itraq-reagent (accessed June 15, 2015).

17. Verano-Braga, T., Schwämmle, V., Sylvester, M., Passos-Silva, D.G., Peluso, A.A.B., Etelvino, G.M., Santos, R.A.S., Roepstorff, P. (2012) Time-resolved Quantitative

Phosphoproteomics: New Insights into Angiotensin-(1-7) Signaling Networks in Human Endothelial Cells. J. Proteome Res. 11: 3370–3381.

18. Yuan, J., Bennett, B.D., Rabinowitz, J.D. (2008) Kinetic Flux Profiling for Quantitation of Cellular Metabolic Fluxes. Nature Protoc. 3: 1328–1340.

19. Bennett, B.D., Yuan, J., Kimball, E.H., Rabinowitz, J.D. (2008) Absolute Quantitation of Intracellular Metabolite Concentrations by an Isotope Ratio-based Approach. Nature Protoc. 3: 1299–1311.

20. Ting, H., Hu, J.-B., Hsieh, K.-T., Urban, P.L. (2014) A Pinch-Valve Interface for Automated Sampling and Monitoring of Dynamic Processes by Gas Chromatography–Mass Spectrometry. Anal. Meth. 6: 4652–4660.

21. Fritzsche, S., Ohla, S., Glaser, P., Giera, D.S., Sickert, M., Schneider, C., Belder D. (2011) Asymmetric Organocatalysis and Analysis on a Single Microfluidic Nanospray Chip. Angew. Chem. Int. Ed. 50: 9467–9470.

22. Vikse, K.L., Ahmadi, Z., Manning, C.C., Harrington, D.A., McIndoe, J.S. (2011) Powerful Insight into Catalytic Mechanisms through Simultaneous Monitoring of Reactants, Products, and Intermediates. Angew. Chem. Int. Ed. 50: 8304–8306.

23. Agilent Technologies. (2012) Agilent 6200 Series TOF and 6500 Series Q-TOF LC/MS System. Concepts Guide, http://www.agilent.com/cs/library/usermanuals/public/G3335-90173_TOF_Q-TOF_Concepts.pdf (accessed September 16, 2015).

24. Hu, L., Jiang, G., Xu, S., Pan, C., Zou, H. (2006) Monitoring Enzyme Reaction and Screening Enzyme Inhibitor Based on MALDI-TOF-MS Platform with a Matrix of Oxidized Carbon Nanotubes. J. Am. Soc. Mass Spectrom. 17: 1616–1619.

9

Data Treatment in Time-resolved Mass Spectrometry

9.1 Overview

The main goal of data treatment in conventional mass spectrometry (MS) is to facilitate identification and quantification of analytes. The focus of time-resolved mass spectrometry (TRMS) is to track variations of identities and quantities of analytes and products over time. In many chemical reactions, the concentrations of reactants decrease and those of products increase with time. In more complex reactions, reaction intermediates exist and their concentrations may increase and decrease within certain periods of time. The evolution of reaction intermediates is distinct from that of reactants and products. TRMS provides an insight into the progress of reactions by identifying molecules based on their mass-to-charge (m/z) ratios. It also determines concentrations of molecules based on signal intensities of ions. Thus, it is important for TRMS to interpret MS data on highly complex and dynamic systems correctly. This chapter will first introduce definitions of various technical terms, and then discuss how to predict molecular compositions of complex mixtures based on the information contained in mass spectra.

One should emphasize that the stability of MS instruments is critical for obtaining reliable TRMS data. In the course of chemical reactions, molecular identities and quantities are highly dynamic. Thus, it is difficult to judge whether variation in signal intensity is due to the progression of a chemical reaction or changes in conditions within an instrument. The user needs to be constantly aware and in control of instrument conditions to ensure no significant fluctuations in sensitivity. In many cases, it is advantageous to use internal standard compounds for mass and ion abundance calibrations (see Chapter 8).

Time-Resolved Mass Spectrometry: From Concept to Applications, First Edition.
Pawel Lukasz Urban, Yu-Chie Chen and Yi-Sheng Wang.
© 2016 John Wiley & Sons, Ltd. Published 2016 by John Wiley & Sons, Ltd.

9.2   Definition of Terms

Mass spectra reveal a lot of information on the analyzed samples. Apart from the m/z ratios, signal intensity, isotopic patterns, or even differences between spectral patterns reveal useful information on samples. In most cases, it is important to compare observed spectral features with theoretical predictions. The following terms are fundamental to correct interpretation of mass spectra. The definitions introduced below are applicable to molecules, atoms, and ions:

- *Exact mass:* Mass calculated based on the empirical formula of a chemical species. Table 9.1 lists exact masses of important atoms and their stable isotopes [1, 2].
- *Accurate mass:* The mass value obtained experimentally.
- *Average mass:* The calculated mass using the weighted average mass of all isotopes of every atom.
- *Nominal mass:* The mass calculated using the integer mass of all components (it is listed in Table 9.1).
- *Mass defect:* The difference between nominal and accurate mass.
- *Monoisotopic mass:* The mass calculated using the principal (most abundant) isotope of each component. The monoisotopic mass of molecules does not necessarily correspond to the most abundant isotope in an isotopic distribution.
- *Base peak:* The most intense spectral feature in a mass spectrum with a finite mass range.
- *Mass accuracy:* The error of mass determination, usually expressed in parts per million (ppm). For details please refer to Section 5.3.2.
- *Mass resolving power:* The ratio of the mass of an ion to the full-width-at-half-maximum (FWHM) of its spectral feature in a mass spectrum, or $m/\Delta m$. Notably, the definitions of resolving power and resolution were reversed for historical reasons. Since in modern MS, one usually reports only $m/\Delta m$, in this book we use the terms "mass resolving power" and "resolution" to refer to the same properties of mass spectra. For details please refer to Section 5.3.1.

9.3   Spectral Patterns

Identification of molecules present in a mass spectrum is fundamental to MS data interpretation. Successful molecular identification relies on the quality of the mass spectra. In addition to displaying spectral features with adequate signal-to-noise (S/N) ratios, a high quality spectrum also features high mass-resolving power and accuracy. Even when using a high-end instrument, it is critical to assure adequate maintenance of the instrument and to perform correct calibration for optimal mass-resolving power and accuracy.

After high quality mass spectra are obtained, the next step in data interpretation is to deduce the atomic composition of mass spectral features. Since mass spectra relate ion abundances with their m/z values, in order to find the molecular weight (m) of a compound, it is necessary to predict the charge state (z) of recorded ions. The number of electric charges acquired by ions in the gas phase is influenced by the ionization method. This problem used to be less important because "old" ionization techniques [e.g., electron ionization (EI), chemical ionization (CI), fast atom bombardment (FAB), secondary ion mass

Table 9.1 *The exact and nominal masses of important atoms, and their natural abundances expressed as a percentage share in the sum of all isotopes [1, 2]*

Isotope	Exact mass	Nominal mass	Natural isotopic abundance (%)
^{1}H	1.00783	1	100.0
^{2}H[a]	2.01410	2	0.0016
^{12}C	12.00000	12	98.9
^{13}C	13.00335	13	1.1
^{14}N	14.00307	14	99.6
^{15}N	15.00011	15	0.4
^{16}O	15.99492	16	99.8
^{18}O	17.99916	18	0.2
^{23}Na	22.98980	23	100.0
^{31}P	30.97376	31	100.0
^{32}S	31.97207	32	95.0
^{33}S	32.97146	33	0.8
^{34}S	33.96786	34	4.2
^{35}Cl	34.96885	35	75.8
^{37}Cl	36.99999	37	24.2
^{39}K	38.96371	39	93.3
^{41}K	40.96183	41	6.7
^{54}Fe	53.93961	54	5.8
^{56}Fe	55.93494	56	91.8
^{57}Fe	56.93539	57	2.1
^{58}Fe	57.93328	58	0.3

[a]Frequently used in quantitative analysis.

spectrometry (SIMS)] mostly produced singly charged species. Determination of z became more problematic after the introduction of electrospray ionization (ESI), especially in the analysis of large biological molecules. Ionization of such species by ESI leads to fewer singly charged species. Therefore, interpretation of spectra representing singly and multiply charged species is more complicated.

9.3.1 Accurate Mass

Although a spectral feature reveals the mass of an ion, the value is not necessarily accurate enough for reliable identification of the corresponding molecule. In low-resolution mass spectra, the isotopic pattern of an ion may be buried within a single spectral feature. The measurement of m/z may result in an observation closer to the average mass than the exact mass of the ion (as discussed in Section 5.3.2). Even in the case of a well-resolved isotopic pattern, mass measurement may still be affected by the width and shape of the spectral feature. Deconvolution of spectral features "contaminated" with multiple ions of similar mass is important for accurate molecular identification but beyond the scope of this book. Here we will focus on the determination of accurate masses of single-component samples.

There are several ways to determine accurate mass, including spectral local maxima, centroids, and least-squares curve-fitting [3, 4]. We will first introduce the related terminology.

9.3.1.1   Local Maxima

The most straightforward method of spectral-feature assignment is finding accurate masses based on local maxima. In this way, m/z values can be obtained even during data acquisition. On the other hand, other methods require dedicated computational tools and suitable algorithms. However, this method is the least accurate because peak maxima may shift for various reasons. First, the peak position of an analyte can fluctuate when unwanted chemicals (e.g., metastable ions) or background noise is present. This problem is especially apparent when the S/N ratio of a peak is low, and background noise interference becomes more evident. Thus, local maxima are utilized mainly when signal intensity is high enough to minimize the influence of signal fluctuations in the proximity of the peak. In many cases, smoothing of adjacent data points is helpful for the extraction of m/z values. Second, the observed local maxima of peaks may deviate considerably from the exact peak position if the density of data points is insufficient to describe a spectral feature correctly. Normally, accurate peak positions can be determined when the envelope of the peak contains 10 or more data points. On the other hand, mass errors may exist if spectral features with dissimilar shapes are present in one mass spectrum, because the local maxima of an asymmetric spectral feature are different from those of a symmetric feature.

9.3.1.2   Centroids

The centroid mass (m_{cen}) of a spectral feature is derived from the average of its integrated intensity. It can be calculated using the equation:

$$m_{cen} = \sum_i m_i I_i / \sum_i I_i \tag{9.1}$$

where m_i and I_i represent the mass and the intensity of the ith data point, respectively. The range of i should cover the entire envelope of the spectral feature. The centroid of a symmetric feature is simply its center, but it deviates from the center in the case of asymmetric features. The centroid method is the most frequently used method for accurate mass determination because it is relatively simple, insensitive to the S/N ratio and shape of the spectral feature. The disadvantage of the centroid method is that it may result in serious systematic errors if spectral features are highly asymmetric; for example, when the energy distribution of ions is affected by imperfect voltage pulses and stray electric fields. In this case, the centroid method is the least accurate of all the m/z determination methods discussed here.

9.3.1.3   Least-squares Curve-fitting

Least-squares curve-fitting enables determination of accurate mass by finding the maximum of a simulated curve that best describes an observation. The Gaussian function is one of the most frequently used functions to describe features in mass spectra. Fitting peaks is done with software tools which rely on the least-squares method. Other functions can also be used to fit spectral features, such as Lorentzian and polynomial. Notably, the accurate mass determined using this method depends on the fitting function. Therefore, the selection of a suitable fitting function is important, and every spectral feature in a mass spectrum should be fitted with the same fitting function to minimize errors. Least-squares curve-fitting provides the most accurate mass, but it relies on extensive computation and is normally performed after completion of data acquisition.

9.3.2   Mass Calibration

Calibration of peak position for accurate mass determination can be performed internally or externally to minimize systematic errors. Internal calibration can be conducted when compounds with known molecular weight (called calibration compounds or calibrants) are mixed with the sample prior to the introduction into the ion source. This calibration can be performed, for example, by adding the calibrant to the liquid-phase sample while diluting it prior to analysis. The best result is achieved when multiple calibration signals are used to interpolate the m/z of ions within the range of interest. In proteomics, a tryptic digest of albumin from horse heart is typically used as the calibrant because it covers a wide m/z range (e.g., m/z 800–3000) that is ideal for mass calibration of low- to medium-sized peptides. In external calibration, the calibrants are analyzed before the analysis of real samples. The peaks of the calibrants are used to create and set the calibration equation in the data acquisition software. This method provides less mass accuracy because the instrument condition may still vary between the calibration and analyses of real samples. However, external calibrations save time and calibration compounds, and such methods also make analyses of analytes free from interferences caused by calibrants.

9.3.3   Singly Charged Molecules

Singly charged ions encompass radical ions, protonated/deprotonated molecules, products of alkali ion additions, or complex ions with other charge carriers. In the case of singly charged radical ions, the molecular weight of an analyte molecule approximately equals to the m/z value of that ion (one electron affects the measurement by only 0.00055 u). In the case of protonated or deprotonated molecules, the m/z values are expressed as $m + 1$ or $m - 1$, respectively. Alkali metal adducts are also commonly observed in MS; for example, $m + 23$ (sodium adducts) or $m + 39$ (potassium adducts). The alkali ions are mostly contaminants, which are very difficult to remove from sample vials, solvents, or sample plates. However, some analytes such as carbohydrates can only be ionized by association with alkali ions [5, 6].

Singly charged molecules generally produce spectral features that are easy to assign. This characteristic is advantageous because different compounds with distinct masses do not interfere with each other in the spectrum. Multiply charged species often result in broad distributions of spectral features that can easily overlap with the signals from other species. Analyzing singly charged species is also less demanding on a high-resolution mass spectrometer because the isotopic features of a singly charged molecule are mostly 1 mass unit apart. Most commercial mass spectrometers are able to resolve adjacent features separated by 1 u up to an m/z limit of 1000 units. Therefore, a high mass-accuracy mass spectrometer is suitable for identifying singly charged small molecules. If the mass resolving power of the instrument is unable to resolve isotopic features, the mass accuracy also decreases, and the observed masses are close to the average mass of the ion. The phenomenon is discussed in Section 5.3.2.

The isotopic distribution of ions can be predicted based on the empirical formula and natural abundances of every isotope (listed in Table 9.1). The mass of a species is the sum of the mass of every isotope, whereas the abundance is the product of the natural abundance of every isotope multiplied by the number of all possible structural combinations (n) of the given molecule. For example, the monoisotopic mass of a molecule with an atomic composition of $C_aH_bN_cO_d$ is:

$$M(C_aH_bN_cO_d) = a \times 12.000 + b \times 1.008 + c \times 14.003 + d \times 15.995 \qquad (9.2)$$

while its natural abundance (A, in percentage) is given by:

$$A(C_aH_bN_cO_d) = [(98.9\%)^a \times (100.0\%)^b \times (99.6\%)^c \times (99.8\%)^d] \times n \qquad (9.3)$$

Consider protonated arginine ($[C_6H_{15}N_4O_2]^+$) as an example: its monoisotopic mass is 175.122 Da while the natural abundance of its most abundant isotopic ion is 91.7%. The second most abundant isotopic ion of arginine is $[^{13}C^{12}C_5H_{15}N_4O_2]^+$, with one of the ^{12}C (12.000 Da) atoms replaced by ^{13}C (13.003 Da). It has a molecular weight of:

$$M(^{13}C^{12}C_5H_{15}N_4O_2)$$

$$= 1 \times 13.003 + 5 \times 12.000 + 15 \times 1.008 + 4 \times 14.003 + 2 \times 15.995 = 176.125 \qquad (9.4)$$

Notably, the abundance of this isotopic ion is 6.1% of the most abundant isotopic ion:

$$A(^{13}C^{12}C_5H_{15}N_4O_2) = [1.1\% \times (98.9\%)^5 \times (100.0\%)^{15} \times (99.6\%)^4 \times (99.8\%)^2] \times 6 \qquad (9.5)$$

The factor of 6 in the last term is the number of all possible carbon atom positions which can be replaced by ^{13}C. The third most abundant isotopic form of protonated arginine is $[^{12}C_6H_{15}{}^{15}N^{14}N_4O_2]^+$, which has a mass of 176.119 and an abundance of 1.5% that of the most abundant isotope. Notably, most mass spectrometers cannot resolve 176.125 and 176.119, resulting in a single combined feature near 176.122 and an abundance of roughly 7.6%. Abundances of the other isotopic ions can be estimated in the same way.

The monoisotopic mass of a small ion is typically its most abundant isotopic feature, but the relative abundance of monoisotopic ions decrease as the ion masses increase. For macromolecules, the abundance of a monoisotopic ion may be too low to be detected. Figure 9.1 depicts the simulated spectra of protonated arginine, angiotensin I, and cytochrome c, with their corresponding monoisotopic features marked with inverted triangles. The monoisotopic masses of neutral arginine, angiotensin I, and cytochrome c are 174.1, 1295.7, and 12375.3, respectively. In the low-mass range, the monoisotopic feature of protonated arginine dominates the mass spectrum (Figure 9.1a). The second isotopic feature has very low abundance. In the middle-mass range, the second isotope of protonated angiotensin I is roughly 80% that of the monoisotopic feature (Figure 9.1b). The increase in the abundance of heavier isotopes is due to the increasing probability of heavy isotopes being present as the number of atoms increases. In the high-mass

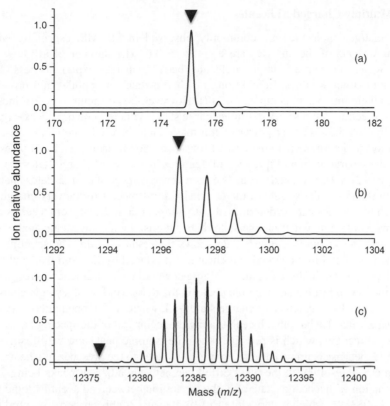

Figure 9.1 *Simulated spectrum of protonated arginine (a), angiotensin I (b), and cytochrome c (c). The position of the monoisotopic feature of the corresponding ion is indicated by the inverted triangle. In the low- and middle-mass range, such as in the case of protonated arginine and angiotensin I, the monoisotopic feature is also the most abundant feature in the spectra. In the high-mass region, the monoisotopic feature is too weak to be observed*

range, as in the case of protonated cytochrome c, the most abundant feature is no longer the monoisotopic feature (Figure 9.1c). In fact, the monoisotopic feature is no longer detectable due to its very low abundance relative to others (0.02% that of the most abundant feature). As mentioned above, the abundance of the monoisotopic feature is very low because the probability that no heavy isotopes are present in such a large molecule is very low.

The relative abundances of isotopic peaks are important for derivation of atomic compositions. It is especially convenient if the studied molecules contain atoms with unique isotopic distributions, such as when irregular changes of isotopic abundance are present. For example, as listed in Table 9.1, the abundances of sulfur isotopes follow the order: $^{32}S > ^{34}S > ^{33}S$. In the case of potassium it is $^{39}K > ^{41}K$, in the case of chlorine it is $^{35}Cl > ^{37}Cl$, and in the case of iron it is $^{56}Fe > ^{54}Fe > ^{57}Fe > ^{58}Fe$.

9.3.4 Multiply Charged Molecules

Multiply charged molecules are commonly observed in ESI-MS, especially when the molecular weights of the analytes are above ~ 500 u. The number of available charge states is greater in the case of larger molecules because charges experience less Coulomb repulsions in larger volumes. In addition, the charge states of peptides and proteins also depend on their amino acid compositions [7]. However, it is difficult to determine m from the m/z of spectral features with multiple charge states without knowing the exact z values.

Depending on the resolving power of the mass analyzer used, one can choose one of the two ways to determine z. In the case of low-resolution instruments, z can be estimated based on the separation of multiple spectral features in a series of charge states (n), such as n, $n+1$, $n+2$, $n+3$, $\ldots$, and so on. The correct assignment of charge states enables calculation of the molecular weight of the original neutral species from every charge-related feature. Thus, z values can be determined with a higher confidence from a series of spectral patterns containing wider charge state distributions. Such charge state analysis is also called *charge deconvolution*. Figure 9.2a shows the result of fitting the spectral features corresponding to multiply charged cytochrome c produced by ESI and detected using an IT mass spectrometer. In this case, the resolving power of the instrument was roughly 1200, which is too low to resolve any isotopic peaks in the mass window of a single charge state. The insets show the peak shape of charge state +13, where no isotope pattern can be seen.

The z values can also be found by calculating the reciprocal of the spacing between adjacent isotopic features, which is possible when the isotopic peaks are well-resolved. This possibility is because isotopic patterns are mostly related to the presence of a heavy isotope of carbon (^{13}C), resulting in the mass difference between adjacent peaks being 1.00335. Thus, for an ion with a charge state of n, the separation between adjacent isotopic features becomes $1.00335/n$, which is very close to $1/n$. Because of this property, spectral features corresponding to multiply charged species can be charge-deconvoluted without analyzing spectral features of other charge states. Since this method relies on the resolving power of the instrument, it is only feasible using high-resolution instruments operated in the low m/z range. Figure 9.2b shows a high-resolution spectrum of cytochrome c obtained using an FT-ICR-MS instrument. Since the instrument has a resolving power of $>1\,000\,000$, the isotopic pattern of cytochrome c carrying 13 additional protons can still be resolved (see the inset in Figure 9.2b).

9.4 Mass Accuracy

The mass accuracy of a mass spectrometer is the most important factor for atomic composition determination. It is more important than the mass resolving power because accurate mass directly affects the correctness of the molecular weight used to deduce atomic composition. In contrast, mass resolving power is less important for analyzing pure samples or when analyzing light ions (e.g., $m/z < 400$). Mass resolving power can be improved by reducing scanning speed (as mentioned in Chapter 5). Although increasing mass resolving power improves mass accuracy, it is difficult to further improve the mass accuracy when distinct spectral peaks are already recorded.

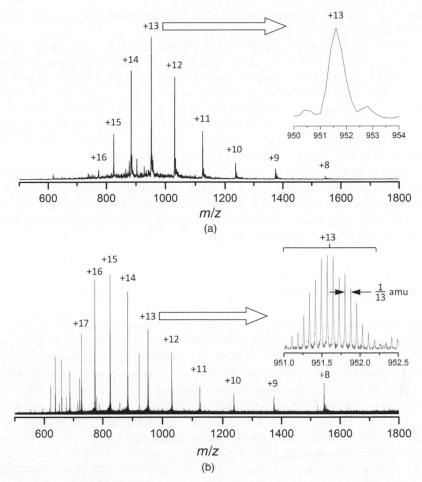

Figure 9.2 *Spectra of ESI-generated cytochrome c ions obtained with (a) an ion trap (IT) and (b) a Fourier transform ion cyclotron resonance (FT-ICR) mass spectrometer. The mass resolving power of the IT is low, so the charge state distribution has to be derived from a series of peaks at different charge states. The mass resolving power of the FT-ICR is high enough to resolve the isotopic pattern of one charge state. The insets display the isotopic patterns of the charge state +13*

Mass accuracy is affected mainly by the precision of machining and assembly of the electrodes of mass analyzers as well as the stability of their electronic systems. Imperfections in the construction of mass analyzers originate from mechanical errors during manufacturing and assembly of components. It is difficult to ensure an overall machining and assembly error of $<10\,\mu m$ even for a small mass analyzer, such as IT. Assuming one dimension of a mass analyzer is roughly $6\,cm$, a $10\,\mu m$ error corresponds to 0.017% relative error. The stability of the electronic system is another important parameter affecting mass accuracy and was discussed in Section 5.3.2.

The number of candidates available for determination of atomic composition relies on mass accuracy. This reliance is especially apparent in the case of molecules composed of simple elements such as C, H, O, and N atoms. For example, it has been estimated that to achieve unambiguous determination of atomic composition of an ion with a m/z of 118, one needs a mass accuracy of 34 ppm [8]. When the m/z of the targeted ion increases to 750, the mass accuracy needs to be 0.018 ppm in order to obtain single atomic composition. Unfortunately, even for the most accurate mass spectrometer presently available, such as FT-MS, average mass accuracy of 0.02 ppm is still unachievable.

Since identification of small ions requires a lower mass accuracy than that for large ions, confidence in the identification of large ions can be improved by breaking down large ions into low-mass fragments by tandem MS. That is, the deduction of the atomic composition of low-mass fragments is easier than that for intact ions. This option shows an advantage of using tandem MS in structural identification.

9.5   Structural Derivation

The atomic composition of an ion can be predicted with good confidence if accurate mass is obtained with a reasonable accuracy. Most chemicals and biological samples contain several high-abundance elements: C, H, O, N, S, P, and Na. Therefore, structural characterization of molecules can be based on certain assumptions regarding elemental composition. In fact, identification of analytes with known elemental compositions is much easier than identification of analytes with unknown elemental compositions. Because mass spectrometers detect ions, it is important to determine the type of charge ions have (i.e., positive *vs.* negative). For example, the charge may be due to the addition or removal of electrons, protons, or other ionic species. Thus, the first step in structural derivation is to analyze charge-related spectral features in order to obtain the masses of uncharged molecules. Precise determination of molecular structure can also be carried out based on fragmentation patterns obtained by tandem MS. However, fragmentation patterns should be interpreted with caution because some molecular structures may undergo random rearrangements after ionization [9].

Upon ionization, the number of electrons in a molecule may change. For example, EI causes the loss of a valence electron from a neutral molecule (see Chapter 2). Because neutral molecules contain even numbers of electrons, EI produces odd-electron ions from neutral molecules. The odd-electron ions, denoted by $M^{+\bullet}$, are radical cations. The dot symbol indicates the presence of an unpaired electron. In contrast to the radical cations, electron capture ionization produces radical anions, denoted by $M^{-\bullet}$. $M^{+\bullet}$ and $M^{-\bullet}$ are generally known as molecular ions.

In addition to molecular ions which contain an odd number of electron, ions may also contain even numbers of electrons. Even-electron ions are typically produced from molecular ions by the decomposition of a radical, such as a hydrogen radical ($H^{\bullet}$). Examples of reaction mechanisms producing even-electron ions are the generation of primary ions and secondary reagent ions discussed in Section 2.2. Another source of even-electron ions is the addition or removal of protons to or from a neutral molecule.

Some structural information about molecules is embedded in the empirical formula derived from accurate mass. Typical saturated organic molecules have simple atomic compositions, such as C_xH_{2x+2} in the case of hydrocarbons and $C_xH_{2x+2}O_y$ (where $y < 2x + 2$) in the case of alcohols with linear hydrocarbon chains. There exist a few basic principles which facilitate structural derivation of other types of compounds.

9.5.1 Unsaturation and Ring Moieties

Unsaturated hydrocarbons and cyclic hydrocarbons have characteristic empirical formulae. The general formula of hydrocarbons is $C_xH_{2x+2-2m}$ while that of alcohols is $C_xH_{2x+2-2m}O_y$ (where m represents the number of rings and unsaturated bonds in the molecular structure). An important characteristic of molecules composed of C, H, O, and S is that their presence results in an even number molecular weight. This characteristic, however, disappears at the presence of a nitrogen atom, which relates to the so-called "nitrogen rule".

9.5.2 Nitrogen Rule

The nitrogen rule reflects the unique mass and valence electron parities of the nitrogen atom, which has an even atomic number and an odd number of valence electrons. It is unique because other atoms have mass and electron parity. Thus, replacing a hydrogen atom from a hydrocarbon by a nitrogen atom results in the addition of two additional hydrogen atoms to the nitrogen atom (due to the octet rule). Thus, a single replacement of such a functional group changes mass of a hydrocarbon from even to odd number; for instance, from C_xH_{2x+2} to $C_xH_{2x+1}NH_2$ or $C_xH_{2x+3}N$. The mass changes again from odd to even number if a second nitrogen atom is present. Thus, the nitrogen rule states that a molecule has an odd mass number if it contains an odd number of nitrogen atoms, otherwise it should have an even mass number. A molecule has an even mass number if it contains no nitrogen atom or an even number of nitrogen atoms.

9.5.3 Functional Groups

Even when the atomic composition is unambiguously determined, molecules with the same composition can still have various molecular structures. In MS, fragmentation of functional groups during or after ionization produces characteristic spectral features revealing structural information of ions. Fragmentation follows typical processes, such as the decomposition of the weakest bonds after ions have been internally activated. The readers can find a comprehensive discussion on these processes and the use of tandem MS techniques in Section 3.3.2. Fragmentation patterns are also frequently observed when conducting analyses with EI, CI, and matrix-assisted laser desorption/ionization (MALDI). They are less pronounced in the case of ESI and atmospheric pressure chemical ionization (APCI) ion sources because those ionization methods are softer. The spectral signatures of various functional groups are contained in the corresponding neutral loss peaks. Table 9.2 lists the commonly observed decomposition channels. More complete lists can be found in other works [9, 10].

Table 9.2 *Important functional groups and corresponding mass (rounded to integer numbers)*

Functional group	Chemical formula	Mass
Methyl	$-CH_3$	15
Amine	$-NH_2$	16
Hydroxy	$-OH$	17
Hydrate	$-OH_2$	18
Ethyl	$-C_2H_5$	25
Carbonyl	$-CO$	28
Nitroso	$-NO$	30
Carboxy	$-CO_2H$	45
Sulfonic acid	$-SO_3H$	81

9.6 Molecule Abundance

The signal intensity of spectral features contains quantitative information but it is highly sensitive to changes in experimental conditions. Therefore, special precautions need to be taken to extract quantitative information from mass spectra. Discussion of quantitative data treatment is also included in Chapter 8. In this section, we will focus on the basic features of mass spectra that carry information on ion abundances.

9.6.1 Signal Intensity

The signal intensity of a spectral feature depends on multiple factors. The first determining factor is the concentration of analyte before ionization, which is also the amount of the molecule of interest. However, signal intensity depends even more on ionization efficiency – a parameter that is very hard to predict and control. Signal intensity also relies on the transmission efficiency of ions from ion sources to detectors, and the detection efficiency of detectors towards the mass and kinetic energy of incoming ions [11, 12]. Since the influence of every parameter on signal intensity is unknown, mass spectra only reveal relative abundances of ions. These relative abundances can be used to estimate absolute abundances and concentrations of analytes in the analyzed samples.

Relative molecule abundance is obtained from the signal intensity of a spectral feature with respect to another feature in the same spectrum, or the same feature in reference spectra. In the case of a single spectrum, the relative signal intensity of the target feature divided by a reference feature can be done readily. Variation of this relative intensity in a series of spectra can be assessed to determine the relative abundance change of the target molecule. Quantity estimation can be achieved with higher precision if reference spectra are available. In this case, the abundance of the molecule of interest can be estimated by comparing its signal intensity with the corresponding peak intensity in reference spectra obtained from samples with known concentration. This method is, in fact, used in quantitative analyses based on calibration curves.

Before going any further, it is necessary to define various terms relating to ion abundance. Amplitude, peak area, and the S/N ratio all link to ion abundances. However, they have different purposes and are used under different circumstances. The general guidelines for usage of these concepts are outlined below:

Amplitude: The amplitude of a spectral feature is its largest intensity value. It is a convenient measure of ion abundance in semi-quantitative comparisons. The criterion for using amplitude is that amplitude should not be affected by background noise. This characteristic is especially true at low amplitudes which can be affected considerably by the fluctuation of background noise. Notably, the amplitude of a spectral feature changes with spectral width because the peak area of an ion should, in principle, be constant for a given sample. Therefore, a broader feature results in a lower amplitude, and vice versa. This important property makes relative quantification using amplitude difficult. Except for quadrupole mass spectrometers, most mass spectrometers give rise to peaks with distinct widths at different m/z. Thus, amplitude is only suitable for relative and rough abundance comparisons of peaks situated within a narrow mass range.

Area: The area of a spectral feature is the integral of the intensity of that feature above the baseline. Area can be used to conduct relatively accurate quantification because it is free from the uncertainty caused by changes in spectral width. However, it has several limitations. First, its calculation is based on different formulae in the case of data obtained on different mass spectrometers. This difference is due to the unequal density of data points across the m/z range (typical for most mass spectrometers). Secondly, the accuracy of the result is good for high-resolution spectra, but the area may be considerably overestimated in low-resolution spectra. This overestimation is because a single feature may reflect the sum of abundances of multiple ions. To overcome these limitations, thorough analysis of spectral features with stand-alone fitting software is typically necessary.

S/N ratio: The S/N ratio is the ratio of signal amplitude to the root-mean-square (RMS) of background noise. It is widely used when background noise has non-negligible impact on the determination of peak amplitude. This impact is especially apparent when analyzing diluted samples producing very low intensity peaks. S/N ratio does not take changes in spectral width into account, so the comparison of relative quantities between spectral features with different spectral widths may cause considerable errors and should be performed with care. S/N ratio is less useful in the characterization of intense spectral features because, in this case, the low accuracy of noise estimations can significantly affect results. Comparisons of S/N ratios in different spectra are normally done when the two spectra are recorded with identical acquisition cycles. This way of data comparison is necessary because the S/N ratio increases with the increasing number of accumulated spectra.

9.6.2 Quantity Calibration

Similar to mass calibration, quantity calibration can be done internally or externally by using appropriate calibration compounds. The calibration curve covers the intensity range corresponding to the intensities of analyte peaks. In order to obtain the highest accuracy, the calibrants used in quantity calibration should have similar atomic composition, molecular structure, and mass as analytes. This requirement is because ionization efficiency depends on molecular structure, and detection efficiency is mass-dependent. In fact, ion detectors rely on secondary electrons produced by the impact of primary ions with detector surfaces, including microchannel plates, electron multipliers, and many others.

The most reliable way to perform quantity calibration is to label the compound of interest with a heavy isotope, and use such an isotope-labeled compound as a calibration reference. Because the molecular weight as well as physical and chemical properties of such isotope-labeled compounds are very similar to those of the analyte, the ionization and detection efficiency of the mass spectrometer for analytes and the calibrants are almost the same. An important example of quantitative measurement with isotope-labeled compounds is isotope-coded affinity tags (ICATs) in quantitative proteomics (see also Chapter 8).

9.6.3   Dynamic Range

Dynamic range describes the ratio of the highest to lowest intensity feature in a mass spectrum. The lowest intensity feature needs to provide meaningful information on an ion, such as with the S/N ratio > 3. Dynamic range is an important factor when mass spectrometers are used to analyze samples with large concentration differences; for example, when one of the sample molecules has an abundance several thousand times lower than others. Dynamic range is distinct from detection limit of a mass spectrometer because detection limit concerns the smallest detectable sample quantity without considering the interference of other species.

The dynamic range of a mass spectrometer is determined by several instrumental characteristics. First, the ion capacity of mass analyzers limits the dynamic range of the instrument. For example, the practical number of ions that can be trapped in a three-dimensional (3D) IT mass analyzer is in the range of 1000 before reaching the space-charge limit. Based on this limitation, the largest abundance ratio between two different ions that can be simultaneously accommodated in the 3D IT mass analyzer and be detected is 999:1. This value, 999, in this case, represents the dynamic range of 3D IT-MS. The dynamic range of two-dimensional (2D) IT-MS is higher (roughly 10^4) owing to its roughly 10 times larger trapping volume.

The performance of the data acquisition system or digitizer of a mass spectrometer is another parameter affecting its dynamic range. The performance encompasses sampling rate and vertical resolution. These aspects determine the density of data point in mass (normally x axis) and signal intensity (normally y axis) coordinates of mass spectra, respectively. The sampling rate of the digitizer determines how many data points can be recorded per second. It is important to use a digitizer with adequately fast sampling rate to record enough data points to display the correct shape of a particular spectral feature, or resolve spectral features with similar m/z values. It is especially important for TOF-MS because mass spectra span only several microseconds, so a digitizer with a sampling rate in the range of 10^9 samples per second is normally used. If the sampling rate of the digitizer is too low, the shape of spectral features may not be determined correctly, and this defect will result in a reduction in the dynamic range of the instrument. Vertical resolution, on the other hand, reflects the density of data points from zero to the highest signal intensity that can be detected. For an 8-bit digitizer, signal intensity is only divided into 2^8 levels in every scan, or 0–255 units. Therefore, the dynamic range of a spectrum for a single acquisition event using an 8-bit digitizer is 256, which is too low to analyze highly complex samples quantitatively. To improve dynamic range without using higher vertical resolution digitizers (e.g., 10-, 12-, or 16-bit), accumulating mass spectra is a simple and effective solution.

In principle, the dynamic range can be increased 10-fold by accumulating 10 mass spectra to yield the final spectrum.

The dynamic range of TOF-MS is approximately $10^4 - 10^5$; in sector (S)-MS, it is $10^4 - 10^5$; 3D IT-MS, $10^2 - 10^3$; 2D IT-MS, $10^3 - 10^4$; while in FT-MS, $10^5 - 10^6$. The high dynamic range of FT-MS is mainly due to its large ion capacity and high-performance digitizer.

9.7   Time-dependent Data Treatment

A unique feature of TRMS is that both identities and quantities of ions may be studied in the time domain. For an engineer, the relationship between conventional MS measurements and TRMS is like the relationship between multimeters and oscilloscopes: multimeters only indicate average electric properties while oscilloscopes measure and record fast changes in electrical signals over time. By analogy, TRMS uses a high-performance chemical detector (mass spectrometer) to study the dynamic phenomena of atoms, molecules, and ions during reactions.

In most TRMS approaches, mass spectrometers are utilized to monitor changes in molecules involved in physical, chemical, and biochemical reactions (see Chapters 10–13). Time-dependent data acquisition is exemplified by selected reaction monitoring (SRM) or multiple reaction monitoring (MRM) methods. For instance, the current of fragment ions produced in fragmentation reactions can be monitored by using triple quadrupole mass spectrometers. In the SRM and MRM modes, the reactant ions are selected by the first quadrupole mass filter (QMF), and introduced into a collision-induced dissociation (CID) cell. The newly generated fragment ions are further analyzed by a second QMF. In this measurement, the extent of the reaction which occurred in the collision cell is obtained from the ion current plot of fragment ions.

Variations in the amounts of ions are typically displayed along the time axis. For example, total ion abundance from an ion source can be monitored over a period of time to obtain information on all analytes. Results for total ion abundance are plotted against time in a total-ion current (TIC) chromatogram. Ion chromatograms are widely used with continuous ion sources such as ESI, EI, CI, APCI and atmospheric pressure photoionization (APPI), because these ion sources are typically used with chromatography systems. A TIC chromatogram is typically used to indicate the correlation between ion signal and retention times in the chromatographic system. Figure 9.3a shows a TIC chromatogram of polypeptides produced from trypsin digestion of bovine serum albumin (BSA), analyzed by liquid chromatography (LC)-MS equipped with an ESI interface system. The ion signal starts from the 6th minute and lasts until the 23rd minute. It is convenient and useful to trace the signal for subsequent data analysis after measurement. However, TIC only provides general sample information because the ion current produced from individual analytes is indistinguishable from others. To obtain the ion current chromatogram of a specific ion, one can operate the mass spectrometer in selected ion monitoring (SIM), or SRM/MRM modes.

It is also possible to plot the ion current of certain ions recorded over time. In such measurements, a large data file is typically recorded with hyphenated MS methods, such as gas chromatography (GC)- or LC-MS analyses; in those experiments, $10^3 - 10^5$ mass spectra are recorded over several minutes to hours. Thus, the current of any ion of interest can be

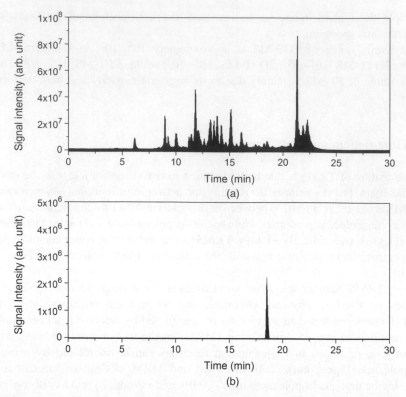

Figure 9.3 *Ion current chromatogram of a BSA digest analyzed by LC-MS. (a) Total-ion current chromatogram; and (b) extracted-ion current chromatogram of a spectral feature with m/z = 1163*

extracted from every spectrum and displayed as a "chromatogram", which is known as an extracted-ion current (EIC) chromatogram. Figure 9.3b shows such a diagram displaying the ion chromatogram of the signal of a peptide, with m/z of 1163, extracted from the same data file as the one shown in Figure 9.3a. The signal starts at 18.3 min and ends at 18.8 min. The peak apex is at 18.5 min. The same feature is also evident in Figure 9.3a because the data are extracted from the same data file. EIC chromatograms provide important information on the progress of reactions of interest. The TIC and EIC chromatograms are widely used in TRMS, as discussed in subsequent chapters.

References

1. Commission on Isotopic Abundances and Atomic Weights. (2014) Isotopic Abundances, http://www.ciaaw.org/isotopic-abundances.htm (accessed September 15, 2015).
2. Wieser, M.E., Coplen, T.B. (2011) Atomic Weights of the Elements 2009 (IUPAC Technical Report). Pure Appl. Chem. 83: 359–396.
3. Yang, C., He, Z.Y., Yu, W.C. (2009) Comparison of Public Peak Detection Algorithms for MALDI Mass Spectrometry Data Analysis. BMC Bioinformatics 10: 4.

4. Savitzky, A., Golay, M.J.E. (1964) Smoothing and Differentiation of Data by Simplified Least Squares Procedures. Anal. Chem. 36: 1627–1639.
5. Harvey, D.J. (1999) Matrix-assisted Laser Desorption/Ionization Mass Spectrometry of Carbohydrates. Mass Spectrom. Rev. 18: 349–450.
6. Harvey, D.J. (2000) Electrospray Mass Spectrometry and Fragmentation of n-linked Carbohydrates Derivatized at the Reducing Terminus. J. Am. Soc. Mass Spectrom. 11: 900–915.
7. Aebersold, R., Goodlett, D.R. (2001) Mass Spectrometry in Proteomics. Chem. Rev. 101: 269–295.
8. Gross, M.L. (1994) Accurate Masses for Structure Confirmation. J. Am. Soc. Mass Spectrom. 5: 57–57.
9. McLafferty, F.W., Tureécek, F. (1993) Interpretation of Mass Spectra, 4th Edition. University Science Books, Mill Valley, CA.
10. Chapman, J.R. (1993) Practical Organic Mass Spectrometry: A Guide for Chemical and Biochemical Analysis, 2nd Edition. John Wiley & Sons, Ltd, Chichester.
11. Chen, X.Y., Westphall, M.S., Smith, L.M. (2003) Mass Spectrometric Analysis of DNA Mixtures: Instrumental Effects Responsible for Decreased Sensitivity with Increasing Mass. Anal. Chem. 75: 5944–5952.
12. Hellsing, M., Karlsson, L., Andren, H.O., Norden, H. (1985) Performance of a Microchannel Plate Ion Detector in the Energy Range 3–25 keV. J. Phys. E: Sci. Instrum. 18: 920–925.

10

Applications in Fundamental Studies of Physical Chemistry

10.1   Overview

A mass spectrometer is a convenient tool for investigating chemical reactions and physical processes in real-time. It provides information-rich output, enabling accurate and precise measurements based on only a small amount of sample. There is no other analytical device offering such advantages simultaneously. For example, optical absorption spectroscopy provides some structural information but requires a large amount of sample, fluorescence spectroscopy is sensitive but is suitable only for fluorescent molecules, and nuclear magnetic resonance spectroscopy provides important structural information but is unsuitable for analyzing small amounts or complex samples. In addition, many spectroscopy techniques are unsuitable for time-resolved analysis since they may not be able to provide sufficient information due to low resolving power and low speeds. Such analytical capabilities are vital for physical chemistry studies because reaction kinetics – the core of physical chemistry research – concerns variations and changes in molecular quantities and identities in a real-time manner.

Most physical chemistry molecular reactions have distinct temporal properties. These reaction times range from femtoseconds, such as in the case of ultrafast quantum chemical reactions, to seconds, such as in the case of slow chemical equilibria. Quantum chemical reactions typically involve excitation of the internal energy of molecules from the femtosecond to picosecond range via laser excitation or ion–molecule collisions. This excitation induces the transition of a molecule from its original quantum state, normally a lower state (e.g., ground state), to a higher state (e.g., excited state). After excitation, the excess energy will dissipate or relax via various reaction channels. The energy-relaxation time depends on the quantum states involved in the process. For gaseous molecules in the absence of collision, electronic relaxation normally takes few femtoseconds to sub-picoseconds. This relaxation involves the transition of electrons from higher to lower electronic states. Electronically excited molecules typically have lifetimes in the sub-picosecond to low picosecond range, but fluorescent and phosphorescent molecules

Time-Resolved Mass Spectrometry: From Concept to Applications, First Edition.
Pawel Lukasz Urban, Yu-Chie Chen and Yi-Sheng Wang.
© 2016 John Wiley & Sons, Ltd. Published 2016 by John Wiley & Sons, Ltd.

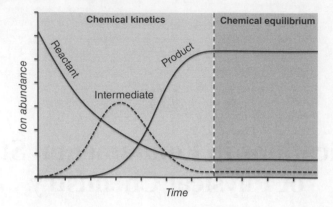

Figure 10.1   *Correlation between chemical kinetics and chemical equilibrium in TRMS*

have longer lifetimes, typically spanning from few nanoseconds to tens of nanoseconds and from few sub-microseconds to minutes, respectively. Intramolecular vibrational relaxation of molecules may range from few sub-picoseconds to sub-nanoseconds, depending on the relaxation process. The selection of suitable time-resolved mass spectrometry (TRMS) approach for the study of certain reactions needs to take into account the spectrum acquisition time (SAT) of the mass spectrometer; the SAT needs to be shorter than duration of the reaction of interest.

This chapter addresses the basic aspects of TRMS in physical chemistry studies. A few important representative examples, which highlight the usefulness of TRMS in solving various physical chemistry problems, are presented and discussed. Based on timescales, physical chemistry problems can be grouped qualitatively into two categories: chemical kinetics and chemical equilibrium. Chemcial kinetics can also be divided into quantum chemistry and reaction kinetics since quantum chemistry, as mentioned above, normally concerns ultrafast reactions. Most ultrafast quantum chemical reactions occur in a single molecule. Reaction dynamics, on the other hand, normally concerns reactions involving two or more molecules in a longer timescale than quantum chemical reactions. Special care should be taken to understand the correlation between chemical kinetics and equilibrium. As a chemical reaction proceeds, the abundance of a reactant reduces while that of the product increases as a function of time, as illustrated in Figure 10.1. Intermediates are also present in the reaction mechanism of complex reactions. At the end of the reaction, the abundance of the reactants, intermediates, and products reach a steady state and no further change is observed, as shown in the region to the right from the vertical dashed line in Figure 10.1. In this case, chemical kinetics concerns the rate of a reaction, whereas chemical equilibrium concerns the resultant steady state.

10.2   Chemical Kinetics

10.2.1   Quantum Chemistry

As mentioned in Section 5.2, every mass spectrometer has a characteristic temporal property and SAT. Although in most cases mass spectrometers are unable to provide time resolution in the low picosecond to femtosecond range, ultrafast chemical reactions can still be studied if the mass spectrometer is equipped with suitable ion sources. For instance, pump–probe laser spectroscopy is widely used for studying quantum chemical properties and the dynamic behavior of atoms or molecules [1–6]. Ultrafast pump–probe experiments use two ultrafast laser beams to excite the internal energy of atoms or molecules involved in the reaction. The two laser beams excite the sample consecutively with a variable delay to perform time-resolved analysis. The correlation between delay time and the reaction products reveals quantum chemical properties of the sample being studied. The two laser beams can be the output from two separate lasers or a single laser after splitting.

Time-resolved spectroscopy, or pump–probe spectroscopy, is widely used to study the lifetime of electronically excited states of molecules [7–9]. Figure 10.2a shows the typical experimental setup involving pump–probe measurements. The pump laser normally irradiates the sample with a fixed optical path, while the probe laser passes through an adjustable mirror system (mirror I and mirror II in Figure 10.2a). The adjustable mirror system can move simultaneously to increase the laser beam path resulting in longer delay times between the pump and probe lasers. The lifetime of the excited state molecules can then be obtained by comparing the delay time with the presence or absence of reaction products [10]. Figure 10.2b shows an example of the consecutive excitation of a molecule by pump and probe ultraviolet lasers, which results in ionization of the molecule. This method, also known as *multiphoton ionization* (MPI), is used to study the ionization energy of gaseous molecules. In MPI reactions, the first laser excites the molecule from its electronic ground state (S_0) to the first excited state (S_1). The second laser irradiates the molecules after the first laser, with a delay from zero to tens of picoseconds. Because the excited state lifetime of a molecule is short, the molecule may return to the ground state via relaxation in the sub-picosecond range. Thus, monitoring the presence or absence of the analyte ion by scanning the delay time of the second laser with respect to the first one provides useful information on the excited state lifetime of the molecule.

In MPI, ionization occurs only if the second laser interacts with molecules excited by the first laser before they return to the S_0 state or transfer to other electronic states. Ions are produced when a molecule is excited to the ionization continuum. It can happen when the probe laser irradiates the molecule before the electron leaves the S_1 state. On the other hand, ion production is not achievable if the delay time between the first and the second laser is longer than the S_1 lifetime of the molecule.

Most gaseous pump–probe experiments require custom-made mass spectrometers [9, 11, 12]. Neutral analytes are typically introduced into the ionization region of the vacuum chamber as a gas stream through a small opening. The ionization region contains an electric field to push ions toward the mass analyzer. Pump–probe laser beams intersect with the gas stream within this ionization region. Since no charged species are produced before ionization occurs, the electric field in the ion source region does not affect the motion of neutral analytes. When ions are produced, they are accelerated toward the mass analyzer immediately. Here, commonly used mass analyzers are time-of-flight (TOF), quadrupole mass filter (QMF) and sector mass analyzers. In order to maintain vacuum conditions inside the instrument, the ion source region is typically a large vacuum chamber operated with high-speed pumping systems.

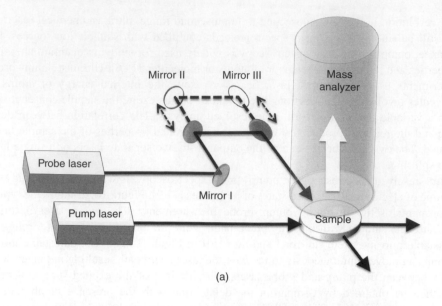

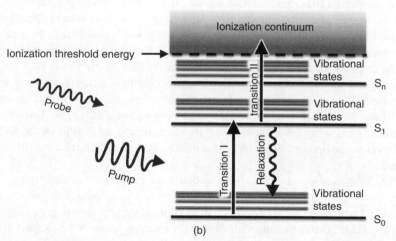

(b)

Figure 10.2   *Time-resolved spectroscopy involving pump–probe lasers. (a) Schematic of the experimental setup. The pump laser irradiates the sample with a fixed beam path. The timing of the probe laser is controlled by using a moveable mirror (mirrors I and II) system. (b) The pump laser excites molecules from the S_0 state to the S_1 state, and the probe laser excites the molecule from the S_1 state to the ionization continuum. Molecules in the S_1 state can undergo relaxation back to the S_0 state*

Figure 10.3 illustrates an application of a pump–probe system in a study of the excited states dynamics of DNA and RNA bases: cytosine, guanine, thymine, and uracil [8]. The experiment was conducted using a TOF mass analyzer equipped with a photoionization ion source. The samples were heated and desorbed using an oven. The vapor was carried into the ionization region using a supersonic argon jet. The pump and probe lasers

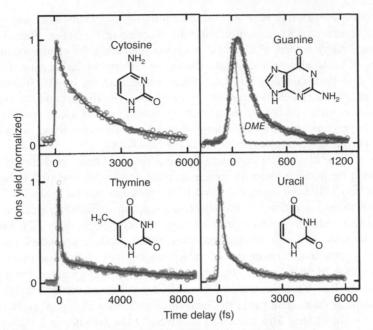

Figure 10.3   *Result of time-resolved pump–probe spectroscopy in combination with TOF-MS. The graphs show femtosecond ionization signals for four DNA and RNA bases, namely cytosine, guanine, thymine, and uracil. The transient of dimethylether (DME) was displayed along with the transient of guanine for comparison [8]. Reprinted with permission from Canuel, C., Mons, M., Piuzzi, F., Tardivel, B., Dimicoli, I., Elhanine, M. (2005) Excited States Dynamics of DNA and RNA Bases: Characterization of a Stepwise Deactivation Pathway in the Gas Phase. J. Chem. Phys. 122: 074316. Copyright (2005) AIP Publishing LLC*

were 267 nm and 400 nm femtosecond laser beams, respectively. By adjusting the time delays between the pump and probe lasers, the femtosecond transient ionization signals were detected using the TOF mass analyzer with a time resolution of 80 fs. The transients were fitted to convolutions of a biexponential decay and a Gaussian function. According to this result, the target molecules showed a fast decay component ranging from 105 to 160 fs, and a slow decay component ranging from 0.36 to 5.12 ps. In conjunction with theoretical calculations, the method enables investigation of energy transfer mechanisms within such biologically essential molecules, and changes in such mechanisms due to variations of molecular structure.

10.2.2   Reaction Kinetics

Real-time MS provides information on the change of identity and quantity of compounds delivered to its interface (ion source). It is an ideal tool for studying the progress of chemical reactions. In many cases, changes in reactant and product concentrations can be monitored continuously. This information is necessary for kinetic characterization of chemical reactions.

Reaction kinetics is the study of the mechanisms and progress of chemical reactions before an equilibrium state is reached [13]. Studying reaction kinetics is usually more important than the determination of reaction equilibria, especially when addressing fundamental questions, or when manipulating the outcome of a reaction. Reaction kinetics encompasses many aspects of chemical reactions, including the change of identity and abundance of molecules, which in turn determine the trend of a reaction. Notably, the species involved in a chemical reaction, which can be analyzed by MS, include not only reactants and products, but also reaction intermediates [14–16]. A common experimental method used in such studies is to monitor the evolution of mass spectra over certain time intervals. Depending on the specific purpose and type of process being investigated, the timescale can range from femtoseconds to hours [1, 17, 18].

Ionization is one of the most important chemical reactions associated closely with time. In addition to the ultrafast ionization reactions mentioned in Section 10.2.1, the evolution of reaction products from other ionization techniques can also be studied with MS. For instance, chemical ionization (CI) involves chain reactions initiated by electron ionization of reagent molecules to produce primary reagent ions (see Section 2.2). The primary reagent ions undergo ion–molecule reactions which produce secondary reagent ions. Ion–molecule reactions between the secondary reagent ions and analyte molecules result in the formation of analyte ions. The evolution of CI process can be monitored using TRMS. Figure 10.4 illustrates the evolution of CI products of diethyl ethylphosphonate using methane as the reagent gas [19]. The result shows that CH_3^+ was produced first and diminished within 10 ms. This trend was followed by the presence of CH_5^+ and $C_2H_5^+$, which decayed in 35 and 60 ms, respectively. The abundance of protonated diethyl

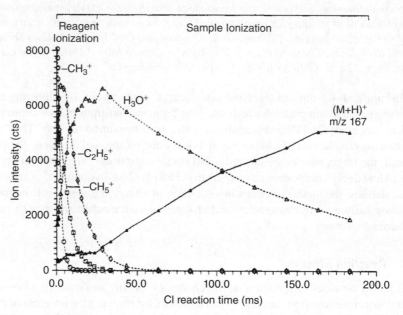

Figure 10.4 *Evolution of chemical ionization products of diethyl ethylphosphonate using methane as reagent gas [19]. Reprinted with permission from Johnson, J.V., Yost, R.A., Kelley, P.E., Bradford, D.C. (1990) Tandem-in-space and Tandem-in-time Mass-spectrometry – Triple Quadrupoles and Quadrupole Ion Traps. Anal. Chem. 62: 2162–2172. Copyright (1990) American Chemical Society*

ethylphosphonate increased as the reaction time increased. These results resemble the CI mechanism introduced in Section 2.2. Unimolecular decomposition is another important type of reaction that has been studied with TRMS. An example is the monitoring of photolysis or photodissociation of gaseous molecules which are excited by light sources (e.g., lasers or flash lamps), in which the evolution of precursor and decomposition products are used to deduce reaction rate constants [20–22].

Another simple chemical reaction involves the transformation of reactants to reaction products with a reaction rate constant K_1:

$$aA + bB \xrightarrow{k_1} cC + dD \qquad (10.1)$$

where A and B are reactants while C and D are products. The constants a, b, c, and d are the stoichiometric coefficients of the corresponding compounds. The reaction rate (r) of this reaction is defined as the rate of reduction of reactant concentration divided by its stoichiometric number, or the rate of increase in product concentration divided by the stoichiometric number:

$$r = -\frac{1}{a}\frac{d[A]}{dt} = -\frac{1}{b}\frac{d[B]}{dt} = \frac{1}{c}\frac{d[C]}{dt} = \frac{1}{d}\frac{d[D]}{dt} \qquad (10.2)$$

Thus, the reaction rate can be derived from the rate of signal intensity change of any of the components. The rate is normally expressed in moles per liter per second $(\text{mol } l^{-1} \, s^{-1} \text{ or } M \, s^{-1})$.

Although Equation 10.2 represents the reaction rate over a time span, the instantaneous reaction rate is more often expressed in a form that takes into account the concentration of reactants [13]:

$$r = K_1[A]^x[B]^y \qquad (10.3)$$

where x and y are not necessarily the same as a and b because many complex reaction schemes involve intricate forward and reverse reactions. Since the reaction rate depends on temperature and activation energy (E_a), K_1 can be further expressed as:

$$K_1 = Ae^{-\frac{E_a}{RT}} \qquad (10.4)$$

where A is a pre-exponential factor, R is the ideal gas constant, while T is temperature. Equation 10.4 is known as the *Arrhenius equation* [23]. The reaction rate can be derived from the change of reactant concentration, which is determined based on the signal intensity of the corresponding mass spectral feature. Hence, the K_1 value can readily be obtained. If the reaction temperature can be precisely measured or estimated, the activation energy of the reaction can also be computed.

Both on-line and off-line sampling techniques can be used for the mechanistic study of chemical reactions [24, 25]. As mentioned previously in Chapter 2, on-line sampling methods, such as ambient ionization techniques, offer advantages over conventional ionization techniques for TRMS. It is because chemical reactions can be studied in a real-time manner under ambient condition. With off-line sampling methods, molecules are normally extracted and transferred manually to the ion source of a mass spectrometer [26, 27]. Examples of mechanistic studies with TRMS include catalytic reactions [28], radical formation [21, 29], and many others [30–33].

A popular example of chemical kinetics studied by means of TRMS is the determination of formation or dissociation dynamics and the reactivity of molecules [34–36] or non-covalent complexes [32, 37–39]. It is usually conducted using ion trapping devices, which can monitor changes in ion abundance over time. The advantage of using trapping devices is that the reaction rate constants can be determined with high accuracy. However, the method is only suitable for studying reactions of gaseous ions.

Examples of formation and dissociation dynamics of complexes have been demonstrated using Fourier transform ion cyclotron resonance mass spectrometry (FT-ICR-MS) [37–41]. Using FT-ICR-MS for studying reaction dynamics is advantageous because the ICR cell traps ions in ultrahigh vacuum conditions that minimizes unwanted ion–molecule reactions. Non-covalent complexes are typically produced by ESI or an external high-pressure ion source facilitating supersonic expansion [37, 39–41]. The ionic complexes are then trapped in the ICR cell, and excited by ion–molecule collisions or electromagnetic radiation. The complexes may also gradually build up their internal energy and dissociate owing to the absorption of the blackbody infrared radiation emitted from the ICR cell [33], as discussed in Section 3.3.2.1. Figure 10.5 shows the evolution of solvated magnesium and magnesium monohydroxide ions trapped in an ICR mass analyzer [39]. Progressive de-solvation of single water molecules can be observed by the decrease of a specific cluster ion as well as the increase of the corresponding product in the mass spectra. For example, the dissociation of $Mg(OH)^+(H_2O)_6$ to form $Mg(OH)^+(H_2O)_5$ is shown in Figure 10.5a. The successive changes from $Mg^+(H_2O)_{19}$ to $Mg^+(H_2O)_{18}$ and from $Mg^+(H_2O)_{18}$ to $Mg(OH)^+(H_2O)_{17}$ are shown in Figure 10.5b, and $Mg^+(H_2O)_{37}$ to $Mg^+(H_2O)_{36}$ in Figure 10.5c. Because ion activation via blackbody infrared radiation is inefficient, the reaction timescale is in the range of tens of milliseconds up to several seconds in the case of dissociation of non-covalent bonds. It is unsuitable for dissociation of covalent bonds in molecules because, in this method, the ion trapping device needs to be heated up to more than 450 K, and ion activation takes hundreds of seconds [36, 42]. Such experimental conditions are impractical for routine analysis.

In contrast to blackbody infrared radiation, energy transfer via ion–molecule collisions is much more efficient. Rapid energy transfer can be facilitated by storing ions in a radio frquency trapping device, in which buffer gases are used for thermalization of both the internal and translational energies. In this approach, ions are thermalized with their surroundings (e.g., buffer gas, electrodes of ion trap) via numerous ion–molecule and molecule–surface collisions. Therefore, the ion temperature can be determined precisely. Since the ion temperature is known, thermodynamic properties of complexes can be obtained by recording changes of the signal intensity of reactants and products.

The binding energy or dissociation energy (E_{diss}) of non-covalent complexes can be calculated by:

$$E_{diss} = E_a + \langle E_{vib} \rangle + a K_B T \tag{10.5}$$

where $\langle E_{vib} \rangle$ is the average vibrational energy while $a k_B T$ is a correction term accounting for the conversion efficiency of thermal energy from the surroundings to complex ions. The a factor is in the range of 1.2–1.8 [43, 44]. Herein, the E_a represents the activation energy of dissociation, which can be determined by Equation 10.4. The average vibrational energy $\langle E_{vib} \rangle$ can be estimated from the vibrational frequencies predicted by

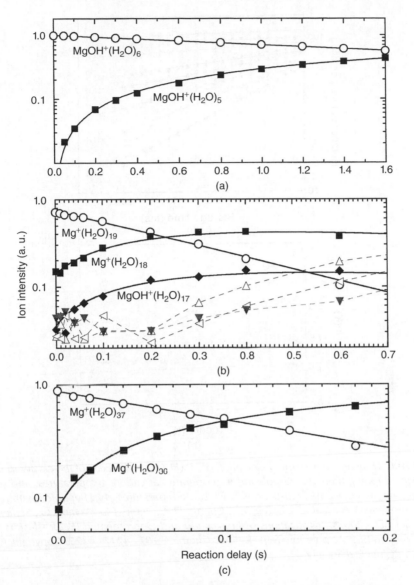

Figure 10.5 *Evolution of solvated magnesium and magnesium hydroxide ions as a function of reaction delay (trapping time) in FT-ICR. (a) Desolvation of a single water molecule from* $Mg(OH)^+(H_2O)_6$ *to* $Mg(OH)^+(H_2O)_5$*; (b) successive changes from* $Mg^+(H_2O)_{19}$ *to* $Mg^+(H_2O)_{18}$ *to* $Mg(OH)^+(H_2O)_{17}$*; and (c) dissociation from* $Mg^+(H_2O)_{37}$ *to* $Mg^+(H_2O)_{36}$ *[39]. Reprinted from Chem. Phys., 239, Berg, C., Beyer, M., Achatz, U., Joos, S., Niedner-Schatteburg, G., Bondybey, V. E., Stability and Reactivity of Hydrated Magnesium Cations, 379–392. Copyright (1998), with permission from Elsevier*

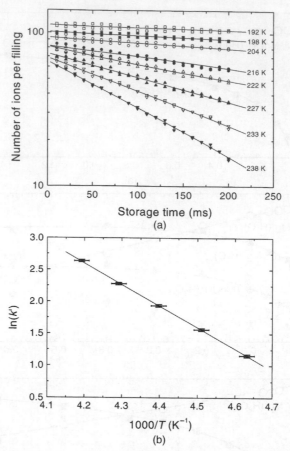

Figure 10.6 *Kinetic study of the dissociation of $H^+(H_2O)_5$. (a) Decay of the cluster ions with increasing trapping time in a 22-pole ion trap operated at various temperatures; and (b) the dissociation rate constant of cluster ions at various temperatures [43]. Reprinted with permission from Wang, Y.-S., Tsai, C.-H., Lee, Y.T., Chang, H.-C., Jiang, J.C., Asvany, O., Schlemmer, S., Gerlich, D. (2003) Investigations of Protonated and Deprotonated Water Clusters Using a Low-temperature 22-pole Ion Trap. J. Phys. Chem. A 107: 4217–4225. Copyright (2003) American Chemical Society*

quantum chemistry calculations [43, 45–47]. Since the ion temperature is carefully controlled, the dissociation energy E_{diss} can be computed. Figure 10.6 shows the dissociation characteristics of $H^+(H_2O)_5$ studied in a 22-pole ion trap maintained at various temperatures [45]. The cluster ions were produced by using corona discharge followed by supersonic expansion into a vacuum chamber. The $H^+(H_2O)_5$ cluster ions were selected by a QMF and stored in the ion trap to monitor the change of ion abundance as a function of trapping time. Because the evaporation of water molecules from the clusters is faster at higher ion temperatures, the decay of $H^+(H_2O)_5$ abundance at higher temperatures is faster (Figure 10.6a). The main product of the evaporation reaction was $H^+(H_2O)_4$, but the product ion could again undergo evaporation. The results can be used to obtain an Arrhenius plot (Figure 10.6b) to deduce E_a of evaporation of $H^+(H_2O)_5$. In this case, the dissociation time of non-covalent complexes is in the range of a few hundreds of milliseconds because energy transfer via collisions is more efficient than photon absorption.

If intense light source (lasers) or continuous ion–molecule collisions are implemented, this generic method can be used to determine the dissociation energy of covalent bonds in macromolecules [48, 49]. A continuous infrared laser beam is widely used because most macromolecules exhibit good infrared absorbance over a wide wavelength range (as discussed in Chapter 3). When the laser irradiation of trapped ions to induce multiphoton absorption lasts long enough, a sufficient amount of energy may be absorbed to induce decomposition. This kind of effect has been demonstrated with trapped small ions [50, 51] or biomolecular ions [52, 53]. In addition to photodissociation, intense light sources and lasers are also widely used to induce chemical reactions of molecules in mechanistic studies. Important examples include photodesorption dynamics of molecules from a surface [54, 55], ozonolysis in the presence of other compounds [56, 57], and others [58]. Many of such reactions were conducted using TOF-MS because these reactions occurred in the low microsecond range or shorter. These reaction times are too short to be analyzed by other types of mass analyzer. In such measurements, photoexcitations were commonly performed in the ion source region of TOF-MS [59, 60]. Such devices are similar to the devices used in pump–probe laser spectroscopy.

10.3 Chemical Equilibrium

In contrast to chemical kinetics, which focuses on the rates of chemical reactions, *chemical equilibrium* focuses on the final state of reactions. It is defined as the state in which the concentrations of all the components reach a steady-state condition and no further changes occur macroscopically. That means there is no tendency toward changes in molecular concentrations or ion abundances, although they do change microscopically (dynamic equilibrium state). One should keep in mind that chemical equilibria must involve both forward and reverse reactions whereas chemical kinetics only concerns forward reactions. Contrary to the simple reaction represented in Equation 10.1, the equilibrium state involves both the forward and the reverse processes, for example:

$$aA + bB \underset{K_1'}{\overset{K_1}{\rightleftharpoons}} cC + dD \tag{10.6}$$

where K_1 and K_1' are rate constants of the forward and reverse reactions, respectively. The equilibrium constant (K_{eq}) of the whole reaction is then expressed as:

$$K_{eq} = \frac{K_1}{K_1'} = \frac{[A]^a[B]^b}{[C]^c[D]^d} \tag{10.7}$$

In the case of processes occurring in a reactor, the molecules can be delivered directly via transfer line systems or picked up by carrier gases to the inlet of a mass spectrometer. If the reaction occurs in ambient conditions, the molecules can be sampled by atmospheric pressure ion sources. If the ionization, transmission, and detection efficiencies of the reactants and the products of Equation 10.6 are available, their concentrations can be estimated from signal intensities in mass spectra. For instance, QMF can be used to track the abundance of molecules involved in oxidation/reduction reactions. The results can be used to determine the equilibrium constant of the studied reactions [61].

Once the equilibrium constant is obtained, the standard Gibbs free energy (ΔG^0) of the reaction can be computed using the following equation:

$$\Delta G^0 = -RTln(K_{eq}) \tag{10.8}$$

Notably, the change in K_{eq} with temperature can also be estimated by monitoring spectral features as a function of reaction temperature, T. Since $\Delta G^0 = \Delta H^0 - T\Delta S^0$, the standard enthalpy change (ΔH^0) and standard entropy change (ΔS^0) of the studied reaction can be derived:

$$ln(K_{eq}) = -\frac{\Delta H^o}{RT} + \frac{\Delta S^o}{R} \tag{10.9}$$

According to Equation 10.9, the ΔH^0 and ΔS^0 of the reaction can be obtained from the slope and the intercept by plotting $ln(K_{eq})$ against $1/T$, respectively. Such plots are called *van't Hoff plots*. Figure 10.7a and b show the van't Hoff plots of an endothermic and an exothermic reaction, respectively. Since the reaction enthalpy of endothermic reactions is a positive value, the slope of the van't Hoff plot is negative; in the case of an exothermic reaction, the slope is positive. Notably, van't Hoff plots are different from Arrhenius plots although they all represent the relationships between temperature and constants associated with stoichiometric reactions. The difference, however, lies in that the Arrhenius plots concern the reaction rate constants in elementary reactions, whereas the van't Hoff plots concern equilibrium constants, which comprise forward and reverse reaction rate constants.

The distinction between chemical kinetics and equilibria can be clearly seen with the evolution of cluster ions. The dissociation processes of cluster ions discussed in Figure 10.5 and Figure 10.6 are typical chemical kinetic problems because the trapped cluster ions can only undergo dissociation of solvated molecules. To reach a chemical equilibrium state, the association of solvent molecules must occurs in the reverse elementary reaction to dissociation.

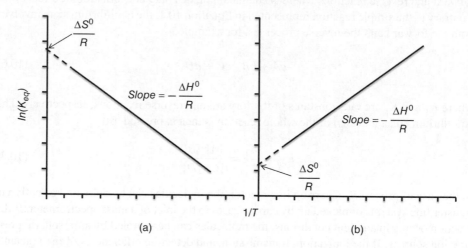

Figure 10.7   *The van't Hoff plots of endothermic (a) and exothermic (b) reactions*

The association of solvation molecules to an ion is called *nucleation reactions*. It is normally conducted under low vacuum conditions to facilitate ion–molecule interactions.

Nucleation is an important reaction in environmental chemistry because clustering of molecules on ions can provide vital information on the molecular dynamics of the Earth's atmosphere [62, 63]. Mass spectrometers can readily characterize stepwise development of cluster ions. Such measurements can be conducted using an ion source installed in a nucleation chamber or a low vacuum chamber filled with gaseous solvation molecules, with a pressure of typically tens to thousands of pascal [62–65]. Such an interface is coupled with a mass analyzer to analyze reaction products. Another example is related to the study of nucleation reactions of trapped ions in an ultra-cold multipole ion trap [66]. In these experiments, high mass accuracy and resolving power are not critical because the dissociation and association of a solvation molecule exhibit significant mass changes, e.g. 18 u for H_2O and 32 u for CH_3OH.

Figure 10.8 shows a temperature-controlled ion flow reactor equipped with a QMF for studying the association and dissociation of water molecules to form $H^+(H_2SO_4)_s(H_2O)_w$ [63]:

$$H^+(H_2SO_4)_s(H_2O)_{w-1} + H_2O + He \rightleftharpoons H^+(H_2SO_4)_s(H_2O)_w + He \qquad (10.10)$$

where s and w are the stoichiometric numbers of H_2SO_4 and H_2O, respectively. To produce the primary ion, water vapor was carried by helium gas and was introduced upstream of the ion flow reactor (right-hand side, Figure 10.8). Stable primary ions, such as $H^+(H_2O)$, were produced by an electron source next to the water vapor supplier. These ions were associated with H_2SO_4 supplied from the side of the ion flow reactor, which is downstream of the H_2O supplier. Additional water molecules attached to the ions to form various combinations of

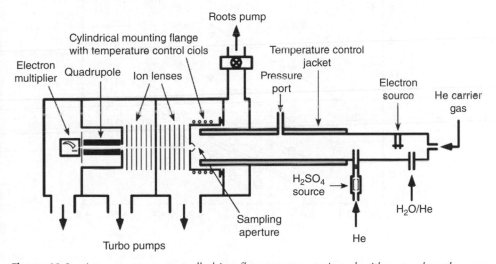

Figure 10.8 *A temperature-controlled ion flow reactor equipped with a quadrupole mass filter for studying cluster ions [63]. Reprinted with permission from Froyd, K.D., Lovejoy, E.R. (2003) Experimental Thermodynamics of Cluster Ions Composed of H_2SO_4 and H_2O. 1. Positive Ions. J. Phys. Chem. A 107: 9800–9811. Copyright (2003) American Chemical Society*

cluster ions as they traveled downstream to the sampling aperture (middle, Figure 10.8) of the mass analyzer. The ion flow reactor was maintained at various temperatures (200–400 K) to facilitate the growth of the cluster ions. Using this setup, the mass spectra of cluster ions obtained at various temperatures could be recorded. Figure 10.9 shows the van't Hoff plots of the clusters with various stoichiometric numbers (s, w). Based on these results, the ΔH and ΔS values as well as other thermodynamic properties of clusters could be derived.

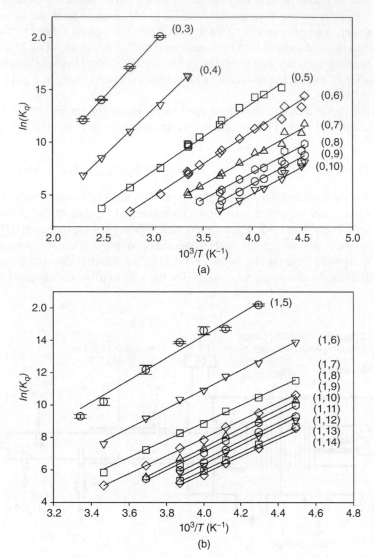

Figure 10.9   *The van't Hoff plots of $H^+(H_2SO_4)_s(H_2O)_w$ clusters with various stoichiometric numbers (s, w). (a) Association of H_2O with $H^+(H_2O)_{3-10}$; (b) association of H_2O with $H^+(H_2SO_4)(H_2O)_{5-14}$ [63]. Reprinted with permission from Froyd, K.D., Lovejoy, E.R. (2003) Experimental Thermodynamics of Cluster Ions Composed of H_2SO_4 and H_2O. 1. Positive Ions. J. Phys. Chem. A 107: 9800–9811. Copyright (2003) American Chemical Society*

Determination of the *proton affinity* of molecules is another important group of chemical equilibrium studies using TRMS [67, 68]. In *equilibrium methods*, relative proton affinity is the ΔH of the following general reaction [69, 70]:

$$B_1H^+ + B_2 \rightleftharpoons B_1 + B_2H^+ \qquad (10.11)$$

in which B_1 and B_2 are two basic molecules. Such an experiment is typically performed in an ion trapping instrument, in which the gaseous basic molecules are introduced to react with the ions [71, 72]. The competing reaction is monitored over a long time until the equilibrium state is reached to determine the equilibrium constant. The result is subsequently used to deduce the ΔG. The ΔG represents the relative gas phase basicities of B_1 and B_2. Once ΔS is available through theoretical calculation, relative proton affinity can be obtained from Equation 10.9. The method has been used to determine the proton affinity of biomolecules [73].

Notably, the relative proton affinity of molecules can also be studied with *kinetic methods* [68]. This kind of analysis can be done by monitoring competition between B_1 and B_2 for a proton in the following general reaction scheme [74]:

$$[B_1 \cdots H \cdots B_2]^+ \begin{array}{c} \overset{K_1}{\rightarrow} [B_1H]^+ + B_2 \\ \underset{K2}{\rightarrow} B_1[B_2H]^+ \end{array} \qquad (10.12)$$

By comparing the abundance of $[B_1H]^+$ and $[B_2H]^+$ and the corresponding dissociation rates, relative proton affinities can be estimated. In this method, protonated pseudo-dimer (e.g., the reactant ion of Equation 10.12) is produced by the ion source and stored in the ion trapping device. Because there is no ion–molecule reaction involved in this reaction, it is unnecessary to introduce gaseous basic molecules into the trapping device. The method has been used to determine the proton affinity of amino acids [75–77].

Other applications of TRMS for chemical equilibrium investigations also involve the estimation of the internal energy of ions [78]. The internal energy of ions can be estimated by measuring the time required to thermalize ions in ion trapping devices. TRMS is also utilized to study chemical reactions that occur at interfaces, such as redox reactions occurring under ambient conditions [79, 80] or in a vacuum [81]. In these experiments, the sampling time represents the time variable of the interfacial chemical reactions. In order to monitor the reaction with the highest temporal resolution, the spectrum acquisition is optimized to the highest speed. Alternatively, selected ion monitoring/selected reaction monitoring/multiple reaction monitoring (SIM/SRM/MRM) mode is used to ensure the highest duty cycle (up to 100%) is available. Such experiments are typically conducted using custom-made mass spectrometers or commercial instruments equipped with specially customized ion sources.

References

1. Zhong, D.P., Zewail, A.H. (1998) Femtosecond Real-time Probing of Reactions. 23. Studies of temporal, Velocity, Angular, and State Dynamics from Transition States to Final Products by Femtosecond-resolved Mass Spectrometry. J. Phys. Chem. A 102: 4031–4058.

2. Fuss, W., Schmid, W.E., Trushin, S.A. (2000) Time-resolved Dissociative Intense-Laser Field Ionization for Probing Dynamics: Femtosecond Photochemical Ring Opening of 1,3-Cyclohexadiene. J. Chem. Phys. 112: 8347–8362.
3. Minegishi, Y., Morimoto, D., Matsumoto, J., Shiromaru, H., Hashimoto, K., Fujino, T. (2012) Desorption Dynamics of Tetracene Ion from Tetracene-doped Anthracene Crystals Studied by Femtosecond Time-resolved Mass Spectrometry. J. Phys. Chem. C 116: 3059–3064.
4. Studzinski, H., Zhang, S., Wang, Y., Temps, F. (2008) Ultrafast Nonradiative Dynamics in Electronically Excited Hexafluorobenzene by Femtosecond Time-resolved Mass Spectrometry. J. Chem. Phys. 128: 164314.
5. Drescher, M., Hentschel, M., Kienberger, R., Uiberacker, M., Yakovlev, V., Scrinzi, A., Westerwalbesloh, T., Kleineberg, U., Heinzmann, U., Krausz, F. (2002) Time-resolved Atomic Inner-shell Spectroscopy. Nature 419: 803–807.
6. Rosker, M. J., Dantus, M., Zewail, A.H. (1988) Femtosecond Real-time Probing of Reactions. 1. The Technique. J. Chem. Phys. 89: 6113–6127.
7. Luhrs, D.C., Viallon, J., Fischer, I. (2001) Excited State Spectroscopy and Dynamics of Isolated Adenine and 9-Methyladenine. Phys. Chem. Chem. Phys. 3: 1827–1831.
8. Canuel, C., Mons, M., Piuzzi, F., Tardivel, B., Dimicoli, I., Elhanine, M. (2005) Excited States Dynamics of DNA and RNA Bases: Characterization of a Stepwise Deactivation Pathway in the Gas Phase. J. Chem. Phys. 122: 074316.
9. Kang, H., Lee, K.T., Kim, S.K. (2002) Femtosecond Real Time Dynamics of Hydrogen Bond Dissociation in Photoexcited Adenine–Water Clusters. Chem. Phys. Lett. 359: 213–219.
10. Baumert, T., Grosser, M., Thalweiser, R., Gerber, G. (1991) Femtosecond Time-resolved Molecular Multiphoton Ionization – the Na_2 system. Phys. Rev. Lett. 67: 3753–3756.
11. Meijer, G., Devries, M.S., Hunziker, H.E., Wendt, H.R. (1990) Laser Desorption Jet-cooling of Organic Molecules – Cooling Characteristics and Detection Sensitivity. Appl. Phys. B 51: 395–403.
12. Carrasquillo, E., Zwier, T.S., Levy, D.H. (1985) The Multiphoton Ionization Spectrum of Complexes of Benzene and Acetylene. J. Chem. Phys. 83: 4990–4999.
13. Noggle, J.H. (1989) Physical Chemistry, 2nd Edition. Scott, Foresman, and Co., Glenview, IL.
14. Santos, L.S. (2009) Reactive Intermediates MS Investigations in Solution. Wiley-VCH Verlag GmbH, Weinheim.
15. Zhu, W.T., Yuan, Y., Zhou, P., Zeng, L., Wang, H., Tang, L., Guo, B., Chen, B. (2012) The Expanding Role of Electrospray Ionization Mass Spectrometry for Probing Reactive Intermediates in Solution. Molecules 17: 11507–11537.
16. Moody, C.J., Whitham, G.H. (1992) Reactive Intermediates. Oxford University Press, Oxford.
17. Rowan, S.J., Cantrill, S.J., Cousins, G.R.L., Sanders, J.K.M., Stoddart, J.F. (2002) Dynamic Covalent Chemistry. Angew. Chem. Int. Ed. 41: 898–952.
18. Vonderlinde, D., Danielzik, B. (1989) Picosecond Time-resolved Laser Mass Spectroscopy. IEEE J. Quantum Electron. 25: 2540–2549.

19. Johnson, J.V., Yost, R.A., Kelley, P.E., Bradford, D.C. (1990) Tandem-in-space and Tandem-in-time Mass Spectrometry – Triple Quadrupoles and Quadrupole Ion Traps. Anal. Chem. 62: 2162–2172.

20. Meyer, R.T. (1968) Flash Photolysis and Time-resolved Mass Spectrometry. 2. Decomposition of Methyl Iodide and Reactivity of $I(^2P_{1/2})$ atoms. J. Phys. Chem. 72: 1583–1591.

21. Meyer, R.T. (1967) Flash Photolysis and Time-resolved Mass Spectrometry. I. Detection of Hydroxyl Radical. J. Chem. Phys. 46: 967–972.

22. Meyer, R.T. (1967) Apparatus for Flash Photolysis and Time Resolved Mass Spectrometry. J. Sci. Instrum. 44: 422–426.

23. Laidler, K.J. (1996) A Glossary of Terms Used in Chemical Kinetics, Including Reaction Dynamics. Pure Appl. Chem. 68: 149–192.

24. Santos, L.S., Knaack, L., Metzger, J.O. (2005) Investigation of Chemical Reactions in Solution Using API-MS. Int. J. Mass Spectrom. 246: 84–104.

25. Ma, X.X., Zhang, S.C., Zhang, X.R. (2012) An Instrumentation Perspective on Reaction Monitoring by Ambient Mass Spectrometry. Trends Anal. Chem. 35: 50–66.

26. Wilson, S.R., Perez, J., Pasternak, A. (1993) ESI-MS Detection of Ionic Intermediates in Phosphine-mediated Reactions. J. Am. Chem. Soc. 115: 1994–1997.

27. Aliprantis, A.O., Canary, J.W. (1994) Observation of Catalytic Intermediates in the Suzuki Reaction by Electrospray Mass Spectrometry. J. Am. Chem. Soc. 116: 6985–6986.

28. Santos, L.S. (2008) Online Mechanistic Investigations of Catalyzed Reactions by Electrospray Ionization Mass Spectrometry: A Tool to Intercept Transient Species in Solution. Eur. J. Org. Chem. 2008: 235–253.

29. Meyer, S., Metzger, J.O. (2003) Use of Electrospray Ionization Mass Spectrometry for the Investigation of Radical Cation Chain Reactions in Solution: Detection of Transient Radical Cations. Anal. Bioanal. Chem. 377: 1108–1114.

30. Kaltashov, I.A., Eyles, S.J. (2005) Mass Spectrometry in Biophysics : Conformation and Dynamics of Biomolecules. John Wiley & Sons, Inc., Hoboken.

31. Dunbar, R.C. (1989) Time-resolved Unimolecular Dissociation of Styrene Ion – Rates and Activation Parameters. J. Am. Chem. Soc. 111: 5572–5576.

32. Dunbar, R.C., McMahon, T.B. (1998) Activation of Unimolecular Reactions by Ambient Blackbody Radiation. Science 279: 194–197.

33. Dunbar, R.C. (1994) Kinetics of Thermal Unimolecular Dissociation by Ambient Infrared Radiation. J. Phys. Chem. 98: 8705–8712.

34. Louris, J.N., Cooks, R.G., Syka, J.E.P., Kelley, P.E., Stafford, G.C., Todd, J.F.J. (1987) Instrumentation, Applications, and Energy Deposition in Quadrupole Ion-trap Tandem Mass Spectrometry. Anal. Chem. 59: 1677–1685.

35. Stephenson, J.L., Booth, M.M., Shalosky, J.A., Eyler, J.R., Yost, R.A. (1994) Infrared Multiple-photon DIssociation in the Quadrupole Ion-trap Via a Multipass Optical Arrangement. J. Am. Soc. Mass Spectrom. 5: 886–893.

36. Price, W.D., Schnier, P.D., Williams, E.R. (1996) Tandem Mass Spectrometry of Large Biomolecule Ions by Blackbody Infrared Radiative Dissociation. Anal. Chem. 68: 859–866.

37. Thölmann, D., Tonner, D.S., McMahon, T.B. (1994) Spontaneous Unimolecular Dissociation of Small Cluster Ions, $(H_3O^+)L_n$ and $Cl^-(H_2O)_n$ ($n=2-4$), under Fourier Transform Ion Cyclotron Resonance Conditions. J. Phys. Chem. 98: 2002–2004.
38. Ho, Y.P., Yang, Y.C., Klippenstein, S.J., Dunbar, R.C. (1997) Binding Energies of Ag^+ and Cd^+ Complexes from Analysis of Radiative Association Kinetics. J. Phys. Chem. A 101: 3338–3347.
39. Berg, C., Beyer, M., Achatz, U., Joos, S., Niedner-Schatteburg, G., Bondybey, V.E. (1998) Stability and Reactivity of Hydrated Magnesium Cations. Chem. Phys. 239: 379–392.
40. Schindler, T., Berg, C., Niednerschatteburg, G., Bondybey, V.E. (1994) Solvation of Hydrochloric Acid in Protonated Water Clusters. Chem. Phys. Lett. 229: 57–64.
41. Wang, Y.-S., Sabu, S., Wei, S.C., Kao, C.M.J., Kong, X., Liau, S.C., Han, C.C., Chang, H.-C., Tu, S.Y., Kung, A.H., Zhang, J.Z.H. (2006) Dissociation of Heme from Gaseous Myoglobin Ions Studied by Infrared Multiphoton Dissociation Spectroscopy and Fourier Transform Ion Cyclotron Resonance Mass Spectrometry. J. Chem. Phys. 125: 133310.
42. Schnier, P.D., Price, W.D., Jockusch, R.A., Williams, E.R. (1996) Blackbody Infrared Radiative Dissociation of Bradykinin and its Analogues: Energetics, Dynamics, and Evidence for Salt-bridge Structures in the Gas Phase. J. Am. Chem. Soc. 118: 7178–7189.
43. Wang, Y.-S., Tsai, C.-H., Lee, Y.T., Chang, H.-C., Jiang, J.C., Asvany, O., Schlemmer, S., Gerlich, D. (2003) Investigations of Protonated and Deprotonated Water Clusters Using a Low-temperature 22-Pole Ion Trap. J. Phys. Chem. A 107: 4217–4225.
44. Lovejoy, E.R., Bianco, R. (2000) Temperature Dependence of Cluster Ion Decomposition in a Quadrupole Ion Trap. J. Phys. Chem. A 104: 10280–10287.
45. Wang, Y.-S., Jiang, J.-C., Cheng, C.L., Lin, S.H., Lee, Y.T., Chang, H.-C. (1997) Identifying 2- and 3-Coordinated H_2O in Protonated Ion Water Clusters by Vibrational Pre-dissociation Spectroscopy and Ab Initio Calculations. J. Chem. Phys. 107: 9695–9698.
46. Jiang, J.-C., Wang, Y.-S., Chang, H.C., Lin, S.H., Lee, Y.T., Niedner-Schatteburg, G., Chang, H.-C. (2000) Infrared Spectra of $H^+(H_2O)_{5-8}$ clusters: Evidence for Symmetric Proton Hydration. J. Am. Chem. Soc. 122: 1398–1410.
47. Wang, Y.-S., Chang, H.C., Jiang, J.-C., Lin, S.H., Lee, Y.T., Chang, H.-C. (1998) Structures and Isomeric Transitions of $NH_4^+(H_2O)_{3-6}$: From Single to Double Rings. J. Am. Chem. Soc. 120: 8777–8788.
48. McLuckey, S.A., Goeringer, D.E. (1997) Slow Heating Methods in Tandem Mass Spectrometry. J. Mass Spectrom. 32: 461–474.
49. Schafer, M., Schmuck, C., Heil, M., Cooper, H.J., Hendrickson, C.L., Chalmers, M.J., Marshall, A.G. (2003) Determination of the Activation Energy for Unimolecular Dissociation of a Non-covalent Gas-phase Peptide: Substrate Complex by Infrared Multiphoton Dissociation Fourier Transform Ion Cyclotron Resonance Mass Spectrometry. J. Am. Soc. Mass Spectrom. 14: 1282–1289.
50. Cui, W.D., Hadas, B., Cao, B.P., Lifshitz, C. (2000) Time-resolved Photodissociation (TRPD) of the Naphthalene and Azulene Cations in an Ion Trap/Reflectron. J. Phys. Chem. A 104: 6339–6344.

51. Dunbar, R.C. (1987) Time-resolved Photodissociation of Chlorobenzene Ion in the ICR Spectrometer. J. Phys. Chem. 91: 2801–2804.
52. Flora, J.W., Muddiman, D.C. (2004) Determination of the Relative Energies of Activation for the Dissociation of Aromatic Versus Aliphatic Phosphopeptides by ESI-FTICR-MS and IRMPD. J. Am. Soc. Mass Spectrom. 15: 121–127.
53. Marzluff, E.M., Campbell, S., Rodgers, M.T., Beauchamp, J.L. (1994) Low-energy Dissociation Pathways of Small Deprotonated Peptides in the Gas Phase. J. Am. Chem. Soc. 116: 7787–7796.
54. Chuang, T.J., Hussla, I. (1984) Time-resolved Mass-spectrometric Study on Infrared-laser Photodesorption of Ammonia from Cu(100). Phys. Rev. Lett. 52: 2045–2048.
55. Kools, J.C.S., Baller, T.S., Dezwart, S.T., Dieleman, J. (1992) Gas-flow Dynamics in Laser Ablation Deposition. J. Appl. Phys. 71: 4547–4556.
56. Welz, O., Savee, J.D., Osborn, D.L., Vasu, S.S., Percival, C.J., Shallcross, D.E., Taatjes, C.A. (2012) Direct Kinetic Measurements of Criegee Intermediate (CH_2OO) formed by Reaction of CH_2I with O_2. Science 335: 204–207.
57. Smith, G.D., Woods, E., DeForest, C.L., Baer, T., Miller, R.E. (2002) Reactive Uptake of Ozone by Oleic Acid Aerosol Particles: Application of Single-particle Mass Spectrometry to Heterogeneous Reaction Kinetics. J. Phys. Chem. A 106: 8085–8095.
58. Van Breemen, R.B., Snow, M., Cotter, R.J. (1983) Time-resolved Laser Desorption Mass Spectrometry. 1. Desorption of Preformed Ions. Int. J. Mass Spectrom. Ion Proc. 49: 35–50.
59. Osborn, D.L., Zou, P., Johnsen, H., Hayden, C.C., Taatjes, C.A., Knyazev, V.D., North, S.W., Peterka, D.S., Ahmed, M., Leone, S.R. (2008) The Multiplexed Chemical Kinetic Photoionization Mass Spectrometer: A New Approach to Isomer-resolved Chemical Kinetics. Rev. Sci. Instrum. 79: 104103.
60. Baeza-Romero, M.T., Blitz, M.A., Goddard, A., Seakins, P.W. (2012) Time-of-flight Mass Spectrometry for Time-resolved Measurements: Some Developments and Applications. Int. J. Chem. Kinet. 44: 532–545.
61. Schweich, D., Villermaux, J. (1979) Measurement of Chemical-equilibrium Constants by Continuous Mass Spectrometry of Transient Compositions at the Outlet of a Pulsed Reactor. Anal. Chem. 51: 77–79.
62. Froyd, K.D., Lovejoy, E.R. (2003) Experimental Thermodynamics of Cluster Ions Composed of H_2SO_4 and H_2O. 2. Measurements and Ab Initio Structures of Negative Ions. J. Phys. Chem. A 107: 9812–9824.
63. Froyd, K.D., Lovejoy, E.R. (2003) Experimental Thermodynamics of Cluster Ions Composed of H_2SO_4 and H_2O. 1. Positive Ions. J. Phys. Chem. A 107: 9800–9811.
64. Cunningh, A.J., Payzant, J.D., Kebarle, P. (1972) Kinetic Study of Proton Hydrate $H^+(H_2O)_n$ Equilibria in Gas Phase. J. Am. Chem. Soc. 94: 7627–7632.
65. Castleman, A.W., Bowen, K.H. (1996) Clusters: Structure, Energetics, and Dynamics of Intermediate States of Matter. J. Phys. Chem. 100: 12911–12944.
66. Gerlich, D. (1995) Ion-neutral Collisions in a 22-Pole Trap at Very Low Energies. Phys. Scripta T59: 256–263.
67. Harrison, A. (1997) The Gas-phase Basicities and Proton Affinities of Amino Acids and Peptides. Mass Spectrom. Rev. 16: 201–217.

68. Cooks, R.G., Patrick, J.S., Kotiaho, T., McLuckey, S.A. (1994) Thermochemical Determinations by the Kinetic Method. Mass Spectrom. Rev. 13: 287–339.
69. Hunter, E.P.L., Lias, S.G. (1998) Evaluated Gas Phase Basicities and Proton Affinities of Molecules: An Update. J. Phys. Chem. Ref. Data 27: 413–656.
70. Lias, S.G., Liebman, J.F., Levin, R.D. (1984) Evaluated Gas-phase Basicities and Proton Affinities of Molecules – Heats of Formation of Protonated Molecules. J. Phys. Chem. Ref. Data 13: 695–808.
71. Kebarle, P. (1992) Ion Molecule Equilibria, How and Why. J. Am. Soc. Mass Spectrom. 3: 1–9.
72. Brodbeltlustig, J.S., Cooks, R.G. (1989) Determination of Relative Gas-phase Basicities by the Proton-transfer Equilibrium Technique and the Kinetic Method in a Quadrupole Ion-trap. Talanta 36: 255–260.
73. Aue, D.H., Webb, H.M., Bowers, M.T. (1976) Quantitative Proton Affinities, Ionization Potentials, and Hydrogen Affinities of Alkylamines. J. Am. Chem. Soc. 98: 311–317.
74. Cooks, R.G., Kruger, T.L. (1977) Intrinsic Basicity Determination Using Metastable Ions. J. Am. Chem. Soc. 99: 1279–1281.
75. Burlet, O., Gaskell, S.J. (1993) Decompositions of Cationized Heterodimers of Amino Acids in Relation to Charge Location in Peptide Ions. J. Am. Soc. Mass Spectrom. 4: 461–469.
76. Bojesen, G. (1987) The Order of Proton Affinities of the 20 Common L-α-Amino Acids. J. Am. Chem. Soc. 109: 5557–5558.
77. Cheng, X.H., Wu, Z.C., Fenselau, C. (1993) Collision Energy-dependence of Proton-bound Dimer Dissociation – Entropy Effects, Proton Affinities, and Intramolecular Hydrogen-bonding in Protonated Peptides. J. Am. Chem. Soc. 115: 4844–4848.
78. Konn, D.O., Murrell, J., Despeyroux, D., Gaskell, S.J. (2005) Comparison of the Effects of Ionization Mechanism, Analyte Concentration, and Ion "Cool-times" on the Internal Energies of Peptide Ions Produced by Electrospray and Atmospheric Pressure Matrix-assisted Laser Desorption Ionization. J. Am. Soc. Mass Spectrom. 16: 743–751.
79. Grimm, R.L., Hodyss, R., Beauchamp, J.L. (2006) Probing Interfacial Chemistry of Single Droplets with Field-induced Droplet Ionization Mass Spectrometry: Physical Adsorption of Polycyclic Aromatic Hydrocarbons and Ozonolysis of Oleic Acid and Related Compounds. Anal. Chem. 78: 3800–3806.
80. Abonnenc, M., Qiao, L.A., Liu, B.H., Girault, H.H. (2010) Electrochemical Aspects of Electrospray and Laser Desorption/Ionization for Mass Spectrometry. Annu. Rev. Anal. Chem. 3: 231–254.
81. Zhou, L., Piekiel, N., Chowdhury, S., Zachariah, M.R. (2010) Time-resolved Mass Spectrometry of the Exothermic Reaction between Nanoaluminum and Metal Oxides: The Role of Oxygen Release. J. Phys. Chem. C 114: 14269–14275.

11

Application of Time-resolved Mass Spectrometry in the Monitoring of Chemical Reactions

Mass spectrometry (MS) is implemented in the monitoring of chemical reactions to answer one or both of the following two questions:

1. What intermediates and products are generated during the reactions?
2. What mechanism do the reactions/kinetics follow?

In some cases, these two questions may be answered with the aid of the acquired MS signals. If the reactions are slow, and the intermediates are stable, the intermediates and the products can be gradually recorded using a temporal MS scan. Thus, the analytical method should be stable over time. The ability to monitor slow reactions is important because the kinetic information can assist investigation of the reaction mechanism. However, some reactions proceed fast while their intermediates have short life times (<1 s) such reactions require suitable ionization techniques that are capable to record intermediates observing the intermediate with short life times. Fortunately, most ionization techniques described in Chapter 2 provide short ionization times. Therefore, it is often feasible to capture short-lived intermediates with millisecond timescale. The main challenge is to decrease the lag time between sampling and ionization. Many ionization techniques have been used to enable monitoring of various chemical reactions. In this chapter, we will provide examples of studies where MS was implemented in the monitoring of various reactions including organic reactions, catalytic reactions, and photoionizations. All of them have quite different time spans. Monitoring slow (minutes to hours), fast (milliseconds to seconds), and ultra-fast reactions (sub-milliseconds) by MS is described in the chapter. Applications of MS in the monitoring of biochemical reactions are covered in Chapter 13.

Time-Resolved Mass Spectrometry: From Concept to Applications, First Edition.
Pawel Lukasz Urban, Yu-Chie Chen and Yi-Sheng Wang.
© 2016 John Wiley & Sons, Ltd. Published 2016 by John Wiley & Sons, Ltd.

11.1   Organic Reactions

Most organic reactions are carried out in the liquid phase. Ionization techniques operated under atmospheric pressure – in many cases – enable direct infusion of liquid samples to the MS interfaces (e.g., electrospray ionization, ESI [1–19]). Such interfaces have commonly been used to couple chemical reactors with MS. In some cases, both qualitative and quantitative chemical information can be obtained during the reaction monitoring. Reactants can be monitored either off-line or on-line (see also Chapter 4). Off-line monitoring can be conducted by collecting reaction solution at a given sampling frequency. This mode of analysis is suitable for the reactions with low rates. When using MS to monitor chemical reactions, the consecutive mass spectra may represent different patterns that reflect changes in the investigated system. As pointed out in Chapter 1, single-point measurements provide snapshots of the investigated reactions at given time points. When off-line monitoring of chemical reactions is conducted, simple separations of reaction species by thin-layer chromatography (TLC) can be conducted to avoid the ion suppression effects that occur during the MS analysis [20–22]. Highly dynamic systems sometimes need to be quenched before introduction to the ion source. Real-time monitoring minimizes the time required to obtain each temporal data point. The time interval must be negligible relative to the time span of the investigated process. To avoid losing any important information about short-lived reaction intermediates, one can implement on-line monitoring approaches [1–19]. Moreover, unstable or air-sensitive intermediates can be protected by inert gas or liquid nitrogen during MS analysis. Low-temperature liquid secondary ion mass spectrometry and atmospheric pressure ionization have been used to observe unstable intermediates in a Witting reaction. The reactants were cooled with liquid nitrogen [23]. Furthermore, when combining ion sources operated in atmospheric pressure with tandem MS (MS/MS), structural information can be obtained [24]. A large amount of information about the mechanisms of chemical reactions can be extracted from MS data. Such measurements enabled by appropriate ionization techniques combined with tandem mass analyzers. Additionally, ion mobility MS using ESI as the ion source has been used to monitor reactions in real time [25]. The advantage of using this technique is that the product ions can be separated in gas phase in the ion mobility chamber prior to MS analysis, minimizing ion suppression effects.

On-line derivatization and monitoring of reaction products by MS can be conducted to improve ionization efficiency of analytes with poor ionization capability (e.g. aldehydes) [26]. A charge is incorporated to the target analytes through derivatization. Therefore, the ionization efficiency can be improved. Furthermore, target analytes in complex samples can be selectively reacted with the derivatization reagents and become visible in the mass spectra without conducting time-consuming extraction procedures. For example, 4-(2-(trimethylammonio)ethoxy)benzenaminium halide is a good derivatization reagent for aldehyde-type analytes [26]. The resulting derivatized products have much higher ionization efficiency than the original analytes prior to derivatization [26]. Although the reaction species were analyzed by MS off-line by injecting the reaction solution through an autosampler with a given time interval [26], it is also possible to monitor reactions in real time by MS.

By the end of the 1980s, a sound setup had been proposed for monitoring chemical reactions in real time and investigating reaction kinetics [1]. A reaction chamber equipped with a temperature regulator (using an electric heater and thermostat) was directly connected to an ESI emitter. The ion source used in this system was then called *ion spray* (pneumatically assisted electrospray). It implemented a high speed (~ 300 m s^{-1}) nebulizing gas (e.g., N_2) supplied co-axially to the ESI emitter. The use of the high pressure gas can improve the ionization efficiency. In the above-mentioned prototype, the reaction solution was continuously transferred from the reaction chamber toward the tip of the ESI emitter for ionization. This approach was suitable for monitoring enzymatic and non-enzymatic organic reactions [1]. Radical chain reactions involving short-lived intermediates could also be monitored by ESI-MS [2, 27, 28]. Transient reactants in radical chain reactions could be observed in the ESI mass spectra. This method helped to clarify the mechanism of the chain reaction.

Recently, a sub-millisecond resolution time-resolved mass spectrometry (TRMS) approach for monitoring chemical reactions, implementing desorption electrospray ionization (DESI), has been proposed [29]. Two reactant solutions were mixed rapidly. The mixing was followed by DESI. Because of the high velocity of the liquid jet, high temporal resolution (~ 300 μs) could be achieved. Additionally, a fused-droplet approach for mixing high-speed reactant liquid droplets was demonstrated [30]. It enables recording the kinetics of liquid-phase chemical reactions within microseconds. Upon merger of the two streams of micrometer-sized droplets generated by electrosprays, reactions were initiated and recorded by MS with microsecond temporal resolution. This observation was possible because the time of the formation of the gas-phase ions from nanoliter-sized droplets was previously estimated to be in the order of few tens of microseconds (Table 2.3). Further, fundamental studies are required to give insights on the timing of such complex systems. Whether or not the mixing of two droplet streams is limited by the electric field in proximity to the MS orifice, or within the merged droplets, remains an open question.

In some cases, the reaction solution is transferred from the reaction flask by a series of liquid chromatography (LC) pumps to the mass spectrometer (Figure 11.1). The transfer line incorporates various steps such as quenching, sequential dilution, and addition of a make-up solution to promote ionization in the ion source [10]. If the flow rate is too high, excess solution is removed by a flow splitter before delivering the reaction solution to the ion source (Figure 11.1). This procedure is similar to the handling of high flow rate effluents during LC-MS analysis (see Chapter 6). On the other hand, if the flow rate is in the range of microliters [31] down to nanoliters [32] per minute, splitting the flow is not necessary (see also Chapter 6).

The Venturi easy ambient sonic-spray ionization (V-EASI) interface has also been used to deliver reaction solution from a flask to the mass spectrometer's inlet [33]. It uses the so-called *Venturi effect* to generate hydrodynamic flow in the sampling capillary (see also Chapter 4). This ion source is normally operated by supplying compressed nitrogen at a pressure of ~ 10 bar ($\sim 10^6$ Pa) [33] although versions with lower pressure have been proposed [34–38], in which the typical flow rate of the sample is ~ 15 μl min^{-1}. Although a syringe pump is not used, the flow rate is satisfactorily stable. V-EASI has been used in long-term (e.g., ~ 2 h) monitoring of the Morita–Baylis–Hillman reaction [33], which is a slow reaction involving carbon–carbon bond formation between the α-position of an activated alkene and a carbon electrophile such as an aldehyde. The reaction intermediates could be recorded in real time as the reaction proceeded during a 2 h time span. This ionization and interface system is straightforward because it does not require any electric contact

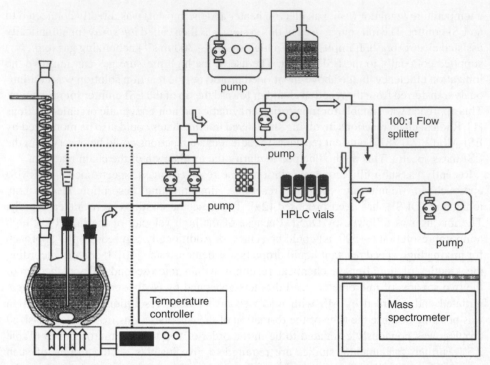

Figure 11.1 *Schematic diagram of the apparatus for on-line MS interrogation of a chemical reaction. Pump no. 1 samples the reaction solution; pump no. 2 provides ambient-temperature methanol to quench the reaction, and to carry out primary dilution; pump no. 3 is used to meter solution to a flow splitter; and pump no. 4 adds dilute HCl to enable solute protonation [10]. Reprinted with permission from Dell'Orco, P., Brum, J., Matsuoka, R., Badlani, M., Muske, K. (1999) Monitoring Process-scale Reactions Using API Mass Spectrometry. Anal. Chem. 71: 5165–5170. Copyright (1999) American Chemical Society*

between the sample emitter and power supply. The stream of nitrogen or air is sufficient to deliver reaction solution and facilitate nebulization of liquid effluent at the V-EASI emitter, so that reactants can be ionized. Nevertheless, the pressure of the gas stream in the V-EASI setup needs to be optimized to obtain high-quality data.

Contrary to V-EASI, in the case of polarization induced (PI)-ESI [e.g., contactless atmospheric pressure ionization (C-API) and ultrasonication-assisted spray ionization (UASI)] [39–41], one does not need to supply nebulizing gas, which makes the monitoring system even more simplistic. In C-API, a glass capillary (~ 20 cm, inner diameter: 50 μm) was inserted to the reaction vial to direct the reaction solution toward the mass spectrometer. The tapered end of this capillary was placed in proximity (~ 1 mm) to the inlet of the mass spectrometer (Figure 11.2a). The flow of sample was mainly due to capillary action and the influence of the electric field near the mass spectrometer, which may induce some electroosmotic flow (EOF) (see Chapter 6). In this approach, the sample flow rate is low (~ 100 nl min^{-1}). The sampling time can be reduced by using a shorter capillary as the sampling tube. Similarly to V-EASI, the capillary combines

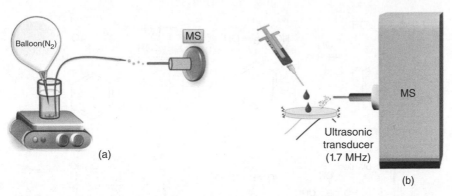

Figure 11.2   *Two setups for monitoring chemical reactions by C-API. (a) Cartoon diagram of the on-line monitoring configuration of an organic reaction by C-API-MS [39]. [Online Monitoring of Chemical Reactions by Contactless Atmospheric Pressure Ionization Mass Spectrometry. Hsieh, C.-H., Chao, C.-S., Mong, K.-K.T., Chen, Y.-C. J. Mass Spectrom. 47: 586–590. Copyright (2012) John Wiley and Sons]. (b) Schematic diagram of the miniature UASI–MS setup [41]. [Real Time Monitoring of Accelerated Chemical Reactions by Ultrasonication-assisted Spray Ionization Mass Spectrometry. Lin, S.-H., Lo, T.-J., Kuo, F.-Y., Chen, Y.-C. J. Mass Spectrom. 49: 50–56. Copyright (2014) John Wiley and Sons]*

the role of sampling device and ion source component. The tapered capillary outlet facilitates accumulation of electric charges due to the electric polarization induced by the electric field in the vicinity of the mass spectrometer's orifice. If the studied reaction requires oxygen-free conditions, the reaction vial can be sealed, and the gas in its headspace replaced by an inert gas (e.g., delivered from a rubber balloon; Figure 11.2a). Furthermore, injecting inert gas can induce hydrodynamic flow in the sampling capillary (see Section 7.4, Equation 7.1) – increasing the flow rate up to $\sim 200\,nl\,min^{-1}$. If additional energy is required to accelerate the reaction, the reaction vial can be heated on a hot plate or exposed to ultrasound [40]. On the other hand, short-lived reaction intermediates (lifetime: approximately milliseconds) can be recorded by UASI-MS using a miniature ultrasonic transducer (Figure 11.2b), which can be used to accelerate organic reactions and to promote formation of ultrafine droplets that can readily be desolvated [41]. The substrates are deposited directly on the surface of the ultrasonic transducer. The reaction products and intermediates are immediately recorded by the mass spectrometer.

The concept of reactive MS has been extensively used in conjunction with DESI [42–45]. In DESI, an electrospray stream carries charged microdroplets on the surface of the sample to dissolve and desorb fine droplets containing analytes (see Chapter 2). If the spray solvent carries a reagent, it can react with co-reactants present in the sample. An interesting example [44] is the rapid and reversible covalent interaction of phenylboronic acid [PhB(OH)$_3$] with diols to form cyclic boronates in a basic aqueous medium (Figure 11.3a). Here, phenylboronic acid (Figure 11.3b) is used to selectively react with analytes with *cis*-diols including fructose and glucose to form the products detected at m/z 301 (Figure 11.3c and d). Product ions will not be recorded by the mass spectrometer unless the carbohydrate analytes are present in the reaction.

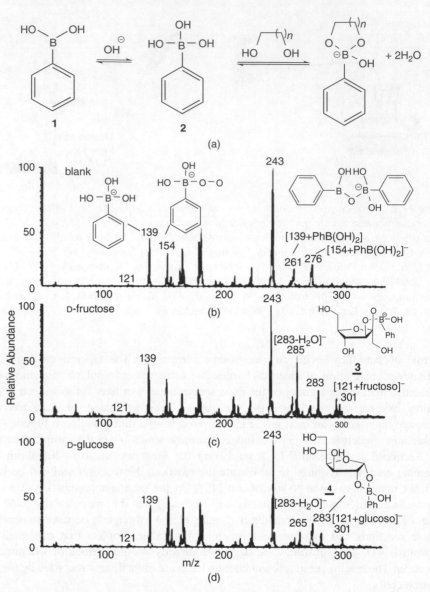

Figure 11.3 *Reactive DESI. (a) Reaction of phenylboronic acid with cis-diols. n = 1 or 2. Reactive-DESI mass spectra showing the ionic species generated from PhB(OH)₃ anions upon interaction with (b) a blank surface, (c) D-fructose on the surface, and (d) D-glucose absorbed in a cotton tip [44]. Reproduced from Chen, H., Cotte-Rodríguez, I., Cooks, R.G. (2006) with permission of the Royal Society of Chemistry*

Charged droplets formed in the electrospray can be used to study the course of chemical reactions and to identify reaction intermediates [46–53]. The Hantzsch synthesis of symmetric 1,4-dihydropyridine derivatives was used as the model reaction in a clever experimental design [46]. Droplets containing the reagents (ethyl acetoacetate, ammonium acetate, and an aromatic aldehyde), were directed to the inlet of the mass spectrometer in a way that the flight distance could be varied. The effects of the distance between the sprayer and the inlet of the mass spectrometer on the progress of the reaction and the degree of desolvation of the droplets were examined. It is quite interesting that more intermediate ions and product ions could be observed as the length of the transfer tube was extended. The results indicated that longer travel distance/time can allow the reaction to proceed on the way to the mass spectrometer's inlet. Furthermore, if finer droplets are present, the reaction can reach its completion more readily.

Extractive electrospray ionization (EESI) – a modified version of ESI – is also useful in the on-line monitoring of chemical reactions characterized by low or high rates [14, 54–56]. For example, the monitoring of the Michael addition reaction of phenylethylamine and acrylonitrile in ethanol conducted at room temperature has been demonstrated [14]. EESI-MS was used to track the appearance of the reaction product, phenylethylaminopropionitrile. The signal of the product appeared after ~1 s. The reaction was monitored for ~ 1 h. Intermediate and side product ions were also recorded in the mass spectra. The side product, 3-[(2-cyanoethyl)phenylethyl-amino]propionitrile appeared after 40 min. In other work, a simple reaction of base hydrolysis of ethyl salicylate (m/z 165) to salicylic acid (m/z 137) via an intermediate, which is methyl salicylate (m/z 151), was used as a model reaction to emphasize the potential of EESI-MS in real-time reaction monitoring (Figure 11.4a–c) [54]. This EESI interface consisted of a grounded nebulizer, which produced an analyte aerosol, and a Venturi pump used to transfer the sample to the ionization region (Figure 11.4d). The reaction was monitored for ~1 h. Initially, the mass spectrum was dominated by deprotonated ethyl salicylate ([ES-H]$^-$) (Figure 11.4a). After 25 min, the intermediate methyl salicylate ([MS-H]$^-$) became the base peak (Figure 11.4b). Finally, the peak at m/z 137 (derived from the deprotonated salicylic acid) dominated mass spectra (Figure 11.4c) [54]. Reaction intermediates can readily be recorded using this approach. Reactions conducted in viscous solutions could also be monitored on-line using EESI-MS [55]. Reactants were directed up to the liquid surface with the aid of bubbles and delivered to the ion source (see Section 2.7.1.3). Bubble rupture occurs generating of microdroplets through a mechanism known as microjetting [55]. Slow reactions such as fructose dehydration were monitored for 1 h to record the appearance of intermediate and product ions. Additionally, reactive EESI has also been demonstrated in the detection of acetonitrile [56]. Because of its low proton affinity of acetonitrile (779 kJ mol^{-1}) [56], it is difficult to be observe acetonitrile peaks in ESI mass spectra. When the ESI spray in EESI contains silver ions, silver adduct ions of acetonitrile can be observed in the mass spectrum after fusing the gas phase acetonitrile with the silver ions [56]. Determination of trace amounts of acetonitrile present in breath is important because acetonitrile has been recognized as a breath marker of lung cancer and smoking habit [56].

In the case of ESI used in reaction monitoring, the components are normally mixed in advance, and such a reaction mixture is loaded to a reservoir. Alternatively, reactant solutions can be mixed in on-line mixers. For monitoring very fast processes, reactive ESI can be employed by mixing two ESI sprays containing different reactants in the air. It has been shown that DESI and EESI approaches can readily be implemented in the monitoring of fast reactions. For example, in the *reactive-DESI*, a reaction will not be initiated until the spray containing one reactant impinges on the sample substrate containing the other reactant. On the other hand, reactions can also be initiated by fusing droplets containing

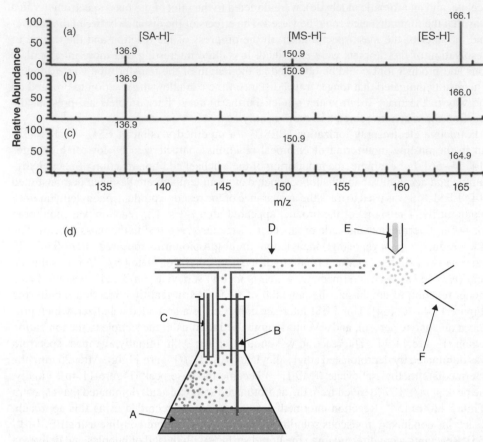

Figure 11.4   *EESI-MS used for reaction montoring. Mass spectra from EESI reaction monitoring of the hydrolysis of ethyl salicylate (ES, m/z 165) to salicylic acid (SA, m/z 151) via methyl salicylate (MS, m/z 137). Spectrum (a) was recorded at the start of the reaction. The reactant (ES) is the base peak. Spectrum (b) was recorded after 25 min. The ES peak has dropped significantly while the intermediate (MS) and product (SA) signals have both increased. Spectrum (c) was recorded after 45 min. The ES peak has reduced in intensity further, and the product peak is now the base peak in the spectrum. (d) Schematic of the EESI interface. Sample is drawn from the reaction mixture (A) via a polyether ether ketone capillary (B) using a high-performance liquid chromatography or peristaltic pump which is split with the major flow returned directly to the reaction mixture (not shown) and the minor flow sprayed back into the reaction vessel using a grounded nebulizer (C). The resulting aerosol is sampled using a Venturi pump (D), the output of which is positioned orthogonal to the ESI nebulizer of the mass spectrometer (E) where the transferred analytes are ionized before entering the mass spectrometer through the sample orifice (F) [54]. [On-line Reaction Monitoring by Extractive Electrospray Ionisation. McCullough, B.J., Bristow, T., O'Connor, G., Hopley, C. Rapid Commun. Mass Spectrom. 25:/1445–1451. Copyright (2011) John Wiley and Sons]*

reactants in EESI [54–56]. Additionally, EESI and conventional ESI are suitable interfaces for continuous monitoring of batch reactions over long periods of time. The unique feature of PI-ESI - a modified version of ESI - is the simplicity of the instrumental setup. Several adaptations of DESI, FDESI, and EESI are suitable for ultra-fast reaction monitoring (sub-milliseconds) while ESI and PI-ESI can readily be employed in the monitoring of fast and slow reactions (milliseconds to minutes).

Probe electrospray ionization (PESI) is a variant of ESI [57]. Here, a solid needle is used as a sampling tool and sample emitter, which is connected to a high voltage (Figure 11.5). The needle is inserted into a sample solution first, and then moved to a location in proximity to the mass spectrometer's inlet (Figure 11.5). The small quantity of sample adsorbed on the needle tip (Figure 11.5) is sufficient for ESI-MS measurement. When the needle is driven up, the adsorbed sample is ionized in front of the mass spectrometer's orifice. Similarly to most previous implementations of ESI, PESI-MS can be used to monitor slower reactions in real time. In the early demonstration of the technique, an air blower was used to drive a drop of solution B slowly toward solution A (Figure 11.5). At the same time, the needle was moving periodically up and down. The frequency of the needle movement was 3 Hz. Thus, mass spectra could be acquired every ~ 0.33 s [57], which enabled constant monitoring of the reaction progress. The model reaction used in this work was the reaction of benzaldehyde and ethanolamine, which gave rise to the formation of a Schiff base. Solution A contained benzaldehyde (0.2 ml, 10 mM) prepared in ethanol/water (7:3, v/v) while solution B contained ethanolamine (0.01 ml, 0.1 M) prepared in methanol/water (7:3, v/v). After 30 s, the protonated Schiff base product was detected, and the reactant ions were decreased to noise level. The reaction can be conducted in small droplets. In this case, reaction solution was not continuously introduced to the mass spectrometer, so that the sample consumption was very small. This way, the possibility to

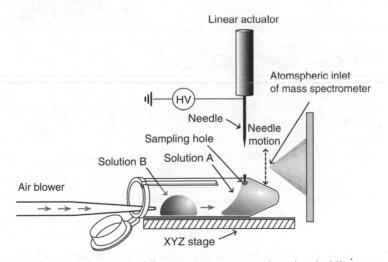

Figure 11.5 *Schematic depiction of the PESI-MS setup combined with diffusive mixing strategy for real-time monitoring [57]. [Real-time Reaction Monitoring by Probe Electrospray Ionization Mass Spectrometry. Yu, Z., Chen, L.C., Erra-Balsells, R., Nonami, H., Hiraoka, K. Rapid Commun. Mass Spectrom. 24: 1507–1513. Copyright (2010) John Wiley and Sons]*

contaminate the mass spectrometer was minimized. Presumably, the amount of the sample consumed during one analysis should be at the sub-microliter level. In fact, low sample consumption can be considered to be the main advantage of PESI in monitoring chemical reactions.

Chemical reactions can also be carried out above the surface of a heated metal probe and – at the same time – monitored by MS (Figure 11.6a). In the so-called paper-assisted thermal ionization [58], substrate solutions are deposited directly on a filter paper positioned on a heated metal probe. The filter paper has triangular form with the sharp angle at one corner, pointing at the mass spectrometer's inlet, to be used as the ESI emitter. Eschweiler–Clarke methylation, which involves a nucleophilic additive–elimination reaction at high temperature, was used as the model reaction to test this system (Figure 11.6b). Here, formaldehyde (HCHO), formic acid (HCOOH), and primary or secondary amines would be methylated to become tertiary amines (Figure 11.6b). The authors of this approach speculated that heating does not only accelerate reactions but it also assists the ionization process.

The Eschweiler–Clarke reaction can also be monitored by DESI [59]. It has been demonstrated that transient intermediates with short lifetimes can be monitored by DESI with little requirement for sample preparation. Additionally, low-temperature plasma (LTP) combined with MS has been used to monitor fast chemical reactions [60, 61]. For example, the

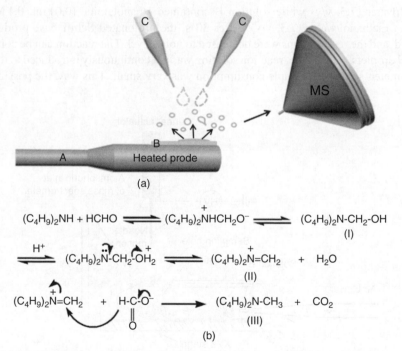

Figure 11.6 *Chemical reactions accelerated under heating for real-time MS monitoring. (a) Schematic diagram of the paper-assisted thermal ionization setup. (A) Heated metal probe, (B) filter paper, and (C) home-made pipette tip. (b) Proposed mechanism of the Eschweiler–Clarke reaction [58]. Reproduced from Pei, J., Kang, Y., Huang, G. (2014) with permission of the Royal Society of Chemistry.*

LTP probe was used to continuously monitor ongoing condensation of ethylenediamine with aldehyde within a few tens of seconds [60]. This approach can monitor reactants and products including polar and non-polar organic compounds. Furthermore, only trace amounts of the reaction species lifted and ionized by the LTP are sufficient for MS analysis. Therefore, the common contamination problems occurring in on-line monitoring MS can be reduced using this approach.

Although most reactions are carried out in liquid phase, some reactions such as *solid-phase peptide synthesis* are conducted directly on solid-phase resins. It is possible to directly characterize reaction products on solid supports using the newly developed ionization techniques. For example, direct analysis in real time (DART) has been successfully used to characterize synthetic peptides on solid supports by MS [62]. In fact, it is possible to monitor various reaction species produced on solid substrates by employing suitable ionization techniques (see Chapter 2).

11.2   Catalytic Reactions

Catalysts boost yields within short periods of time by lowering the activation energy of the process. Analytical methods can support the search for suitable catalysts for various reactions [63–73]. Because of its selectivity, sensitivity, speed, and reliability, ESI-MS has been widely used to monitor catalytic reactions [63–73]. For example, ESI-MS/MS has been implemented in the screening of the Brookhart-type Pd(II) olefin polymerization catalysts [74]. A Brookhart-type precursor, Pd(II) $(\eta^2, \eta^2$-cycloocta-1, 5-diene)chloromethylpalladium(II) ([(cod)Pd(CH$_3$)(Cl)]) can form complexes with eight different ligands (Figure 11.7a), which could be used as catalysts in polymerization reactions. The ESI-MS and ESI-MS/MS were used as detection tools to screen suitable catalysts. The appearance of the product ions in the mass spectra after a few hours revealed the progress of the reaction. Initially, the complexes (**1a–h**) were prepared by mixing equimolar amounts of the eight individually prepared diimine ligands in diethyl ether with one equivalent of [(cod)Pd(CH$_3$)(Cl)] at room temperature. The overnight reaction was followed by evaporation to dryness, washing with hexane, and activation of the solid orange residue in dichloromethane with silver trimethanesulfonate (AgOTf) (Figure 11.7a). Polymerization was conducted by reacting the mixture of the catalysts in dichloromethane with ethylene at −10 °C for 1 h, then quenched by adding to a 100-fold greater volume of 3% dimethyl sulfoxide (DMSO) in CH$_2$Cl$_2$. The resulting solution was analyzed by ESI-MS (Figure 11.7b) and ESI-MS/MS (Figure 11.7c; parent ions $m/z > 2200$). The results show that complex **4c** dominated the tandem mass spectrum, suggesting that complex **1c** was the best catalyst among the eight complexes tested (Figure 11.7a). The MS results of the ESI-MS screening are convergent with those obtained by conventional assays. In fact, using ESI-MS (MS/MS) is an efficient way to screen catalysts of polymerization reactions. Different catalysts can be added to the reaction mixtures, and the resulting product mixtures can be analyzed directly by MS. The polymerization reaction is relatively slow (1 h); thus, it was monitored off-line. Nevertheless, it is also possible to monitor the polymerization reactions in real time by on-line MS.

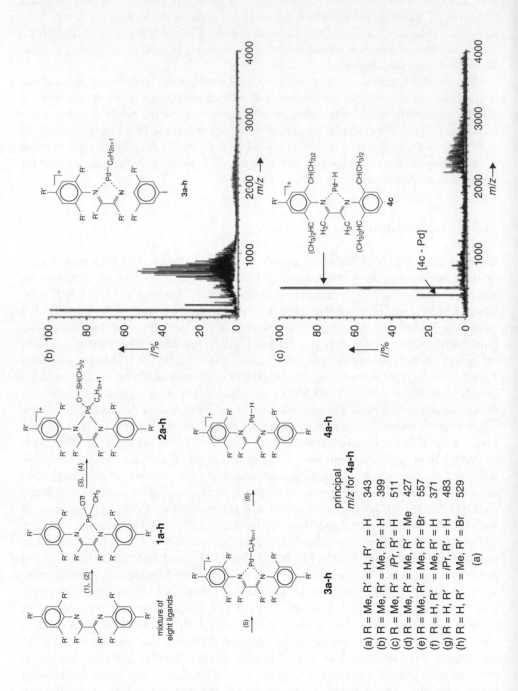

(a)

principal
m/z for **4a-h**

(a) R = Me, R' = H, R'' = H   343
(b) R = Me, R' = Me, R'' = H   399
(c) R = Me, R' = iPr, R'' = H   511
(d) R = Me, R' = Me, R'' = Me   427
(e) R = Me, R' = Me, R'' = Br   557
(f) R = H, R' = Me, R'' = H   371
(g) R = H, R' = iPr, R'' = H   483
(h) R = H, R' = Me, R'' = Br   529

Figure 11.7 *Using ESI-MS to find candidates for catalysts. (a) (1) [(cod)Pd(CH₃)(Cl)], 20 °C,* *10 h; (2) AgSO₃CF₃; (3) excess C₂H₄ , −10 °C, 1 h; (4) DMSO; (5) electrospray under mild* *desolvation conditions to give polymeric ions (and loss of DMSO); (6) reject all ions below a* *certain mass and subject the remaining high-mass ions to collision with Xe. (b) ESI mass spec-* *trum of the mixture of oligomeric/polymeric ions 3a–h after reaction of 1a–h with ethylene* *and quenching with DMSO. (c) After selection of ions with m/z > 2200 and collision-induced* *dissociation with Xe to induce β-hydride elimination, the daughter ion spectrum shows a pre-* *dominant peak at m/z 511 and a smaller peak at m/z 405. The former peak corresponds to 4c* *and the latter to a secondary fragment [4c-Pd] [74]. Reproduced from Hinderling, C., Chen, P.* *(1999) Rapid Screening of Olefin Polymerization Catalyst Libraries by Electrospray Ionization* *Tandem Mass Spectrometry. Angew. Chem. Int. Ed. 38: 2253–2256. with permission from* *John Wiley and Sons*

Short-lived reaction intermediates derived from catalytic reactions can also be detected using various MS approaches. Because of its unique design, DESI-MS is highly suit-able for detection of short-lived intermediates. Reactions are initiated by the merger of all the reagents and catalysts. For example, the DESI spray can contain reactants, while the sample loading substrate can be coated with catalyst [75] (Figure 11.8). The reaction intermediates/products can be immediately detected by MS once the spray impinges on the catalyst to generate splashed charged droplets for MS analysis. The gas-phase ions, derived from the charged droplets, which may contain reaction inter-mediates/products, can be instantly analyzed by MS. An archetypal C-H amination reaction catalyzed by a dirhodium tetracarboxylate complex, was used as the model reaction to demonstrate this approach [75]. The short-lived (lifetime: nanosecond to microsecond regime) transient reactive species were monitored using this method [75]. The C-H amination experiments were performed by spraying a solution containing the sulfamate ester (2) (ROSO₂NH₂; R = CH₂CCl₃) and iodine(III) oxidant (3) at bis[rhodium(α, α, α′, α′-tetramethyl-1, 3-benzenedipropionic acid)] (Rh₂(esp)₂) (1) or a

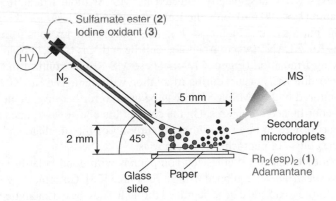

Figure 11.8 *Using DESI-MS to detect short-lived intermediates from catalytic reac-* *tions. DESI-MS setup for intercepting transient intermediates of the Rh₂(esp)₂-catalyzed* *C-H amination of adamantane [75]. Reproduced from Perry, R.H., Cahill III, T.J., Roizen, J.L.,* *Bois, J.D., Zare, R.N. (2012) with permission of PNAS*

mixture of (1) and a hydrocarbon substrate (adamantane) deposited on a paper surface (Figure 11.8). Short-lived reaction intermediates, including two nitrenoid complexes that differ in oxidation state, were detected using this approach, which was proposed earlier without direct evidence. Certainly, the information on the occurrence of specific reaction intermediates can be helpful when clarifying reaction mechanisms.

11.3   Photochemical Reactions

Short-lived reaction intermediates and products resulting from light flash photolysis have been detected using TOF-MS [76]. A flash lamp was used to induce photochemical reactions of the reactant gases in a reaction vessel. The generated species, such as radicals, could be immediately (in approximately milliseconds) detected by TOF-MS [76]. In other work, a laser beam was combined with an ion cyclotron resonance mass analyzer to follow the process of photodissociation [77]. The dissociation rates and branching ratios for naphthalene ion were measured by means of the time-resolved photodissociation approach. The above-mentioned approaches [76, 77] are limited to detection of species generated from gas-phase substrates.

Detecting products of photochemical reactions conducted in the condensed phase by ESI-MS is also straightforward. Monitoring photoionization products generated in the liquid phase with this method was reported in 1995 [6]. That study was the first one to report on the use of ESI-MS in the on-line monitoring of photosubstitution process. The measurement system was quite simple: a syringe was connected to a reaction quartz cell that was connected to the ESI needle. It took ~ 2 min for the sample solution to flow through the quartz cell. During this time, the sample was irradiated by a Xe lamp [6] (Figure 11.9a). It took a few tens of seconds for the sample to arrive at the tip of the ESI emitter where ionization was initiated. The setup enabled the detection of photoproducts with lifetimes of at least several minutes. In this study [6], the solvent (e.g., acetonitrile) coordinated Ru(II) complex ions obtained from the photodissociation of Ru(2, 2'-bipyridine)$_3$Cl$_2$. The related complexes were subsequently detected by MS. Without irradiating the reaction cell by a Xe lamp ($\lambda > 290$ nm), only the RuL$_3{}^{2+}$ ions (where L is 2,2'- bipyridine) were observed in the mass spectra (Figure 11.9b). The Ru(II) complex ions formed with acetonitrile, that is Ru$_2$XLAN$^+$ (where AN is acetonitrile and X is Cl$^-$), appeared in the mass spectra following irradiation (Figure 11.9c). These MS results confirm that dissociation of Ru(2, 2'-bipyridine)$_3$Cl$_2$ occurs during irradiation with light. The Ru$_2$XLAN$^+$ ions can readily be monitored by ESI-MS. MS/MS can be used to further investigate the structure of the solvent-coordinated complexes. On the basis of the above work, one can concluded that ESI-MS has proved to be useful in the monitoring of intermediate and product ions generated in the course of photochemical reactions.

Moreover, photochemically synthesized compounds with good volatility can be monitored by a mass spectrometer equipped with an EI source [78]. Chlorinated volatile organic compounds are quite toxic and can be degraded through TiO$_2$-based catalytic reactions. EI-MS, which can readily be implemented in the direct analysis of volatile species, has been used to monitoring and identification of the volatile products of ultraviolet photodegradation of chlorinated volatile organic compounds in a TiO$_2$-catalyst flow reactor [78]. Intermediates such as phosgene and chlorine were seen in the EI mass spectra recorded

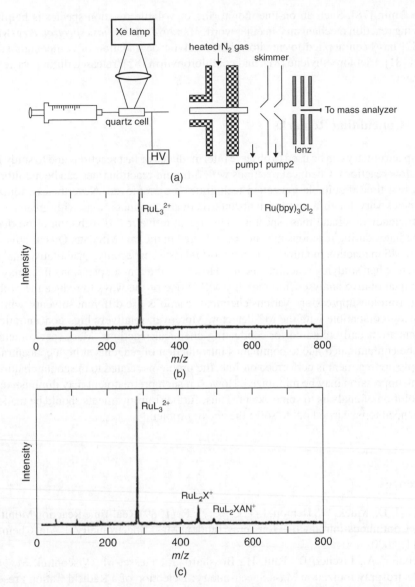

Figure 11.9 *Photochemical reaction monitoring by ESI-MS. (a) Schematic representation of ESI-MS for on-line analysis of photochemical reaction products. The drawing is not to scale. (b, c) ESI mass spectra of Ru(bpy)$_3$Cl$_2$ (bpy = 2,2'-bipyridine) in acetonitrile (0.1 mM): (b) without; and (c) with photoirradiation (λ > 290 nm). L, bpy; AN, acetonitrile; X, Cl$^-$ [6]. Reproduced with permission from Arakawa, R., Jian, L., Yoshimura, A., Nozaki, K., Ohno, T., Doe, H., Matsuo, T. (1995) On-line Mass Analysis of Reaction Products by Electrospray Ionization. Photosubstitution of Ruthenium(II) Diimine Complexes. Inorg. Chem. 34: 3874–3878. Copyright (1995) American Chemical Society*

within 2 min [78]. Such an on-line monitoring of volatile reaction species is helpful for clarifying reaction mechanisms. In other word, *Membrane inlet mass spectrometry* (MIMS) [79–82] has been applied to monitor photocatalytic degradation of compounds such as phenol [81], trichloroethylene [81], and epichlorohydrin [82], released during waste water treatment.

11.4   Concluding Remarks

Mass spectrometry can be used to observe intermediates of fast reactions and to study kinetics of slow reactions. Chemical reactions with different reaction rates can be monitored by MS in real time if suitable ionization techniques are employed. Nevertheless, ionization efficiencies vary for different reactants/intermediates/products. Ions with greater ionization efficiency dominate mass spectra. Thus, quantitative real distribution of products or intermediates during reactions may not be reflected in the raw MS data. Quantitative capability of MS in reaction monitoring is an important issue, especially. Quantitative analysis is important when studying reaction kinetics. However, the ion-suppression effect may complicate quantitative analyses. Ion mobility MS is one of the ways to reduce the problems arising from ionsuppression. Various chemical reactions use different solvents, which are not always compatible with the MS detector. Moreover, relatively high concentrations of reactants are usually required for conducting chemical reactions. Mass spectrometer can easily be contaminated due to continuous introduction of reagents at high concentrations. If sample pretreatment is performed on-line, the problems related to the contamination and the ion-suppression may be minimized. However, such pretreatment may diminish temporal resolution of analysis to some extent. Thus, further improvements should be made, and studies need to be carried out to solve the above problems.

References

1. Lee, E.D., Muck, W., Henion, J.D., Covey, T.R. (1989) Real-time Reaction Monitoring by Continuous-introduction Ion-spray Tandem Mass Spectrometry. J. Am. Chem. Soc. 111: 4600–4604.
2. Schafer, A., Fischer, B., Paul, H., Bosshard, R., Hesse, M., Viscontini, M. (1992) Electrospray Ionization Mass Spectrometry: Detection of a Radical Cation Present in Solution: New Results on the Chemistry of (Tetrahydropteridinone)-Metal Complexes. Helv. Chim. Acta 75: 1955–1964.
3. Kebarle, P., Tang, L. (1993) From Ions in Solution to Ions in the Gas Phase – the Mechanism of Electrospray Mass Spectrometry. Anal. Chem. 65: A972–A986.
4. Aliprantis, A.O., Canary, J.W. (1994) Observation of Catalytic Intermediates in the Suzuki Reaction by Electrospray Mass Spectrometry. J. Am. Chem. Soc. 116: 6985–6986.
5. Arakawa, R., Tachiyashiki, S., Matsuo, T. (1995) Detection of Reaction Intermediates: Photosubstitution of (Polypyridine)ruthenium(II) Complexes Using Online Electrospray Mass Spectrometry. Anal. Chem. 67: 4133–4138.

6. Arakawa, R., Jian, L., Yoshimura, A., Nozaki, K., Ohno, T., Doe, H., Matsuo, T. (1995) On-line Mass Analysis of Reaction Products by Electrospray Ionization. Photosubstitution of Ruthenium(II) Diimine Complexes. Inorg. Chem. 34: 3874–3878.
7. Arakawa, R., Lu, J., Mizuno, K., Inoue, H., Doe, H., Matsuo, T. (1997) On-line Electrospray Mass Analysis of Photoallylation Reactions of Dicyanobenzenes by Allylic Silanes via Photoinduced Electron Transfer. Int. J. Mass Spectrom. Ion Process. 160: 371–376.
8. Zechel, D.L., Konermann, L., Withers, S.G., Douglas, D.J. (1998) Pre-steady State Kinetic Analysis of an Enzymatic Reaction Monitored by Time-resolved Electrospray Ionization Mass Spectrometry. Biochemistry 37: 7664–7669.
9. Espenson, J.H., Tan, H., Mollah, S., Houk, R.S., Eager, M.D. (1998) Base Hydrolysis of Methyltrioxorhenium. The Mechanism Revised and Extended: A Novel Application of Electrospray Mass Spectrometry. Inorg. Chem. 37: 4621–4624.
10. Dell'Orco, P., Brum, J., Matsuoka, R., Badlani, M., Muske, K. (1999) Monitoring Process-scale Reactions Using API Mass Spectrometry. Anal. Chem. 71: 5165–5170.
11. Griep-Raming, J., Meyer, S., Bruhn, T., Metzger J.O. (2002) Investigation of Reactive Intermediates of Chemical Reactions in Solution by Electrospray Ionization Mass Spectrometry: Radical Chain Reactions. Angew. Chem. Int. Ed. 41: 2738–2742.
12. Welz, O., Savee, J.D., Osborn, D.L., Vasu, S.S., Percival, C.J., Shallcross, D.E., Taatjes, C.A. (2012) Direct Kinetic Measurements of Criegee Intermediate (CH_2OO) Formed by Reaction of CH_2I with O_2. Science 335: 204–207.
13. Chen, Y.-C., Urban, P.L. (2013) Time-resolved Mass Spectrometry. Trends Anal. Chem. 44: 106–120.
14. Zhu, L., Gamez, G., Chen, H.W., Huang, H.X., Chingin, K., Zenobi, R. (2008) Real-time, On-line Monitoring of Organic Chemical Reactions Using Extractive Electrospray Ionization Tandem Mass Spectrometry. Rapid Commun. Mass Spectrom. 22: 2993–2998.
15. Regiani, T., Santos, V.G., Godoi, M.N., Vaz, B. G., Eberlin, M.N., Coelho, F. (2011) On the Mechanism of the Aza-Morita–Baylis–Hillman Reaction: ESI-MS Interception of a Unique New Intermediate, Chem. Commun. 47: 6593–6595.
16. Ahmadi, Z., McIndoe, J.S. (2013) A Mechanistic Investigation of Hydrodehalogenation Using ESI-MS. Chem. Commun. 49: 11488–11490.
17. Hsu, F.-J., Liu, T.-L., Laskar, A.H., Shiea, J., Huang, M.-Z. (2014) Gravitational Sampling Electrospray Ionization Mass Spectrometry for Real-time Reaction Monitoring. Rapid Commun. Mass Spectrom. 28: 1979–1986.
18. Sam, J.W., Tang, X.-J., Magliozzo, R.S., Peisachts, J. (1995) Electrospray Mass Spectrometry of Iron Bleomycin 11: Investigation of the Reaction of Fe(III)-Bleomycin with Iodosylbenzene. J. Am. Chem. Soc. 117: 1012–1018.
19. Milagre, C.D.F., Milagre, H.M.S., Santos, L.S., M, Lopes, L.A., Moran, P.J.S., Eberlin, M.N., Rodrigues, J.A.R. (2007) Probing the Mechanism of Direct Mannich-type α-Methylenation of Ketoesters via Electrospray Ionization Mass Spectrometry. J. Mass Spectrom. 42: 1287–1293.
20. Chen, C.-C., Yang, Y.-L., Ou, C.-L. Chou, C.-H., Liaw, C.-C., Lin, P.-C. (2013) Direct Monitoring of Chemical Transformations by Combining Thin Layer Chromatography with Nanoparticle-assisted Laser Desorption/Ionization Mass Spectrometry. Analyst 138: 1379–1385.

21. Smith, N.J., Domin, M.A., Scott, L.T. (2008) HRMS Directly From TLC Slides. A Powerful Tool for Rapid Analysis of Organic Mixtures. Org. Lett. 10: 3493–3496.
22. Cegłowski, M., Smoluch, M., Babij, M., Gotszalk, T., Silberring, J., Schroeder, G., (2014) Dielectric Barrier Discharge Ionization in Characterization of Organic Compounds Separated on Thin-layer Chromatography Plates. PLOS One 9: e106088.
23. Wang, C.-H., Huang, M.-W., Lee, C.-Y., Chei, H.-L., Huang, J.-P., Shiea, J. (1998) Detection of a Thermally Unstable Intermediate in the Wittig Reaction Using Low-temperature Liquid Secondary Ion and Atmospheric Pressure Ionization Mass Spectrometry. J. Am. Soc. Mass Spectrom. 9: 1168–1174.
24. Kotiaho, T., Hayward, M.J., Cooks, R.G. (1991) Direct Determination of Chlorination Products of Organic Amines Using Membrane Introduction Mass Spectrometry. Anal. Chem. 63: 1794–1891.
25. Harry, E.L., Bristow, A.W.T., Wilson, I.D., Creaser, C.S. (2011) Real-time Reaction Monitoring Using Ion Mobility Mass Spectrometry. Analyst 136: 1728–1732.
26. Eggink, M., Wijtmans, M., Ekkebus, R., Lingeman, H., de Esch, I.J.P., Kool, J., Niessen, W.M.A., Irth, H. (2008) Development of a Selective ESI-MS Derivatization Reagent: Synthesis and Optimization for the Analysis of Aldehydes in Biological Mixtures. Anal. Chem. 80: 9042–9051.
27. Meyer, S., Metzger, J.O. (2003) Use of Electrospray Ionization Mass Spectrometry for the Investigation of Radical Cation Chain Reactions in Solution: Detection of Transient Radical Cations. Anal. Bioanal. Chem. 377: 1108–1114.
28. Schäfer, A., Fischer, B., Paul, H., Bosshard, R., Hesse, M. and Viscontinif, M. (1992) Electrospray Ionization Mass Spectrometry: Detection of A Radical Cation Present in Solution: New Results on the Chemistry of (Tetrahydropteridinone)-Metal Complexes. Helv. Chim. Acta. 75: 1955–1964.
29. Miao, Z., Chen, H., Liu, P., Liu, Y. (2011) Development of Submillisecond Time-resolved Mass Spectrometry Using Desorption Electrospray Ionization. Anal. Chem. 83: 3994–3997.
30. Lee, J.K., Kim, S., Nam, H.G., and Zare, R.N. (2015) Microdroplet Fusion Mass Spectrometry for Fast Reaction Kinetics. Proc. Natl. Acad. Sci. USA 112: 3898–3903.
31. Huang, G., Chen, H., Zhang, X., Cooks, R.G., Ouyang, Z. (2007) Rapid Screening of Anabolic Steroids in Urine by Reactive Desorption Electrospray Ionization. Anal. Chem. 79: 8327–8332.
32. Nyadong, L., Hohenstein, E.G., Galhena, A., Lane, A.L., Kubanek, J., Sherrill, C.D., Fernandez, F.M. (2009) Reactive Desorption Electrospray Ionization Mass Spectrometry (DESI-MS) of Natural Products of a Marine Alga. Anal. Bioanal. Chem. 394: 245–254.
33. Santos, V.G., Regiani, T., Dias, F.F.G., Romão, W., Jara, J.L.P., Klitzke, C.F., Coelho, F., Eberlin, M.N. (2011) Venturi Easy Ambient Sonic-Spray Ionization. Anal. Chem. 83: 1375–1380.
34. Hu, J.-B., Chen, T.-R., Chang, C.-H., Cheng, J.-Y., Chen, Y.-C., Urban P.L. (2015) A Compact 3D-printed Interface for Coupling Open Digital Microchips with Venturi Easy Ambient Sonic-spray Ionization Mass Spectrometry. Analyst 140: 1495–1501.
35. Hu, J.-B., Chen, S.-Y., Wu, J.-T., Chen, Y.-C., Urban, P.L. (2014) Automated System for Extraction and Instantaneous Analysis of Millimeter-sized Samples. RSC Adv. 4: 10693–10701.

36. Ting, H., Urban, P.L. (2014) Spatiotemporal Effects of a Bioautocatalytic Chemical Wave Revealed by Time-resolved Mass Spectrometry. RSC Adv. 4: 2103–2108.
37. Chiu, S.-H., Urban, P.L. (2015) Robotics-assisted Mass Spectrometry Assay Platform Enabled by Open-source Electronics. Biosens. Bioelectron. 64: 260–268.
38. Hu, J.-B., Chen, T.-R., Chang, C.-H., Cheng, J.-Y., Chen, Y.-C., Urban P.L. (2015) A Compact 3D-printed Interface for Coupling Open Digital Microchips with Venturi Easy Ambient Sonic-spray Ionization Mass Spectrometry. Analyst 140: 1495–1501.
39. Hsieh, C.-H., Chao, C.-S., Mong, K.-K.T., Chen, Y.-C. (2012) Online Monitoring of Chemical Reactions by Contactless Atmospheric Pressure Ionization Mass Spectrometry. J. Mass Spectrom. 47: 586–590.
40. Chen, T.-Y., Chao, C.-S., Mong, K.-K.T., Chen, Y.-C. (2010) Ultrasonication-assisted Spray Ionization Mass Spectrometry for On-line Monitoring of Organic Reactions. Chem. Commun. 46: 8347–8349.
41. Lin, S.-H., Lo, T.-J., Kuo, F.-Y., Chen, Y.-C. (2014) Real Time Monitoring of Accelerated Chemical Reactions by Ultrasonication-assisted Spray Ionization Mass Spectrometry. J. Mass Spectrom. 49: 50–56.
42. Song, Y., Cooks, R.G. (2007) Reactive Desorption Electrospray Ionization for Selective Detection of the Hydrolysis Products of Phosphonate Esters. J. Mass Spectrom. 42: 1086–1092.
43. Sparrapan, R., Eberlin, L.S., Haddad, R., Cooks, R.G., Eberlin, M.N., Augusti, R. (2006) Ambient Eberlin Reactions via Desorption Electrospray Ionization Mass Spectrometry. J. Mass Spectrom. 41: 1242–1246.
44. Chen, H., Cotte-Rodríguez, I., Cooks, R.G. (2006) cis-Diol Functional Group Recognition by Reactive Desorption Electrospray Ionization (DESI). Chem. Commun. 597–599.
45. Girod, M., Moyano, E., Campbell, D.I., Cooks, R.G. (2011) Accelerated Bimolecular Reactions in Microdroplets Studied by Desorption Electrospray Ionization Mass Spectrometry. Chem. Sci. 2: 501–510.
46. Bain, R.M., Pulliam, C.J., Cooks, R.G. (2015) Accelerated Hantzsch Electrospray Synthesis with Temporal Control of Reaction Intermediates. Chem. Sci. 6: 397–401.
47. Cooks, R.G., Ouyang, Z., Takats, Z., Wiseman, J.M. (2006) Ambient Mass Spectrometry. Science 311: 1566–1570.
48. Monge, M.E., Harris, G.A., Dwivedi, P., Fernández, F.M. (2013) Mass Spectrometry: Recent Advances in Direct Open Air Surface Sampling/Ionization. Chem. Rev. 113: 2269–2308.
49. Ifa, D.R., Wu, C., Ouyang, Z., Cooks, R.G. (2010) Desorption Electrospray Ionization and Other Ambient Ionization Methods: Current Progress and Preview. Analyst 135: 669–681.
50. Badu-Tawiah, A.K., Eberlin, L.S., Ouyang, Z., Cooks, R.G. (2013) Chemical Aspects of the Extractive Methods of Ambient Ionization Mass Spectrometry. Annu. Rev. Phys. Chem. 64: 481–505.
51. Espy, R.D., Wleklinski, M., Yan, X., Cooks, R.G. (2014) Beyond the Flask: Reactions on the Fly in Ambient Mass Spectrometry. Trends Anal. Chem. 57: 135–146.
52. Perry, R.H., Splendore, M., Chien, A., Davis, N.K., Zare, R.N. (2011) Detecting Reaction Intermediates in Liquids on the Millisecond Time Scale Using Desorption Electrospray Ionization. Angew. Chem. Int. Ed. 50: 250–254.

53. Lee, J.K., Kim, S., Namb, H.G., Zare, R.N. (2012) Microdroplet Fusion Mass Spectrometry for Fast Reaction Kinetics. Proc. Natl. Acad. Sci. USA 109: 2246–2250.

54. McCullough, B.J., Bristow, T., O'Connor, G., Hopley, C. (2011) On-line Reaction Monitoring by Extractive Electrospray Ionisation. Rapid Commun. Mass Spectrom. 25: 1445–1451.

55. Law, W.S., Chen, H., Ding, J., Yang, S., Zhu, L., Gamez, G., Chingin, K., Ren, Y., Zenobi, R. (2009) Rapid Characterization of Complex Viscous Liquids at the Molecular Level. Angew. Chem. Int. Ed., 48: 8277 –8280.

56. Li, M., Ding, J., Gu, H., Zhang, Y., Pan, S., Xu, N., Chen, H., Li, H. (2013) Facilitated Diffusion of Acetonitrile Revealed by Quantitative Breath Analysis Using Extractive Electrospray Ionization Mass Spectrometry. Sci. Rep. 3: 1205.

57. Yu, Z., Chen, L.C., Erra-Balsells, R., Nonami, H., Hiraoka, K. (2010) Real-time Reaction Monitoring by Probe Electrospray Ionization Mass Spectrometry. Rapid Commun. Mass Spectrom. 24: 1507–1513.

58. Pei, J., Kang, Y., Huang, G. (2014) Reactive Intermediate Detection in Real Time via Paper Assisted Thermal Ionization Mass Spectrometry. Analyst 139: 5354–5357.

59. Xu, G., Chen, B., Guo, B., He, D., Yao, S. (2011) Detection of Intermediates for the Eschweiler–Clarke Reaction by Liquid-phase Reactive Desorption Electrospray Ionization Mass Spectrometry. Analyst 136: 2385–2390.

60. Zhang, Z., Gong, X., Zhang, S., Yang, H., Shi, Y., Yang, C., Zhang, X., Xiong, X., Fang, X., Ouyang, Z. (2013) Observation of Replacement of Carbon in Benzene with Nitrogen in a Low-temperature Plasma. Sci. Rep. 3: 3481.

61. Ma, X., Zhang, S., Lin, Z., Liu, Y., Xing, Z., Yang, C., Zhang, X. (2009) Real-time Monitoring of Chemical Reactions by Mass Spectrometry Utilizing a Low-temperature Plasma Probe. Analyst 134: 1863–1867.

62. Sanchez, L.M., Curtis, M.E., Bracamonte, B.E., Kurita, K.L., Navarro, G.O., Sparkman, D., Linington, R.G. (2011) Versatile Method for the Detection of Covalently Bound Substrates on Solid Supports by DART Mass Spectrometry. Org. Lett. 13: 3770–3773.

63. Ertl, G. (1990) Elementary Steps in Heterogeneous Catalysis. Angew. Chem. Int. Ed. 29: 1219–1227.

64. Chen, P. (2003) Electrospray Ionization Tandem Mass Spectrometry in High-throughput Screening of Homogeneous Catalysts. Angew. Chem. Int. Ed. 42: 2832–2847.

65. Britovsek, G.J.P., Gibson, V.C., Wass, D.F. (1999) The Search for New-generation Olefin Polymerization Catalysts: Life beyond Metallocenes. Angew. Chem. Int. Ed. 38: 428–447.

66. Marquez, C., Metzger, J.O. (2006) ESI-MS Study on the Aldol Reaction Catalyzed by L-proline. Chem. Commun.1539-1541.

67. dos Santos, M.R., Coriolano, R., Godoi, M.N., Monteiro, A.L., de Oliveira, H.C.B., Eberlin, M.N., Neto, B.A.D. (2014) Phosphine-free Heck Reaction: Mechanistic Insights and Catalysis "on Water" Using a Charge-tagged Palladium Complex. New J. Chem. 38: 2958–2963.

68. Carrasco-Sanchez, V., Simirgiotis, M.J., Santos, L.S. (2009) The Morita-Baylis-Hillman Reaction: Insights into Asymmetry and Reaction Mechanisms by Electrospray Ionization Mass Spectrometry. Molecules 14: 3989–4021.

69. Santos, L., Knaack, S.L., Metzger, J.O. (2005) Investigation of Chemical Reactions in Solution Using API-MS. Int. J. Mass Spectrom. 246: 84–104.
70. Santos, L.S., Pavam, C.H., Almeida, W.P., Coelho, F., Eberlin, M.N. (2004) Probing the Mechanism of the Baylis–Hillman Reaction by Electrospray Ionization Mass and Tandem Mass Spectrometry. Angew. Chem. Int. Ed. 43: 4330–4333.
71. Yan, X., Sokol, E., Li, X., Li, G., Xu, S., Cooks, R.G. (2014) On-line Reaction Monitoring and Mechanistic Studies by Mass Spectrometry: Negishi Cross-coupling, Hydrogenolysis, and Reductive Amination. Angew. Chem. Int. Ed. 53: 5931–5935.
72. Eberlin, M.N. (2007) Electrospray Ionization Mass Spectrometry: A Major Tool to Investigate Reaction Mechanisms in both Solution and Gas Phase. Eur. J. Mass Spectrom. 13: 19–28.
73. Santos, L.S. (2008) Online Mechanistic Investigations of Catalyzed Reactions by Electrospray Ionization Mass Spectrometry: A Tool to Intercept Transient Species in Solution. Eur. J. Org. Chem. 2: 235–253.
74. Hinderling, C., Chen, P. (1999) Rapid Screening of Olefin Polymerization Catalyst Libraries by Electrospray Ionization Tandem Mass Spectrometry. Angew. Chem. Int. Ed. 38: 2253–2256.
75. Perry, R.H., Cahill III,, T.J., Roizen, J.L., Bois, J.D., Zare, R.N. (2012) Capturing Fleeting Intermediates in A Catalytic C-H Amination Reaction Cycle. Proc. Natl. Acad. Sci. USA 109: 18295–18299.
76. Meyer, R.T. (1967) Apparatus for Flash Photolysis and Time Resolved Mass Spectrometry. J. Sci. Instrum. 44: 422–426.
77. Ho, Y.-P., Dunbar, J.R.C., Lifshitz, C. (1995) C-H Bond Strength of Naphthalene Ion. A Reevaluation Using New Time-resolved Photodissociation Results. J. Am. Chem. Soc. 117: 6504–6508.
78. Alberici, R.M., Mendes, M.A., Jardim, W. F., Eberlin, M.N. (1998) Mass Spectrometry On-line Monitoring and MS^2 Product Characterization of TiO$_2$/UV Photocatalytic Degradation of Chlorinated Volatile Organic Compounds. J. Am. Soc. Mass Spectrom. 9: 1321–1327.
79. Lauritsen, F.R., Kotiaho, T. (1996) Advances in Membrane Inlet Mass Spectrometry (MIMS). Rev. Anal. Chem. 15: 237–264.
80. Rios, R.V.R.A., da Rocha, L.L., Vieira, T.G., Lago, R.M., Augusti, R. (2000) On-line Monitoring by Membrane Introduction Mass Spectrometry of Chlorination of Organics in Water. Mechanistic and Kinetic Aspects of Chloroform Formation. J. Mass Spectrom. 35: 618–624.
81. Nogueira, R.F.P., Alberici, R.M., Mendes, M.A., Jardim, W.F., Eberlin, M.N. (1993) Photocatalytic Degradation of Phenol and Trichloroethylene: On-line and Real-time Monitoring via Membrane Introduction Mass Spectrometry. Ind. Eng. Chem. Res. 38: 1754–1758.
82. Johnson, R.C., Koch, K., Cooks, R.G., (1999) On-line Monitoring of Reactions of Epichlorohydrin in Water Using Liquid Membrane Introduction Mass Spectrometry. Ind. Eng. Chem. Res. 38: 343–351.

12

Applications of Time-resolved Mass Spectrometry in the Studies of Protein Structure Dynamics

Most biological processes involve proteins [1]. Molecules of these biopolymers are characterized by multiple descriptors, including empirical formula, amino acid sequence (primary structure), arrangement of polypeptide into helices and sheets (secondary structure), protein folding (tertiary structure), post-translational modifications (e.g., presence of phosphate or carbohydrate moieties), presence of disulfide bridges, formation of multi-unit complexes (quaternary structure), and others. Out of these features, protein folding has recently received considerable attention. Being a dynamic process, protein folding can be defined as "a biased conformational diffusion on a multidimensional energy landscape" [2]. It involves an interplay between entropic and enthalpic contributions to the free energy of the system [1]. The number of possible conformations of a polypeptide chain is extremely large. Various interactions shape physical and chemical properties of proteins, including electrostatic interactions, van der Waals forces, hydrogen interactions, and hydrophobic interactions [3]. Characterizing protein folding can help to explain various molecular events, for example those related to cellular regulation, and assist the design of proteins with novel functions, the utilization of sequence information, and the development of novel therapies [1]. It can contribute to advances in bioengineering and biophysics [3]; for example, production of industrially relevant enzyme proteins. Protein misfolding is linked with several human diseases, including Huntington's disease and Alzheimer's disease [3]. Investigation of protein folding variants is also relevant to human prion diseases among others [4].

Apart from the momentary structure, characterization of structure dynamics is needed for complete understanding of protein function [5]. Protein folding often occurs in time intervals ranging from milliseconds to seconds [3, 6]. To be accomplished within such a short period of time, the folding process must follow predefined pathways rather than random sampling [3]. Folding rates down to $\sim 1\,\mu s^{-1}$ have been recorded for some

Time-Resolved Mass Spectrometry: From Concept to Applications, First Edition.
Pawel Lukasz Urban, Yu-Chie Chen and Yi-Sheng Wang.
© 2016 John Wiley & Sons, Ltd. Published 2016 by John Wiley & Sons, Ltd.

proteins. Precise analysis of the early folding events requires superior kinetic resolution. For example, folding events related to helix–coil transition are believed to occur in the nanosecond and sub-microsecond timescale; duration of the diffusion-limited "collapse" is estimated to be in the order of microseconds, while the events involving intermediates can extend from microseconds to hundreds of seconds [3].

To augment our cognition of the protein folding process, both theoretical and experimental approaches are used synergistically [7]. Crystallographic and nuclear magnetic resonance (NMR)-aided studies led to a large amount of information on the static structures of proteins [8] but they can be extended to time-resolved studies. Two commonly used methods for the studies of protein folding include *stopped flow* measurements and kinetic measurements with *hydrogen exchange* [9]. Conventional circular dichroism spectroscopy can provide temporal resolution within milliseconds [3]. The methods commonly used in the investigation of protein folding at high temporal resolution include ultrafast mixing (>10 µs), temperature jump (1 ns–10 s), optical triggers (100 fs–1 ms), acoustic relaxation (1 ns–1 ms), pressure jump (60 µs–1 s), dielectric relaxation (1 ns–1 s), and NMR line broadening (100 µs–100 ms) [3, 10]. Slower folding events are investigated using stopped-flow techniques (e.g., NMR), and fluorescence- and isotope-labeling [3].

Application of mass spectrometry (MS) to protein conformational studies has recently gained the attention of the research community. Cross-linking is a sample preparation strategy that enables stabilization of protein molecules or their complexes for MS analysis of their three-dimensional structure [11]. Thus, mass spectrometric methods in conjunction with chemical cross-linking are especially adequate for performing "static" analyses of non-covalent complexes [12, 13]. However, MS used along with special quenching methods or soft ionization techniques, can also yield additional information on the dynamic nature of proteins [8, 14, 15]. While some proteins are known to possess two folding states, others fold through transient intermediates [16]. Due to the short lifetimes of folding intermediates and low abundances, studies of folding pathways are complex tasks [16]. Temporal monitoring by MS can provide valuable insights on the complexity of such biochemical processes [17]. While kinetic measurements yield data on protein folding mechanisms, capturing the initial stages of folding is particularly challenging for contemporary analytical chemistry [18].

12.1   Electrospray Ionization in Protein Studies

Electrospray ionization (ESI) mass spectra of proteins are often dominated by peaks representing multiple charge states [19, 20]. Due to this feature, protein samples can be analyzed by means of mass spectrometers with a narrow mas-to-charge (m/z) range, such as ion traps. ESI is soft enough to introduce proteins and nucleic acids into the gas phase without breaking covalent bonds [21]. Moreover, the ESI charge state distribution is sensitive to conformational changes of proteins in the liquid phase [16]. It provides information on different structural states present at equilibrium [5]. In the positive-ion mode, unfolded polypeptide chains acquire a relatively large number of positive charges due to the exposure of many ionizable groups to cations (*e.g.* protons). Thus, due to the increased number of positive charges, they are shifted towards the lower m/z values as compared with the folded species [16]. Transformation of protein tertiary structure may occur on exposure to some solvents. In many cases, these structural changes are reversible. They often occur within the time range of milliseconds to seconds [16].

Many studies of biomolecules carried out by ESI-MS assume that, in the gas-phase conditions, ions retain elements of liquid-phase structures [22]. This assumption is supported by experimental results and molecular dynamics simulations. It has been suggested that solution-phase conformers of some proteins may be preserved for several tens of milliseconds after the ESI [21, 23, 24]. Because of the "soft" character of ESI, some protein–ligand complexes and large protein assemblies may also maintain non-covalent bonding in the gas phase [16, 24–29]. For example, nanospray electrospray ionization (nanoESI)-MS enabled the formation of complexes between a heat shock protein and an unfolding luciferase (client) to be monitored in the course of a few minutes [30].

However, the three-dimensional structure of biomolecules may also undergo changes during the transition to the gas phase (see, e.g., [31–33]). In ESI, large globular species are believed to follow the so-called *charged residue model* while disordered polymers may follow the *chain ejection model* [34]. Introduction of native proteins to the gas phase can induce side-chain collapse, unfolding, and refolding into non-native structures [21]. Desolvation of ions during ESI is a complex process that involves several intricate molecular dynamic events, which may be influenced by the sample matrix composition [35]. For instance, the presence of acetonitrile (27%) did not affect the *compactness* of holo-myoglobin, which was characterized by analytical ultracentrifugation [36]. It was observed that some macromolecular species (e.g., denatured lysozyme) could undergo *compaction* during their transfer into the gas phase (cf. [31, 37]). Thus, careful optimization of experimental conditions, and cross-validation of experimental results is still necessary when implementing ESI-MS in the analysis of intact biomolecules.

On-line ESI-MS is a convenient approach for studying protein folding and unfolding [16]. It should be pointed out that ESI-MS only requires small amounts of proteins for kinetic studies of folding, which is an advantage of this approach compared with other techniques. However, precautions need to be taken that the buffers used to induce folding/refolding are compatible with the ion source [9]. Some of the available methods involve a *quenching step* in order to preserve the molecular state after a short reaction time. However, interpretation of the quench-flow experiments may be complicated because of the necessity to collect kinetic data from numerous single-time-point measurements [16].

This section continues with some representative examples of time-resolved mass spectrometry (TRMS) studies on protein folding dynamics. In an early study, Mirza *et al.* [38] implemented an apparatus that enabled the control of the temperature of protein solutions just before their introduction to ESI. Thus, it was possible to investigate conformational changes of proteins in response to heat. Some of the earlier attempts of protein folding analysis by MS took advantage of stopped flow measurements [39]. However, due to their suitability for studies of fast folding kinetics, continuous flow methods gained greater popularity. Konermann *et al.* [40] developed a method for investigating folding kinetics of proteins which combines continuous flow mixing and ESI-MS (Figure 12.1). Here, the incubation time of the sample and solvent was set by the length of the collective capillary between the T-junction and the ESI emitter. This method was tested on acid-denatured cytochrome *c*. The measured kinetics was described by two lifetimes: 0.17 ± 0.02 and 8.1 ± 0.9 s. They correspond to fast and slow refolding subpopulations of the model protein, respectively.

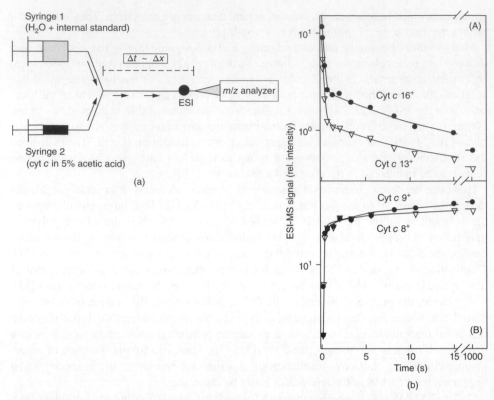

Figure 12.1 *(a) Experimental apparatus for the time-resolved ESI-MS experiments. Arrows indicate the flow directions of the solutions from the two syringes. The flow time Δt in the reaction capillary is proportional to the capillary length Δx. (b) Time course of the signal intensity in the mass spectrum of cytochrome (Cyt) c for charge states 13 + and 16 + (A) and 8 + and 9 + (B). Intensities were measured at different times after changing the pH in the solution from 2.4 to 3.0. The first data point in each curve represents t = 0.1 s; the last one ("t = 1000 s" is taken from the stationary mass spectrum measured at pH 3.0. Data shown here are the average of three independent sets of experiments. The intensities of the data are normalized to the intensity of an internal standard. Solid lines are fits to the experimental data. The ordinates in this figure have a logarithmic scale [40]. Adapted with permission from Konermann, L., Collings, B.A., Douglas, D.J. (1997) Cytochrome c Folding Kinetics Studied by Time-resolved Electrospray Ionization Mass Spectrometry. Biochemistry 36: 5554–5559. Copyright (1997) American Chemical Society*

A similar approach was applied in the study of holo-myoglobin [41]. In both cases, the lifetimes are in good agreement with those obtained by alternative methods. Acid-induced denaturation of holo-myoglobin on a pH jump (6.5 to 3.2) coupled with ESI-MS led to observations of various protein conformations [41]. The MS signals were monitored on a timescale of a few seconds with sub-second to a few seconds intervals. Some of the charge states represent lifetimes shorter than 1 s. The lifetimes obtained by TRMS and stopped-flow measurements of the Soret absorption band are convergent [41]. This work pointed to the sequential two-step mechanism of acid-induced denaturation of holo-myoglobin:

$$(\text{heme-protein})_{\text{native}} \rightarrow (\text{heme-protein})_{\text{unfolded}} \rightarrow \text{heme} + (\text{protein})_{\text{unfolded}} \quad (12.1)$$

It was later noted that holo-myoglobin unfolds through the short-lived intermediate only at acidic pH while under more basic conditions, no intermediate could be observed [42]. The reverse reaction – protein folding following acid-induced denaturation – has been studied using holo-myoglobin as the model analyte [43]. Based on the obtained kinetic data, it was inferred that the reconstitution mechanism cannot be approximated as a simple reversal of acid-induced denaturation [43]. More recently, Wilson *et al.* [44] studied the disassembly and unfolding of a homodimeric protein complex (MW $\approx$ 100 kDa) on a pH jump (7.5 to 2.8) coupled on-line with ESI-time-of-flight (TOF)-MS. The generated protein species differed in ligand binding behavior and charge states. It was observed that denaturation of large multiprotein complexes involved generation of various short-lived intermediates [44].

Simple interfaces can readily be used to follow the dynamics of multimeric proteins. Using the miniaturized version of ESI, nanoESI, combined with a heating device, it was possible to follow assembly and disassembly pathways of the chaperone complex MtGimC [45]. In a representative experiment, solutions containing isolated sub-units were mixed at 37 °C, and analyzed immediately. Following ∼2 min incubation, spectral features corresponding to the intact MtGimC complex could be observed [45]. In a recent study, Cubrilovic *et al.* [46] used native ESI-MS to monitor dissociation of the tumor necrosis factor (TNF)-alpha trimer caused by a small-molecule inhibitor. Partial dissociation of the protein into dimers and monomers occurred upon addition of the inhibitor.

Ionization techniques derived from ESI, such as electrosonic spray ionization (ESSI), are also suitable for studies of fragile molecules. The ESSI-MS approach enables observation of non-covalent complexes of myoglobin, protein kinase A/ATP complex, and other proteins [47]. It can be used to study on-line deprotonation reactions on peptides and proteins [48, 49]. Such analyses can be done by introducing volatile bases between the ion source and mass analyzer. This method is fast: one reference base can be scanned in a time interval of ∼ 1 min. The desorption electrospray ionization (DESI) technique was also implemented in the study of protein conformation in solution [50]. The interaction time between the spray solvent and the protein was estimated to be ∼ 1 ms, and it was suggested that this timescale would be too short for the studied proteins to unfold [50].

12.2 Mass Spectrometry Strategies for Ultra-fast Mixing and Incubation

Modifying the design of the ESI source allows one to reduce the duration of incubation of the reactants prior to the ionization. The continuous flow mixing approach, disclosed by Wilson and Konermann [51], is particularly successful because it enables the duration of reactant incubation to be varied (see Section 4.2.4 and Figure 4.3). Millisecond-scale incubations could readily be accomplished in this way. If even shorter incubations are required, further modification of the standard ESI emitter design may be necessary. For example, Fisher *et al.* [52] used a pulled dual-lumen glass capillary – the so-called *theta capillary* – as nanoESI emitter for short timescale mixing of protein and acid solutions to influence protein charge state distributions without modifying the sample solution. In the theta emitters, reactants are separated by the glass wall until they merge in proximity to the capillary outlet (Figure 4.4; [52–54]). Soon after the merger, the combined solution undergoes desolvation according to the conventional ESI principle. The short mixing time scale enables the study of short-lived unfolding intermediates and higher charge states of non-covalent protein complexes. This method facilitates application of non-volatile super-charging agents shortly before the ionization. Mortensen and Williams [54] used theta emitters to observe conformational changes of acid-denatured cytochrome *c*. The sub-sequent folding was recorded by nanoESI-MS, pointing to a reaction time of 7–25 µs. The time frame during which protein folding or unfolding could occur during nanoESI depended on the initial droplet size and on the solution composition. It is believed that protein folding or unfolding processes that occur within ~ 10 µs intervals can be inves-tigated by means of the rapid mixing enabled by the theta emitters combined with MS [54]. Most recently, kinetic studies on acid-induced unfolding of cytochrome *c* were car-ried out by fusing microdroplets delivered in the plumes of two sprayers positioned in close proximity (Figure 4.5) [55]. This technical improvement also provided a very good tem-poral resolution (in the order of a few microseconds), and should be considered in future discovery-oriented work related to protein folding. One obstacle here is the limited possi-bility to control the incubation time and reaction environment due to the shrinkage of the charged microdroplets.

12.3   Hydrogen/Deuterium Exchange

Exchange of hydrogen isotope atoms is occasionally applied to proteins prior to analysis of their structures by MS. Exposure of a protein to heavy water (D_2O, ^2H_2O) induces rapid exchange between protium and deuterium in the amide moieties at the disordered regions that are not involved in stable hydrogen interactions (cf. [56]). In the earlier years, the hydrogen exchange strategy was implemented in conjunction with radiochemical (using T, ^3H) [57] and NMR detection [58, 59]. It was soon realized that the increase of protein mass can be followed by MS [60]. Hydrogen/deuterium exchange (HDX) ESI-MS soon became a convenient way of investigating conformational changes of proteins in solutions [61].

The HDX method is based on the fact that tightly folded elements of biopolymers are protected; thus, they exhibit slow isotope exchange [56]. HDX MS is sensitive to the back-bone dynamics in solution, and it can help to locate dynamic regions in protein molecules [5]. Intricacies of protein conformation, dynamics, and interactions can be exposed [62]. In particular, HDX MS can assist mapping transiently structured regions of disordered pro-teins [63]. HDX measurements on disordered proteins are particularly applicable when the studied processes are in the millisecond timescale [63]. A thorough understanding of pro-tein function can be gained by combining data obtained from classical methods with those obtained in HDX MS experiments [60].

In one of the pioneering studies, conformational changes were induced in a model pro-tein (bovine ubiquitin) by adding methanol to aqueous acidic solutions of the analyte [61]. Using HDX in combination with ESI-MS, it was possible to follow conformational changes in bovine ubiquitin and in chicken egg lysozyme (due to the reduction of the disulfide cross-linkages by dithiothreitol) [64]. The extent of HDX that occurred in different protein conformers, could be evaluated following preset periods of time. The HDX ESI-MS helped to clarify the existence of alternative pathways in the folding of lysozyme molecules [65]. It also pointed to the co-operative nature of the folding process [66].

In general, one can group the HDX methods into two categories [60]: (i) continuous labeling; and (ii) pulse labeling. The labeled proteins can be analyzed globally or locally (i.e., following fragmentation conducted by means of biochemical or physical methods). Enzymatic digestion of labeled proteins is one of the possible subsequent steps that allows spatial information on the HDX pattern of folding intermediates to be found [2].

In the case of *pulse labeling*, reactants (e.g., a protein solution and a denaturing/refolding solution) are mixed, and subsequently – following a brief incubation time – the labeling reagent is added (cf. [3, 59]). Eventually, the process is quenched – for example by inject-ing a quenching solution (cf. [9]). Buffers with varied acidities are often used to induce folding or refolding. The extent of labeling depends on pH. Thus, acidity of the buffer can be used to initiate and stop the process. At high pH the exchange rate is high while low pH inhibits the process [9]. The extent of labeling can be evaluated based on the appar-ent shifts of peaks in the mass spectra: toward higher or lower m/z values, in the case of $H \rightarrow D$ and $D \rightarrow H$ exchange, respectively. The *pulse HDX* method is one of the useful tools that enables the properties of kinetic intermediates to be studied [2]. For example, in the work conducted by Yang and Smith [67], the kinetics of cytochrome c folding was inves-tigated by high-performance liquid chromatography (HPLC)-ESI-MS following labeling of the folding intermediates (folding times: 5 ms–1 s) and digestion with pepsin. The quench-flow HDX pulse labeling and ESI-MS also enabled the capture of an intermediate in the folding of apo-myoglobin [68].

While conventional HDX protocols use the quenching step, in some cases, by employing special ESI-MS interfaces, operating at high speeds and minimizing delays, it is possible to skip this step to enable studies of short-lived protein conformations [2]. In the setup implemented by Simmons and Konermann [69], a protein sample was electrosprayed after its exposure to the deuterated solvent. The experimental setup consisted of two mixers. The solutions of apo-myoglobin and heme were merged in the first mixer. Subsequently, the mixture was merged with the flow of heavy water for labeling. The final mixture was subsequently infused to a mass spectrometer via an ESI interface without chemical-induced quenching. In this way, reconstitution of holo-myoglobin from unfolded apo-myoglobin and free heme could be followed. Short lifetime intermediates of the protein assembly were observed [69]. HDX ESI-MS performed by mixing solutions of a protein (myoglobin) and heavy water – both containing acetonitrile – enabled observation of the dynamics of partially denatured molecules [36].

The pulse HDX methodology also enables studies of folding and assembly of multi-meric proteins. For example, a study on a protein revealed that each of its monomeric forms consists of two subspecies which differ in the isotope exchange levels [70]. The amount of information obtained here by combining time-resolved ESI-MS and HDX is certainly greater than the information obtained using the individual techniques separately: the existence of four distinct monomeric species, involved in the folding and assembly of the denatured protein, could be confirmed [70]. In addition, the higher-order structure of antibody–drug conjugates and other protein therapeutics can be studied with the HDX MS method [71].

In an interesting work, Liuni *et al.* [72] investigated pre-steady state conformational dynamics in an active enzyme using combined time-resolved ESI-MS and sub-second HDX. Here, biocatalytic processes were monitored as time-dependent intensity changes of reaction intermediates. Conformational dynamics was analyzed by the rate and magnitude of deuterium uptake [72].

HDX MS experiments can also be conducted using microfluidic chips. Liuni *et al.* [73] fabricated a poly(methyl methacrylate) (PMMA) microchip that enables rapid digestion of proteins prior to ESI-MS. It can be used for spatially resolved HDX MS experiments. Along these lines, Rob *et al.* [74] implemented a microfluidic chip to measure HDX in proteins on the millisecond to low second timescale (Figure 12.2). This device facilitated characterization of conformational dynamics in weakly structured regions. It comprises a capillary mixer (with variable time labeling pulse), a static mixer for quenching, a proteolytic microreactor (for protein digestion), and an on-chip ESI source. The labeling times can be adjusted from a few tens of milliseconds up to a few seconds [74]. Activity-linked conformational transitions in proteins can also be revealed with aid of the HDX MS approach taking advantage of microfluidics [75].

Hu *et al.* [76] investigated the kinetic folding of ribonuclease H by HDX (Figure 12.3). The MS pulse-labeling experiment – conducted in this work – was implemented using a setup that consists of four syringes and three mixers. The folding buffer was first mixed with deuterated protein, and the process was allowed to proceed at a pH of 5. Subsequently, the "pulse buffer" was injected to adjust the pH to 10. The exchange of deuterium to protium followed within 10 ms. It was stopped with the aid of the quench buffer injected via the third mixer. The product of the HDX experiment was digested on-line using immobilized proteolytic enzymes, and the resulting peptides were analyzed by HPLC-ESI-MS. The authors concluded that the folding of the studied protein proceeded through distinct intermediates in a stepwise pathway that sequentially incorporates cooperative native-like structural elements to build the native protein [76].

While bottom-up proteomics normally uses protease enzymes to fragment polypeptide chains, the top-down approach implements fragmentation techniques such as electron capture dissociation (ECD) [77] to generate the peptide fragments (see also Section 3.3.2.2). ECD has also been implemented in conjunction with HDX [78, 79]. It could pinpoint locations of protected amides with an average resolution of less than two residues (for horse myoglobin) [79]. The amount of structural information obtained with this method is greater than when using conventional HDX procedures. Pan *et al.* [80] implemented pulsed HDX in conjunction with top-down MS [using ECD and Fourier transform (FT)-MS] in the investigation of short-lived protein folding intermediates (Figure 12.4). On account of the short duration of the acid quenching step (~ 1 s), amide proton back-exchange could be mitigated. Overall, ECD as well as electron transfer dissociation (ETD) minimize the so-called *scrambling* (intramolecular hydrogen migration), thus reducing measurement artifacts [81, 82]. (For a thorough discussion of protein labeling methods combined with MS, also see the excellent reviews published elsewhere [2, 9, 56, 60, 62, 75, 83].)

Protein folding studies generally concern liquid-phase phenomena; however, HDX reactions can also be conducted in the gas phase (e.g., [84–90]; for an overview see [91]). For instance, multiply charged ions of several model proteins reacted with D_2O in a vacuum following pseudo-first-order kinetics [85]. It was also observed that removing solvent significantly increased conformational rigidity. In other work, ions of bovine ubiquitin were

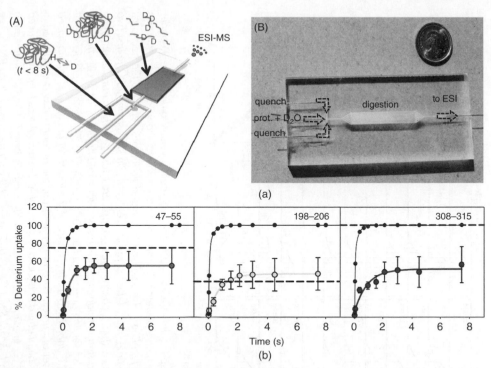

Figure 12.2 Hydrogen/deuterium exchange on a microchip. (a) The microfluidic device. (A) A schematic depiction. Solutions are injected at the "near" side of the device and ESI is carried out at the "far" end. HDX is conducted under native conditions (pD 7.6) in the first phase with adjustable labeling times from ~40 ms to 8 s. This is followed by HDX quenching, which is achieved by dropping the pH to 2.5. The labeled protein is then injected into the proteolytic reactor, generating labeled peptides which can be located on the native structure. (B) A photograph of the device, showing a coin for scale. (b) HDX kinetics for representative peptides from DAHP synthase. (Top) Intrinsic rate profiles $D(t)_{int}$ for each segment are shown in black. (Bottom) The measured profiles are represented by filled circles: right (strong protection); middle (moderate protection); left (low protection). The extent of protection is estimated according to the protection factor defined in this report. For most peptides, there is little correlation between N_{fast} and the number of loop amides in the substrate-bound structure [74]. Adapted with permission from Rob, T., Liuni, P., Gill, P.K., Zhu, S., Balachandran, N., Berti, P.J., Wilson, D.J. (2012) Measuring Dynamics in Weakly Structured Regions of Proteins Using Microfluidics-enabled Subsecond H/D Exchange Mass Spectrometry. Anal. Chem. 84: 3771–3779. Copyright (2012) American Chemical Society

reacted in an ion cyclotron resonance (ICR) cell with D_2O allowing reaction times from 1 s to 1 h [88]. The kinetics of HDX could be followed separately for the ions with different charge states. Apart from the ICR cell, the deuterium labeling could also be conducted in an external ion reservoir [89, 90]. In the most recent work, the mechanism of HDX reactions was studied in the gas phase using ubiquitin and cytochrome c [92]. When a very small amount of heavy water is introduced into the ion trap, bimodal distributions can be recorded for the trapped protein ions. Manipulating protein ions in the gas phase is very

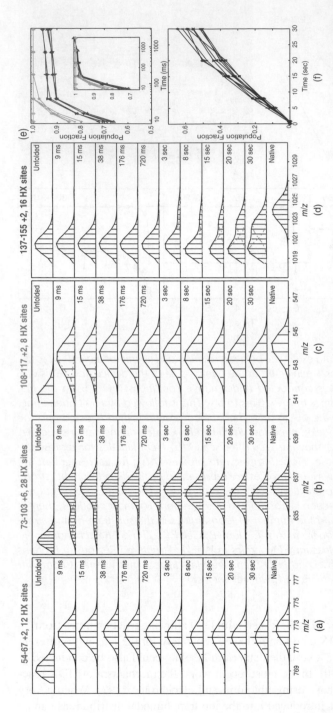

Figure 12.3 *Monitoring the folding process by hydrogen/deuterium pulse labeling. (a–d) Illustrative MS spectra versus folding time. Peptides shown cover each helical segment plus some neighboring sequence in the native protein. The top and bottom frames show control experiments in which the unfolded and native proteins were subjected to the same labeling pulse and analysis. Fitted envelopes separate the fractional populations of the unfolded, intermediate, and native state present at the time of the labeling pulse. Deuterons on side chains and the first two residues of each peptide are lost during sample preparation. The subpeaks within each isotopic envelope are caused by the natural abundance of ^{13}C (~1%) convolved with the carried number of deuterons. A leftward drift in folded peptide mass at long folding times (d) occurs because not-yet-protected sites are exposed to D-to-H exchange during the prepulse folding period (pH 5, 10 °C). (e and f) The time dependence for the formation of the protected state of different protein regions. (Inset) The unblocked folding phase of the lower set of curves is renormalized to 100% to allow direct comparison with the folding time of the upper set of curves. For this comparison, the experiment was conducted in triplicate, and only the highest-precision peptides were used. The segments highlighted in (e) fold in detectably different phases [76]. Reproduced from Hu, W., Walters, B.T., Kan, Z.-Y., Mayne, L., Rosen, L.E., Marqusee, S., Englander, S.W. (2013) Stepwise Protein Folding at Near Amino Acid Resolution by Hydrogen Exchange and Mass Spectrometry. Proc. Natl. Acad. Sci. USA 110: 7684–7689 with permission from PNAS. See colour plate section for colour figure.*

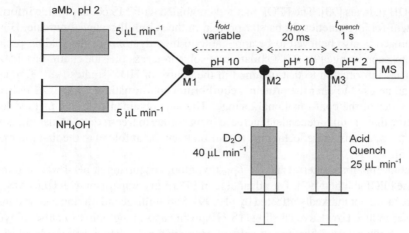

Figure 12.4 *Continuous-flow mixing device for pulsed HDX/MS. Exposure of acid-unfolded protein to ammonium hydroxide solution triggers refolding at mixer M1. After a variable amount of folding time (ranging between 10 ms and 1 s), addition of D_2O initiates the labeling pulse at mixer M2. HDX is quenched by formic acid solution after M3, before the mixture is infused into the electrospray source of the mass spectrometer. Uncorrected pH-meter readings for D_2O-containing solutions are referred to as pH*. t_{fold}, duration of folding; t_{HDX}, duration of labeling; t_{quench}, duration of quenching [80]. Reprinted with permission from Pan, J., Han, J., Borchers, C.H., Konermann, L. (2010) Characterizing Short-lived Protein Folding Intermediates by Top-down Hydrogen Exchange Mass Spectrometry. Anal. Chem. 82: 8591–8597. Copyright (2010) American Chemical Society*

convenient from a technical point of view. Reaction times are not limited by mixing of liquid solutions in the interfaces and ion sources. The labeled protein ions can rapidly be separated and detected. These studies bring insight on the biophysics of proteins. However, considering that the gas phase is not the native environment for proteins, interpretation of the gas-phase data – and drawing conclusions which could enhance our understanding of biological systems – is challenging.

12.4 Photochemical Methods

Utilizing photochemical processes, for example, those induced by laser light, can facilitate ultra-fast quenching of structural transformations of proteins prior to the MS analysis. In the method called *fast photochemical oxidation of proteins* (FPOP), hydroxyl radicals (·OH) can be generated in a nanoliter flow cell by irradiating hydrogen peroxide with a 248 nm KrF excimer laser [93]. The lifetimes of these radicals (in the presence of a scavenger, glutamine) are in the order of ~ 1 µs. During this short period of time, they can label amino acid side chains of proteins [94, 95]. The oxidation sites are located using typical proteomic analysis methods. In the case of apo-myoglobin, the folding/unfolding is a two-step process. Considering the two-stage folding kinetics of the model protein, the first step in unfolding is the breaking of tertiary interactions and admitting water with a relaxation time of ~ 200 µs, which is considered to be shorter than the disappearance time of the ·OH radicals [93]. The FPOP method combined with MS could provide information on protein-folding kinetics for barstar protein in the sub-millisecond timescale. However, the approach is also believed to be applicable to investigations of the folding process in the microsecond range [96, 97]. Oxidation of solvent-accessible side chains provides complementary information to that obtained in the course of HDX studies [98]. Temperature jump can be used to alter the protein's equilibrium conformation, while ·OH radicals act as a reporter of the conformational change. The data obtained with this method led to a conjecture that the time-dependent increase in mass (due to free-radical oxidation) is a measure of the rate constant reflecting the transition from the unfolded to the first intermediate state [96].

The conformational intermediate of apo-myoglobin occurring at pH 4 was also mapped using the FPOP method [99]. The advantage of FPOP, in comparison with HDX MS, is that FPOP data are not markedly affected by pH [99]. Sub-millisecond mixing can be combined with laser-induced oxidative labeling [18]. Exposing apo-myoglobin to a pulse of hydroxyl radical introduced modifications at solvent-accessible side chains, and the level of labeling could be measured by MS. This approach provides spatially resolved measurements of changes in solvent accessibility (Figure 12.5; [18, 100]). Using the above method, it was found that major conformational changes in a model protein are completed within 0.1 s. Overall, implementation of sub-millisecond mixing and slower mixing techniques enables observation of complete folding pathways, from fractions of a millisecond to minutes [18]. In order to extend the capabilities of the FPOP method toward customized kinetic studies of protein folding, it was appealing to know the exact concentration of the generated ·OH radicals, which was recently estimated to be in the order of ~ 1 mM (from 15 mM H_2O_2) [95]. The FPOP methodology has also been used in conjunction with the so-called *multidimensional protein identification technology* (MudPIT) to enhance identification of oxidized peptides in complex systems [101].

For further discussion of oxidative labeling of proteins, the readers are referred to the excellent review papers elsewhere [83, 98, 102–104]. While using ·OH radicals for protein labeling has gained much attention, other reactive species (e.g., $SO_4^-\cdot$) were also used in a variant of the FPOP approach [105]. Such tags have slightly different selectivity toward the target amino acids residues compared with ·OH radicals. Several other labeling methods have been developed which do not rely on photochemical reactions or HDX, for example *pulsed alkylation* [106, 107]. On the other hand, methyl methanethiosulfonate reagent was utilized to label thiol moieties in the course of folding [108].

When discussing photochemical approaches for studying the dynamics of proteins by MS, we should mention yet another method: *ultraviolet photodissociation* (UVPD)-MS [109]. When used in conjunction with *S*-ethylacetimidate labeling, it could assist evaluation of conformational changes as a function of the denaturation process [110]. It is applicable to the analysis of protein sequences in native protein–ligand and protein–protein complexes [111]. Taking advantage of this approach, it was possible to obtain non-covalent fragment ions containing the portion of protein still bound to the ligand. Thus, the result of this kind of measurement provided information on the binding sites [111]. UVPD can be considered to be complementary to other top-down proteomic strategies such as ECD or ETD (see also Section 3.3.2.2).

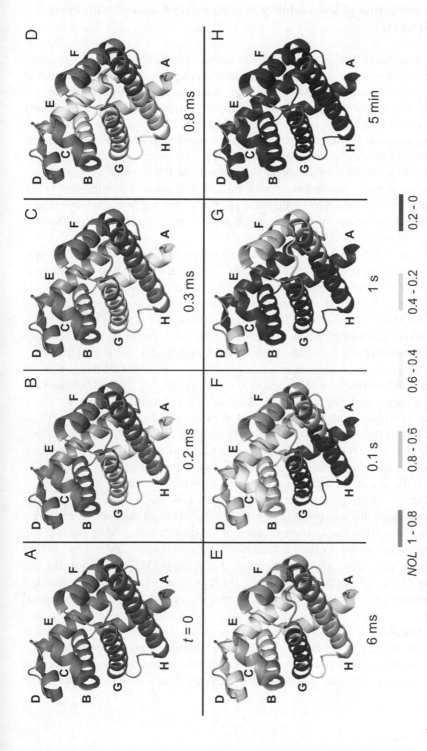

Figure 12.5 *Structural changes during aMb folding as measured by FPOP. Normalized oxidation level (NOL) data were mapped onto the crystal structure of native hMb (PDB code 1WLA) [100] using a five-level color code. Regions for which no structural data are available appear in gray. (A–H) Regions within the aMb polypeptide [18]. Adapted with permission from Vahidi, S., Stocks, B.B., Liaghati-Mobarhan, Y., Konermann, L. (2013) Submillisecond Protein Folding Events Monitored by Rapid Mixing and Mass Spectrometry-Based Oxidative Labeling. Anal. Chem. 85: 8618–8625. Copyright (2013) American Chemical Society. See colour plate section for colour figure*

12.5   Implementation of Ion-mobility Spectrometry Coupled with Mass Spectrometry

Ion-mobility spectrometry (IMS) coupled with MS has recently gained in popularity in studies of intact biomolecules (see Section 6.3). Ions may undergo conformational changes in the gas phase – also during the prolonged storage in an ion trap. This phenomenon was investigated by varying the storage time of ions from ~10 to 200 ms, and subsequently analyzing the stored ions by ion mobility(IM)-TOF-MS [112]. Notably, as pointed out, the studied timescale was shorter than the timescales covered with other instruments using different approaches. In other work, Schenk *et al.* [113] used *trapped ion mobility spectrometry* (TIMS) to investigate the conformational dynamics of the DNA complex (Figure 12.6). Conformational isomer interconversion rates were followed in the course of trapping [113]. Molecular dynamics from the solvent state to the gas-phase state was recorded. The results suggest that the conformations of the interacting peptide are defined by the protonation site, backbone, and side orientations [113].

The combination of IM-MS with HDX enables investigation of conformations of negatively charged peptide and protein ions [114]. Such a combined ion mobility-based method (HDX-TIMS-MS) was implemented to investigate the kinetic intermediates of holo- and apo-myoglobin [115]. It enabled correlation of the ion-neutral collision cross-section and time-resolved H/D back exchange rate. The high mobility resolution of the TIMS cell permitted the observation of multiple bands [115]. The ion-mobility cell of a commercial mass spectrometer can also become the reaction cell for HDX experiments in the gas phase [24]. For example, ND_3 gas can be injected to a *traveling wave ion guide* (TWIG) in front or behind the ion-mobility cell. The HDX process occurs on the way to the TOF analyzer. Fast exchanging sites are labeled within time intervals of $0.1–10$ ms [24].

In other work, cryogenic IM-MS was utilized to probe the structural evolution of an undecapeptide during the final stages of the ESI process [22]. The results demonstrated that a compact dehydrated conformer population can be kinetically trapped on the timescale of several milliseconds, even when an extended gas-phase conformation is energetically favorable. With the aid of IM-MS, Reading *et al.* [116] followed the conformations of membrane proteins, which allowed them to investigate how the surface charge density dictates the stability of folded states (Figure 12.7). The obtained data reveal gas-phase unfolding trajectories. They facilitate observation of the intermediate folding states [116]. Detergent molecules, present in the sample, assist the transition of membrane protein molecules into the gas phase, and their ionization. From Figure 12.7a, it is clear that the temporal peaks of charged species separated by IM span over ~1 ms. The scan speed of the mass spectrometer is sufficient to record several data points across these temporal peaks.

The IM methodology is also suitable for studies of the dynamics of supramolecular complexes. For instance, ESI-IM-MS was implemented to gain insights on amyloid assembly [117]. The dynamics of the oligomers was further investigated by observing subunit exchange after mixing ^{14}N- and ^{15}N-labeled oligomers by means of ESI-MS over a few hours [117].

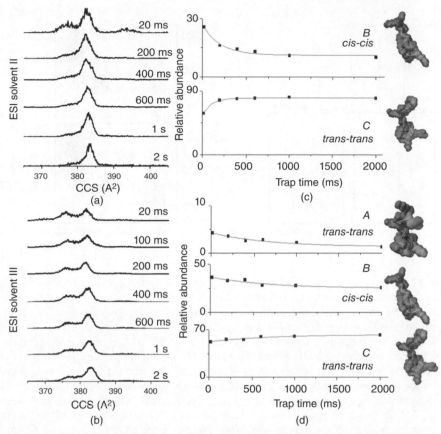

Figure 12.6 *Typical ATHP [M + 2H]$^{2+}$ collision cross section (CCS) profiles for ESI solvent conditions (a) II and (b) III as a function of the trap time (20 ms–2 s). The relative abundances and candidate structures for the most abundant bands (A–C) as a function of the trap time is depicted in (c) and (d) for ESI solvent conditions II and III, respectively. Experimental data have been fitted with exponential decays. Candidate structures are displayed on the far right with the orientations of the proline residues presented. ATHP, "AT hook" decapeptide unit (Lys^1-Arg^2-Pro^3-Arg^4-Gly^5-Arg^6-Pro^7-Arg^8-Lys^9-Trp^{10}) [113]. Reprinted with permission from Schenk, E.R., Ridgeway, M.E., Park, M.A., Leng, F., Fernandez-Lima, F. (2014) Isomerization Kinetics of AT Hook Decapeptide Solution Structures. Anal. Chem. 86: 1210–1214. Copyright (2014) American Chemical Society*

12.6 Concluding Remarks

In summary, MS has enormous potential for the studies of dynamics of protein folding. Conventional sample preparation and ionization techniques have been optimized considering specifics of such measurements. It appears that rapid micromixers (tees, theta capillary emitters, dual emitters) used with ESI, on-line HDX in conjunction with bottom-up

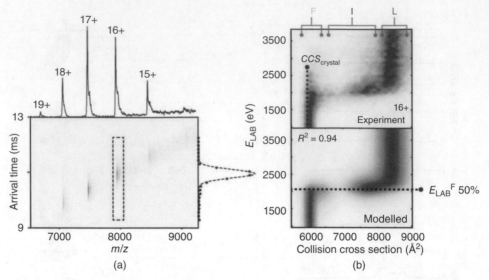

Figure 12.7 *The preservation and unfolding trajectory of folded gas-phase conformations of membrane proteins is dependent on surface charge density and structure. (a) IM-MS spectrum of AmtB in the presence of C8E4. Shown are the arrival time distributions of all ions with the raw data of the 16 + ion (dashed-line box) extracted and fitted with a Gaussian distribution (dashed line). Ion mobility is shown in linear intensity scale. (b) Gas-phase collision-induced unfolding data and modeled fits of AmtB 16 + ion. The folded native-like state (F) is in agreement with the collision cross section (CCS) of the crystal structure (CCS$_{crystal}$). The ions unfold to produce denatured intermediates (I) eventually reaching a maximum CCS for an intermediate population before monomer dissociation (L) [116]. Reproduced from Reading, E., Liko, I., Allison, T.M., Benesch, J.L.P., Laganowsky, A., Robinson, C.V. (2015) The Role of the Detergent Micelle in Preserving the Structure of Membrane Proteins in the Gas Phase. Angew. Chem. Int. Ed. 54: 4577–4581 with permission from John Wiley and Sons*

or top-down proteomic methods, as well as the FPOP method, are the most promising approaches for temporal studies of protein tertiary structures using MS. During the past two decades probing protein conformations in the gas phase has become common practice in biophysics research. Many past studies focused on model protein systems (myoglobin, cytochrome *c*, ubiquitin). In future, it is appealing to look at various other proteins in which misfolding could be associated with disease. The assumption that the gas-phase structures are related to the liquid-phase structures seems to be supported by a substantial amount of experimental evidence obtained for various proteins and experimental conditions. However, taking into account the huge heterogeneity of proteins (amino acid sequence, post-translational modifications, folding patterns), in the authors' opinion, this assumption still needs to be re-verified in future work. Such an additional verification is especially required in the discovery-oriented work that is focused on less studied biomolecules or in one that implements unconventional experimental conditions.

References

1. Dobson, C.M., Šali, A., Karplus, M. (1998) Protein Folding: A Perspective from Theory and Experiment. Angew. Chem. Int. Ed. 37: 868–893.
2. Konermann, L., Simmons, D.A. (2003) Protein-folding Kinetics and Mechanisms Studied by Pulse-Labeling and Mass Spectrometry. Mass Spectrom. Rev. 22: 1–26.
3. Nölting, B. (2006) Protein Folding Kinetics. Biophysical Methods, 2nd Edition. Springer, Berlin.
4. Head, M.W., Ironside, J.W. (2012) Creutzfeldt–Jakob Disease: Prion Protein Type, Disease Phenotype and Agent Strain. Neuropath. Appl. Neurobiol. 38: 296–310.
5. Eyles, S.J., Kaltashov, I.A. (2004) Methods to Study Protein Dynamics and Folding by Mass Spectrometry. Methods 34: 88–99.
6. Wolynes, P.G., Luthey-Schulten, Z., Onuchic, J.N. (1996) Fast-folding Experiments and the Topography of Protein Folding Energy Landscapes. Chem. Biol. 3: 425–432.
7. Plaxco, K.W., Dobson, C.M. (1996) Time-resolved Biophysical Methods in the Study of Protein Folding. Curr. Opin. Struct. Biol. 6: 630–636.
8. Kaltashov, I.A., Eyles, S.J. (2002) Studies of Biomolecular Conformations and Conformational Dynamics by Mass Spectrometry. Mass Spectrom. Rev. 21: 37–71.
9. Kaltashov, I.A., Eyles, S.J. (2012) Chapter 6, Kinetic Studies by Mass Spectrometry. In: Mass Spectrometry in Structural Biology and Biophysics: Architecture, Dynamics, and Interaction of Biomolecules, 2nd Edition. John Wiley & Sons, Inc., Hoboken.
10. Eaton, W.A., Munoz, V., Hagen, S.J., Jas, G.S., Lapidus, L.J., Henry, E.R., Hofrichter, J. (2000) Fast Kinetics and Mechanisms in Protein Folding. Annu. Rev. Biophys. Biomol. Struct. 29: 327–359.
11. Sinz, A. (2006) Chemical Cross-linking and Mass Spectrometry to Map Three-dimensional Protein Structures and Protein–Protein Interactions. Mass Spectrom. Rev. 25: 663–682.
12. Smith, R.D. (2000) Evolution of ESI-Mass Spectrometry and Fourier Transform Ion Cyclotron Resonance for Proteomics and Other Biological Applications. Int. J. Mass Spectrom. 200: 509–544.
13. Lee, Y.J. (2008) Mass Spectrometric Analysis of Cross-linking Sites for the Structure of Proteins and Protein Complexes. Mol. BioSyst. 4: 816–823.
14. Winston, R.L., Fitzgerald, M.C. (1997) Mass Spectrometry as a Readout of Protein Structure and Function. Mass Spectrom. Rev. 16: 165–179.
15. Konermann, L., Vahidi, S., Sowole, M.A. (2014) Mass Spectrometry Methods for Studying Structure and Dynamics of Biological Macromolecules. Anal. Chem. 86: 213–232.
16. Konermann, L., Pan, J., Wilson, D.J. (2006) Protein Folding Mechanisms Studied by Time-resolved Electrospray Mass Spectrometry. BioTechniques 40: 135–141.
17. Zinck, N., Stark, A.-K., Wilson, D.J., Sharon, M. (2014) An Improved Rapid Mixing Device for Time-resolved Electrospray Mass Spectrometry Measurements. ChemistryOpen 3: 109–114.

18. Vahidi, S., Stocks, B.B., Liaghati-Mobarhan, Y., Konermann, L. (2013) Submillisecond Protein Folding Events Monitored by Rapid Mixing and Mass Spectrometry-based Oxidative Labeling. Anal. Chem. 85: 8618–8625.
19. Siu, K.W.M., Gardner, G.J., Berman, S.S. (1989) Multiply Charged Ions in Ionspray Tandem Mass Spectrometry. Org. Mass Spectrom. 24: 931–942.
20. Banerjee, S., Mazumdar, S. (2012) Electrospray Ionization Mass Spectrometry: A Technique to Access the Information beyond the Molecular Weight of the Analyte. Int. J. Anal. Chem. 2012: 282574.
21. Breuker, K., McLafferty, F.W. (2008) Stepwise Evolution of Protein Native Structure with Electrospray into the Gas Phase, 10^{-12} to 10^2 s. Proc. Natl. Acad. Sci. USA 105: 18145–18152.
22. Silveira, J.A., Fort, K.L., Kim, D., Servage, K.A., Pierson, N.A., Clemmer, D.E., Russell, D.H. (2013) From Solution to the Gas Phase: Stepwise Dehydration and Kinetic Trapping of Substance P Reveals the Origin of Peptide Conformations. J. Am. Chem. Soc. 135: 19147–19153.
23. Badman, E.R., Myung, S., Clemmer, D.E. (2005) Evidence for Unfolding and Refolding of Gas-phase Cytochrome *c* Ions in a Paul Trap. J. Am. Soc. Mass Spectrom. 16: 1493–1497.
24. Rand, K.D., Pringle, S.D., Murphy III, J.P., Fadgen, K.E., Brown, J., Engen, J.R. (2009) Gas-phase Hydrogen/Deuterium Exchange in a Traveling Wave Ion Guide for the Examination of Protein Conformations. Anal. Chem. 81: 10019–10028.
25. Loo, J.A. (1995) Observation of Large Subunit Protein Complexes by Electrospray Ionization Mass Spectrometry. J. Mass Spectrom. 30: 180–183.
26. Loo, J.A. (1997) Studying Noncovalent Protein Complexes by Electrospray Ionization Mass Spectrometry. Mass Spectrom. Rev. 16: 1–23.
27. Benesch, J.L.P., Robinson, C.V. (2006) Mass Spectrometry of Macromolecular Assemblies: Preservation and Dissociation. Curr. Opin. Struct. Biol. 16: 245–251.
28. Heck, A.J.R. (2008) Native Mass Spectrometry: A Bridge Between Interactomics and Structural Biology. Nature Meth. 5: 927–933.
29. Hopper, J.T.S., Robinson, C.V. (2014) Mass Spectrometry Quantifies Protein Interactions – From Molecular Chaperones to Membrane Porins. Angew. Chem. Int. Ed. 53: 14002–14015.
30. Stengel, F., Baldwin, A.J., Painter, A.J., Jaya, N., Basha, E., Kay, L.E., Vierling, E., Robinson, C.V., Benesch, J.L.P. (2010) Quaternary Dynamics and Plasticity Underlie Small Heat Shock Protein Chaperone Function. Proc. Natl. Acad. Sci. USA 107: 2007–2012.
31. Gross, D.S., Schnier, P.D., Rodriguez-Cruz, S.E., Fagerquist, C.K., Williams, E.R. (1996) Conformations and Folding of Lysozyme Ions *In Vacuo*. Proc. Natl. Acad. Sci. USA 93: 3143–3148.
32. McLafferty, F.W., Guan, Z., Haupts, U., Wood, T.D., Kelleher, N.L. (1998) Gaseous Conformational Structures of Cytochrome *c*. J. Am. Chem. Soc. 120: 4732–4740.
33. Breuker, K., Oh, H., Horn, D.M., Cerda, B.A., McLafferty, F.W. (2002) Detailed Unfolding and Folding of Gaseous Ubiquitin Ions Characterized by Electron Capture Dissociation. J. Am. Chem. Soc. 124: 6407–6420.
34. Konermann, L., Ahadi, E., Rodriguez, A.D., Vahidi, S. (2013) Unraveling the Mechanism of Electrospray Ionization. Anal. Chem. 85: 2–9.

35. Ahadi, E., Konermann, L. (2011) Ejection of Solvated Ions from Electrosprayed Methanol/Water Nanodroplets Studied by Molecular Dynamics Simulations. J. Am. Chem. Soc. 133: 9354–9363.

36. Simmons, D.A., Dunn, S.D., Konermann, L. (2003) Conformational Dynamics of Partially Denatured Myoglobin Studied by Time-resolved Electrospray Mass Spectrometry with Online Hydrogen–Deuterium Exchange. Biochemistry 42: 5896–5905.

37. Lee, J.W., Kim, H.I. (2015) Solvent-induced Structural Transitions of Lysozyme in an Electrospray Ionization Source. Analyst 140: 3573–3580.

38. Mirza, U.A., Cohen, S.L., Chait, B.T. (1993) Heat-induced Conformational Changes in Proteins Studied by Electrospray Ionization Mass Spectrometry. Anal. Chem. 65: 1–6.

39. Kolakowski, B.M., Konermann, L. (2001) From Small-molecule Reactions to Protein Folding: Studying Biochemical Kinetics by Stopped-flow Electrospray Mass Spectrometry. Anal. Biochem. 292: 107–114.

40. Konermann, L., Collings, B.A., Douglas, D.J. (1997) Cytochrome *c* Folding Kinetics Studied by Time-resolved Electrospray Ionization Mass Spectrometry. Biochemistry 36: 5554–5559.

41. Konermann, L., Rosell, F.I., Mauk, A.G., Douglas, D.J. (1997) Acid-induced Denaturation of Myoglobin Studied by Time-resolved Electrospray Ionization Mass Spectrometry. Biochemistry 36: 6448–6454.

42. Sogbein, O.O., Simmons, D.A., Konermann, L. (2000) Effects of pH on the Kinetic Reaction Mechanism of Myoglobin Unfolding Studied by Time-resolved Electrospray Ionization Mass Spectrometry. J. Am. Soc. Mass Spectrom. 11: 312–319.

43. Lee, V.W.S., Chen, Y.-L., Konermann, L. (1999) Reconstitution of Acid-denatured Holomyoglobin Studied by Time-resolved Electrospray Ionization Mass Spectrometry. Anal. Chem. 71: 4154–4159.

44. Wilson, D.J., Rafferty, S.P., Konermann, L. (2005) Kinetic Unfolding Mechanism of the Inducible Nitric Oxide Synthase Oxygenase Domain Determined by Time-resolved Electrospray Mass Spectrometry. Biochemistry 44. 2276–2283.

45. Fändrich, M., Tito, M.A., Leroux, M.R., Rostom, A.A., Hartl, F.U., Dobson, C.M., Robinson, C.V. (2000) Observation of the Noncovalent Assembly and Disassembly Pathways of the Chaperone Complex MtGimC by Mass Spectrometry. Proc. Natl. Acad. Sci. USA 97: 14151–14155.

46. Cubrilovic, D., Barylyuk, K., Hofmann, D., Walczak, M.J., Gräber, M., Berg, T., Wider, G., Zenobi, R. (2014) Direct Monitoring of Protein–Protein Inhibition Using Nano Electrospray Ionization Mass Spectrometry. Chem. Sci. 5: 2794–2803.

47. Takáts, Z., Wiseman, J.M., Gologan, B., Cooks, R.G. (2004) Electrosonic Spray Ionization. A Gentle Technique for Generating Folded Proteins and Protein Complexes in the Gas Phase and for Studying Ion–Molecule Reactions at Atmospheric Pressure. Anal. Chem. 76: 4050–4058.

48. Touboul, D., Jecklin, M.C., Zenobi, R. (2007) Rapid and Precise Measurements of Gas-phase Basicity of Peptides and Proteins at Atmospheric Pressure by Electrosonic Spray Ionization-Mass Spectrometry. J. Phys. Chem. B Lett. 111: 11629–11631.

49. Touboul, D., Jecklin, M.C., Zenobi, R. (2008) Investigation of Deprotonation Reactions on Globular and Denatured Proteins at Atmospheric Pressure by ESSI-MS. J. Am. Soc. Mass Spectrom. 19: 455–466.

50. Miao, Z., Wu, S., Chen, H. (2010) The Study of Protein Conformation in Solution Via Direct Sampling by Desorption Electrospray Ionization Mass Spectrometry. J. Am. Soc. Mass Spectrom. 21: 1730–1736.

51. Wilson, D.J., Konermann, L. (2003) A Capillary Mixer with Adjustable Reaction Chamber Volume for Millisecond Time-resolved Studies by Electrospray Mass Spectrometry. Anal. Chem. 75: 6408–6414.

52. Fisher, C.M., Kharlamova, A., McLuckey, S.A. (2014) Affecting Protein Charge State Distributions in Nano-electrospray Ionization via In-spray Solution Mixing Using Theta Capillaries. Anal. Chem. 86: 4581–4588.

53. Mortensen, D.N., Williams, E.R. (2014) Theta-glass Capillaries in Electrospray Ionization: Rapid Mixing and Short Droplet Lifetimes. Anal. Chem. 86: 9315–9321.

54. Mortensen, D.N., Williams, E.R. (2015) Investigating Protein Folding and Unfolding in Electrospray Nanodrops upon Rapid Mixing Using Theta-glass Emitters. Anal. Chem. 87: 1281–1287.

55. Lee, J.K., Kim, S., Nam, H.G., Zare, R.N. (2015) Microdroplet Fusion Mass Spectrometry for Fast Reaction Kinetics. Proc. Natl. Acad. Sci. USA 112: 3898–3903.

56. Konermann, L., Pan, J., Liu, Y.-H. (2011) Hydrogen Exchange Mass Spectrometry for Studying Protein Structure and Dynamics. Chem. Soc. Rev. 40: 1224–1234.

57. Kim, P.S., Baldwin, R.L. (1980) Structural Intermediates Trapped During the Folding of Ribonuclease A by Amide Proton Exchange. Biochemistry 19: 6124–6129.

58. Udgaonkar, J.B., Baldwin, R.L. (1988) NMR Evidence for an Early Framework Intermediate on the Folding Pathway of Ribonuclease A. Nature 335: 694–699.

59. Roder, H., Elöve, G.A., Englander, S.W. (1988) Structural Characterization of Folding Intermediates in Cytochrome *c* by H-exchange Labelling and Proton NMR. Nature 335: 700–704.

60. Wales, T.E., Engen, J.R. (2006) Hydrogen Exchange Mass Spectrometry for the Analysis of Protein Dynamics. Mass Spectrom. Rev. 25: 158–170.

61. Katta, V., Chait, B.T. (1991) Conformational Changes in Proteins Probed by Hydrogen-exchange Electrospray-ionization Mass Spectrometry. Rapid Commun. Mass Spectrom. 5: 214–217.

62. Morgan, C.R., Engen, J.R. (2009) Investigating Solution-phase Protein Structure and Dynamics by Hydrogen Exchange Mass Spectrometry. Curr. Protoc. Protein Sci. Suppl. 58: 17.6.1–17.6.17.

63. Keppel, T.R., Weis, D.D. (2015) Mapping Residual Structure in Intrinsically Disordered Proteins at Residue Resolution Using Millisecond Hydrogen/Deuterium Exchange and Residue Averaging. J. Am. Soc. Mass Spectrom. 26: 547–554.

64. Katta, V., Chait, B.T. (1993) Hydrogen/Deuterium Exchange Electrospray Ionization Mass Spectrometry: A Method for Probing Protein Conformational Changes in Solution. J. Am. Chem. Soc. 115: 6317–6321.

65. Miranker, A., Robinson, C.V., Radford, S.E., Aplin, R.T., Dobson, C.M. (1993) Detection of Transient Protein Folding Populations by Mass Spectrometry. Science 262: 896–900.

66. Hooke, S.D., Eyles, S.J., Miranker, A., Radford, S.E., Robinson, C.V., Dobson, C.M. (1995) Cooperative Elements in Protein Folding Monitored by Electrospray Ionization Mass Spectrometry. J. Am. Chem. Soc. 117: 1548–1549.

67. Yang, H., Smith, D.L. (1997) Kinetics of Cytochrome *c* Folding Examined by Hydrogen Exchange and Mass Spectrometry. Biochemistry 36: 14992–14999.

68. Tsui, V., Garcia, C., Cavagnero, S., Siuzdak, G., Dyson, H.J., Wright, P.E. (1999) Quench-flow Experiments Combined with Mass Spectrometry Show Apomyoglobin Folds through an Obligatory Intermediate. Protein Sci. 8: 45–49.

69. Simmons, D.A., Konermann, L. (2002) Characterization of Transient Protein Folding Intermediates during Myoglobin Reconstitution by Time-resolved Electrospray Mass Spectrometry with On-line Isotopic Pulse Labeling. Biochemistry 41: 1906–1914.

70. Pan, J., Rintala-Dempsey, A.C., Li, Y., Shaw, G.S., Konermann, L. (2006) Folding Kinetics of the S100A11 Protein Dimer Studied by Time-resolved Electrospray Mass Spectrometry and Pulsed Hydrogen-Deuterium Exchange. Biochemistry 45: 3005–3013.

71. Pan, L.Y., Salas-Solano, O., Valliere-Douglass, J.F. (2014) Conformation and Dynamics of Interchain Cysteine-linked Antibody–Drug Conjugates as Revealed by Hydrogen/Deuterium Exchange Mass Spectrometry. Anal. Chem. 86: 2657–2664.

72. Liuni, P., Jeganathan, A., Wilson, D.J. (2012) Conformer Selection and Intensified Dynamics during Catalytic Turnover in Chymotrypsin. Angew. Chem. Int. Ed. 51: 9666–9669.

73. Liuni, P., Rob, T., Wilson, D.J. (2010) A Microfluidic Reactor for Rapid, Low-pressure Proteolysis with On-chip Electrospray Ionization. Rapid Commun. Mass Spectrom. 24: 315–320.

74. Rob, T., Liuni, P., Gill, P.K., Zhu, S., Balachandran, N., Berti, P.J., Wilson, D.J. (2012) Measuring Dynamics in Weakly Structured Regions of Proteins Using Microfluidics-enabled Subsecond H/D Exchange Mass Spectrometry. Anal. Chem. 84: 3771–3779.

75. Resetca, D., Wilson, D.J. (2013) Characterizing Rapid, Activity-linked Conformational Transitions in Proteins via Sub-second Hydrogen Deuterium Exchange Mass Spectrometry. FEBS J. 280: 5616–5625.

76. Hu, W., Walters, B.T., Kan, Z.-Y., Mayne, L., Rosen, L.E., Marqusee, S., Englander, S.W. (2013) Stepwise Protein Folding at Near Amino Acid Resolution by Hydrogen Exchange and Mass Spectrometry. Proc. Natl. Acad. Sci. USA 110: 7684–7689.

77. Zubarev, R.A., Kelleher, N.L., McLafferty, F.W. (1998) Electron Capture Dissociation of Multiply Charged Protein Cations. A Nonergodic Process. J. Am. Chem. Soc. 120: 3265–3266.

78. Pan, J., Han, J., Borchers, C.H., Konermann, L. (2008) Electron Capture Dissociation of Electrosprayed Protein Ions for Spatially Resolved Hydrogen Exchange Measurements. J. Am. Chem. Soc. 130: 11574–11575.

79. Pan, J., Han, J., Borchers, C.H., Konermann, L. (2009) Hydrogen/Deuterium Exchange Mass Spectrometry with Top-Down Electron Capture Dissociation for Characterizing Structural Transitions of a 17 kDa Protein. J. Am. Chem. Soc. 131: 12801–12808.

80. Pan, J., Han, J., Borchers, C.H., Konermann, L. (2010) Characterizing Short-lived Protein Folding Intermediates by Top-Down Hydrogen Exchange Mass Spectrometry. Anal. Chem. 82: 8591–8597.

81. Rand, K.D., Adams, C.M., Zubarev, R.A., Jørgensen, T.J.D. (2008) Electron Capture Dissociation Proceeds with a Low Degree of Intramolecular Migration of Peptide Amide Hydrogens. J. Am. Chem. Soc. 130: 1341–1349.

82. Zehl, M., Rand, K.D., Jensen, O.N., Jørgensen, T.J.D. (2008) Electron Transfer Dissociation Facilitates the Measurement of Deuterium Incorporation into Selectively Labeled Peptides with Single Residue Resolution. J. Am. Chem. Soc. 130: 17453–17459.

83. Konermann, L., Tong, X., Pan, Y. (2008) Protein Structure and Dynamics Studied by Mass Spectrometry: H/D Exchange, Hydroxyl Radical Labeling, and Related Approaches. J. Mass Spectrom. 43: 1021–1036.

84. Winger, B.E., Light-Wahl, K.J., Rockwood, A.L., Smith, R.D. (1992) Probing Qualitative Conformation Differences of Multiply Protonated Gas-phase Proteins via H/D Isotopic Exchange with D_2O. J. Am. Chem. Soc. 114: 5897–5898.

85. Suckau, D., Shi, Y., Beu, S.C., Senko, M.W., Quinn, J.P., Wampler III, F.M., McLafferty, F.W. (1993) Coexisting Stable Conformations of Gaseous Protein Ions. Proc. Natl. Acad. Sci. USA 90: 790–793.

86. Campbell, S., Rodgers, M.T., Marzluff, E.M., Beauchamp, J.L. (1994) Structural and Energetic Constraints on Gas Phase Hydrogen/Deuterium Exchange Reactions of Protonated Peptides with D_2O, CD_3OD, CD_3CO_2D, and ND_3. J. Am. Chem. Soc. 116: 9765–9766.

87. Wood, T.D., Chorush, R.A., Wampler III, F.M., Little, D.P., O'Connor, P.B., McLafferty, F.W. (1995) Gas-phase Folding and Unfolding of Cytochrome *c* Cations. Proc. Natl. Acad. Sci. USA 92: 2451–2454.

88. Freitas, M.A., Hendrickson, C.L., Emmett, M.R., Marshall, A.G. (1999) Gas-phase Bovine Ubiquitin Cation Conformations Resolved by Gas-phase Hydrogen/Deuterium Exchange Rate and Extent. Int. J. Mass Spectrom. 185/186/187: 565–575.

89. Hofstadler, S.A., Sannes-Lowery, K.A., Griffey, R.H. (1999) A Gated-beam Electrospray Ionization Source with an External Ion Reservoir. A New Tool for the Characterization of Biomolecules Using Electrospray Ionization Mass Spectrometry. Rapid Commun. Mass Spectrom. 13: 1971–1979.

90. Hofstadler, S.A., Sannes-Lowery, K.A., Griffey, R.H. (2000) Enhanced Gas-phase Hydrogen–Deuterium Exchange of Oligonucleotide and Protein Ions Stored in an External Multipole Ion Reservoir. J. Mass Spectrom. 35: 62–70.

91. Lifshitz, C. (2004) A Review of Gas-phase H/D Exchange Experiments: the Protonated Arginine Dimer and Bradykinin Nonapeptide Systems. Int. J. Mass Spectrom. 234: 63–70.

92. Rajabi, K. (2015) Time-resolved Pulsed Hydrogen/Deuterium Exchange Mass Spectrometry Probes Gaseous Proteins Structural Kinetics. J. Am. Soc. Mass Spectrom. 26: 71–82.

93. Hambly, D.M., Gross, M.L. (2005) Laser Flash Photolysis of Hydrogen Peroxide to Oxidize Protein Solvent-accessible Residues on the Microsecond Timescale. J. Am. Soc. Mass Spectrom. 16: 2057–2063.

94. Gau, B.C., Sharp, J.S., Rempel, D.L., Gross, M.L. (2009) Fast Photochemical Oxidation of Protein Footprints Faster than Protein Unfolding. Anal. Chem. 81: 6563–6571.

95. Niu, B., Zhang, H., Giblin, D., Rempel, D.L., Gross, M.L. (2015) Dosimetry Determines the Initial OH Radical Concentration in Fast Photochemical Oxidation of Proteins (FPOP). J. Am. Soc. Mass Spectrom. 26: 843–846.
96. Chen, J., Rempel, D.L., Gross, M.L. (2010) T-jump and Fast Photochemical Oxidation Probe Sub Millisec Protein Folding. J. Am. Chem. Soc. 132: 15502–15504.
97. Gruebele, M. (2010) Analytical Biochemistry: Weighing up Protein Folding. Nature 468: 640–641.
98. Konermann, L., Pan, Y., Stocks, B.B. (2011) Protein Folding Mechanisms Studied by Pulsed Oxidative Labeling and Mass Spectrometry. Curr. Opin. Struct. Biol. 21: 634–640.
99. Vahidi, S., Stocks, B.B., Liaghati-Mobarhan, Y., Konermann, L. (2012) Mapping pH-induced Protein Structural Changes under Equilibrium Conditions by Pulsed Oxidative Labeling and Mass Spectrometry. Anal. Chem. 84: 9124–9130.
100. Maurus, R., Overall, C.M., Bogumil, R., Luo, Y., Mauk, A.G., Smith, M., Brayer, G.D. (1997) A Myoglobin Variant with a Polar Substitution in a Conserved Hydrophobic Cluster in the Heme Binding Pocket. Biochim. Biophys. Acta 1341: 1–13.
101. Rinas, A., Jones, L.M. (2015) Fast Photochemical Oxidation of Proteins Coupled to Multidimensional Protein Identification Technology (MudPIT): Expanding Footprinting Strategies to Complex Systems. J. Am. Soc. Mass Spectrom. 26: 540–546.
102. Xu, G., Chance, M.R. (2007) Hydroxyl Radical-mediated Modification of Proteins as Probes for Structural Proteomics. Chem. Rev. 107: 3514–3543.
103. Kiselar, J.G., Chance, M.R. (2010) Future Directions of Structural Mass Spectrometry Using Hydroxyl Radical Footprinting. J. Mass. Spectrom. 45: 1373–1382.
104. Konermann, L., Stocks, B.B., Pan, Y., Tong, X. (2010) Mass Spectrometry Combined with Oxidative Labeling for Exploring Protein Structure and Folding. Mass Spectrom. Rev. 29: 651–667.
105. Gau, B.C., Chen, H., Zhang, Y., Gross, M.L. (2010) Sulfate Radical Anion as a New Reagent for Fast Photochemical Oxidation of Proteins. Anal. Chem. 82: 7821–7827.
106. Apuy, J.L., Park, Z.-Y., Swartz, P.D., Dangott, L.J., Russell, D.H., Baldwin, T.O. (2001) Pulsed-alkylation Mass Spectrometry for the Study of Protein Folding and Dynamics: Development and Application to the Study of a Folding/Unfolding Intermediate of Bacterial Luciferase. Biochemistry 40: 15153–15163.
107. Apuy, J.L., Chen, X., Russell, D.H., Baldwin, T.O., Giedroc, D.P. (2001) Ratiometric Pulsed Alkylation/Mass Spectrometry of the Cysteine Pairs in Individual Zinc Fingers of MRE-binding Transcription Factor-1 (MTF-1) as a Probe of Zinc Chelate Stability. Biochemistry 40: 15164–15175.
108. Jha, S.K., Udgaonkar, J.B. (2007) Exploring the Cooperativity of the Fast Folding Reaction of a Small Protein Using Pulsed Thiol Labeling and Mass Spectrometry. J. Biol. Chem. 282: 37479–37491.
109. Bowers, W.D., Delbert, S.S., Hunter, R.L., McIver Jr, R.T. (1984) Fragmentation of Oligopeptide Ions Using Ultraviolet Laser Radiation and Fourier Transform Mass Spectrometry. J. Am. Chem. Soc. 106: 7288–7289.
110. Cammarata, M., Lin, K.-Y., Pruet, J., Liu, H., Brodbelt, J. (2014) Probing the Unfolding of Myoglobin and Domain C of PARP-1 with Covalent Labeling and Top-Down Ultraviolet Photodissociation Mass Spectrometry. Anal. Chem. 86: 2534–2542.

111. O'Brien, J.P., Li, W., Zhang, Y., Brodbelt, J.S. (2014) Characterization of Native Protein Complexes Using Ultraviolet Photodissociation Mass Spectrometry. J. Am. Chem. Soc. 136: 12920–12928.
112. Badman, E.R., Hoaglund-Hyzer, C.S., Clemmer, D.E. (2001) Monitoring Structural Changes of Proteins in an Ion Trap over ~10–200 ms: Unfolding Transitions in Cytochrome *c* Ions. Anal. Chem. 73: 6000–6007.
113. Schenk, E.R., Ridgeway, M.E., Park, M.A., Leng, F., Fernandez-Lima, F. (2014) Isomerization Kinetics of AT Hook Decapeptide Solution Structures. Anal. Chem. 86: 1210–1214.
114. Donohoe, G.C., Khakinejad, M., Valentine, S.J. (2015) Ion Mobility Spectrometry-Hydrogen Deuterium Exchange Mass Spectrometry of Anions: Part 1. Peptides to Proteins. J. Am. Soc. Mass Spectrom. 26: 564–576.
115. Schenk, E.R., Almeida, R., Miksovska, J., Ridgeway, M.E., Park, M.A., Fernandez-Lima, F. (2015) Kinetic Intermediates of Holo- and Apo-myoglobin Studied Using HDX-TIMS-MS and Molecular Dynamic Simulations. J. Am. Soc. Mass Spectrom. 26: 555–563.
116. Reading, E., Liko, I., Allison, T.M., Benesch, J.L.P., Laganowsky, A., Robinson, C.V. (2015) The Role of the Detergent Micelle in Preserving the Structure of Membrane Proteins in the Gas Phase. Angew. Chem. Int. Ed. 54: 4577–4581.
117. Smith, D.P., Radford, S.E., Ashcroft, A.E. (2010) Elongated Oligomers in β_2-Microglobulin Amyloid Assembly Revealed by Ion Mobility Spectrometry-Mass Spectrometry. Proc. Natl. Acad. Sci. USA 107: 6794–6798.

13

Applications of Time-resolved Mass Spectrometry in Biochemical Analysis

Mass spectrometry (MS) has become an essential platform for analysis of biomolecules. It is heavily used by the pharmaceutical industry, especially in the drug discovery process [1]. It can also provide insights on the mechanism of complex dynamic processes [2]. Mass spectrometric systems, incorporating mixing devices and soft ionization, enable kinetic and mechanistic studies on biochemical reactions such as those catalyzed by enzymes [3]. Due to the high selectivity and sensitivity of MS, reaction substrates and products can be measured independently, and – in most cases – without the need for labeling [2]. Moreover, the biotechnology industry takes advantage of the on-line *process analytical technology* (PAT), in which mass spectrometers play the central role. For instance, mass spectrometers are used to monitor the composition of gas generated in bioreactors thus providing vital information on fermentative processes [4]. In the next few paragraphs, we will summarize applications of temporally resolved mass spectrometric measurements in biochemistry, including studies of biocatalytic reactions *in vitro* as well as monitoring the release of biomolecules from biological cells *in vivo* and *ex vivo*.

13.1 Enzymatic Reactions

13.1.1 Requirements of Time-resolved Mass Spectrometry in Biocatalysis

The prerequisite for implementing MS in biochemical analysis is that the analytes are amenable to ionization using one of the available ion sources (cf. Chapter 2). A vast majority of biochemical processes occur in the condensed phase. Most enzymes are active in aqueous solutions while only a few of them can catalyze reactions in organic solvents (e.g., transesterification catalyzed by lipase [5]). Water-based buffers are not always the optimum solvents for the operation of ion sources. In fact, concentrated buffer solutions – which are often used in biochemical assays – can suppress ionization, and lead to contamination of

Time-Resolved Mass Spectrometry: From Concept to Applications, First Edition.
Pawel Lukasz Urban, Yu-Chie Chen and Yi-Sheng Wang.
© 2016 John Wiley & Sons, Ltd. Published 2016 by John Wiley & Sons, Ltd.

surfaces at the mass spectrometer's inlet (cf. Chapter 8). On-line monitoring of enzymatic reactions requires trade-off between the optimum conditions for the reaction and the MS analysis. The use of relatively volatile electrolytes (e.g., solutions of ammonium acetate [6], primary or secondary amines [7]) is generally preferred, as long as they do not significantly deteriorate enzymatic activity.

13.1.2   Off-line and On-line Methods

Experimental design of time-resolved mass spectrometry (TRMS) systems can greatly affect temporal resolution in the analysis of dynamic samples (see also Section 4.2). The two popular MS approaches – laser desorption/ionization (LDI)-MS and electrospray ionization (ESI)-MS – are suitable for studies of enzymatic reactions [8].

Off-line MS methods typically require quenching steps and performing reactions in discrete aliquots. Thus, when the reactions are completed within short periods of time, it is difficult to obtain continuous kinetic profiles. However, off-line methods enable sample fractionation, isolation, or desalting. Off-line methods are represented by LDI-MS (e.g., [9, 10]), in which case aliquots of samples are deposited on sample targets. The reaction medium is normally evaporated before insertion of the sample target to the vacuum compartment of the mass spectrometer. In an elegant methodology, the enzyme-catalyzed reaction was conducted in solution but the reactants were then immobilized onto a self-assembled monolayer, and detected by matrix-assisted laser desorption/ionization(MALDI)-time-of-flight (TOF)-MS [11]. Subsequently, it was found that the time-dependent quantities of reactants obtained by MS correlated well with those obtained by high-performance liquid chromatography (HPLC).

On-line MS methods enable continuous kinetic profiles to be obtained but they cannot easily accommodate complex sample preparation steps. In the 1980s, enzymatic reactions were monitored by a popular – at that time – ionization technique, namely *fast atom bombardment* (FAB)-MS [12, 13]. Heidmann *et al.* [14] used FAB-MS to identify conjugation products of reactive quinones with glutathione by conducting dynamic mass spectral analysis. Soon after the introduction of ESI to MS, its potential in the monitoring of biochemical reactions was recognized, especially in the detection of labile intermediates (cf. [15, 16]). Nowadays ESI and MALDI are prime tools for the analysis of biomolecules. Both techniques are also suitable for the investigation of biocatalytic processes with diverse temporal resolutions [17].

Nanospray electrospray ionization (nanoESI)-MS – the miniaturized version of ESI – can readily be used to obtain structural and dynamic information for protein complexes [18]. An automated robotic nanoESI-MS method for monitoring the dynamics of protein complexes in real time has been developed. Sample solution is deposited into a sample well; reagents are then added to this solution; finally, the robot delivers the final sample to the mass spectrometer. This sampling cycle may be repeated [18]. The time interval between the sampling is >32 s. The method enabled seamless monitoring of the digestion of cytochrome *c* by trypsin (Figure 13.1). Using this method, it was also found that small heat shock proteins, which exist as dodecamers, composed of dimeric building blocks, exchange within minutes, and that the rate of exchange strongly depends on temperature [18].

The *rapid chemical quench* strategy is a prominent approach in the enzymology toolkit, which enables reaction steps with short duration to be studied. It is combined with various detection techniques, including MS [15]. For instance, Clarke *et al.* [19] coupled a quench-flow microreactor with ESI-Fourier transform (FT)-ion cyclotron resonance (ICR)-MS to follow *pre-steady-state* hydrolysis of *p*-nitrophenyl acetate by chymotrypsin (see Section 13.1.3). It was possible to detect the enzyme intermediate involved in this reaction. The initial data points were collected with sub-second resolution. The duration of incubation (from mixing the reactants till quenching) was controlled by varying the flow rates of reactant/enzyme solutions provided by pumps [19]. Although it is an on-line approach, the temporal resolution of the reaction monitoring in this case does not depend on the spectrum acquisition time but it depends on the residence time of the reactants in the continuous flow microreactor (similarly to various other methods which implement fluidic reactors; see, e.g., Section 4.2.4).

While (MA)LDI-MS is a typical off-line technique, it can be used in conjunction with the quench-flow strategy to follow relatively fast (sub-second) phenomena associated with biochemical reactions [20]. For instance, it was employed in the study of pre-steady-state kinetics of a protein-tyrosine phosphatase [20]. The timescale of the studied process extended from ~10 ms to 1 s. The advantage of implementing MALDI-MS is that the obtained spectra are less perturbed by the presence of buffers and other sample additives [20]. A similar method was implemented in the monitoring of dTDP-glucose 4,6-dehydratase [21]. The reaction was quenched using a mixture of guanidine hydrochloride and sodium borohydride, which additionally stabilized the intermediates and the product [21].

Using ESI-MS, enzyme/substrate specificity can be determined in a straightforward multiplexed assay [22]. The high analytical throughput of MS allows multiple enzyme samples to be screened for their biocatalytic activity. Mironov *et al.* [23] presented an automated enzyme assay based on capillary electrophoresis (CE) coupled with ESI-MS (Figure 13.2). It enables multiple injections, incubation, and separation in order to conduct enzyme profiling. It may be used as a tool for identification of enzymes for biocatalytic applications. The retention of the reactant mixture in the capillary enables metering of the reaction time prior to separation and introduction to the ESI-MS system. In other work, Sun *et al.* [24] coupled nanoliter-scale segmented flow reactions with ESI-MS to enable direct analysis of reaction products, and applied this system to a screen of inhibitors. The MS analysis is as short as 1.2 s while the amounts of reactants are in the order of ~1 pmol, which emphasizes high throughput and points to reasonably good analytical sensitivity of the method. Moreover, immobilized enzyme reactors can be coupled with MS to enable proteomic analysis [25]. For example, sequencing peptides could be done by coupling a column with immobilized enzyme with a thermospray ion source mass spectrometer [26].

13.1.3 Time-resolved Mass Spectrometry Studies of Enzyme Kinetics

Mass spectrometry can characterize enzyme kinetics, mechanism, and selectivity [27]. Notably, MS enables kinetic analyses of enzymatic reactions where it is not possible to carry out simple spectrophotometric assays – for example, in the absence of chromogenic substrates [28, 29].

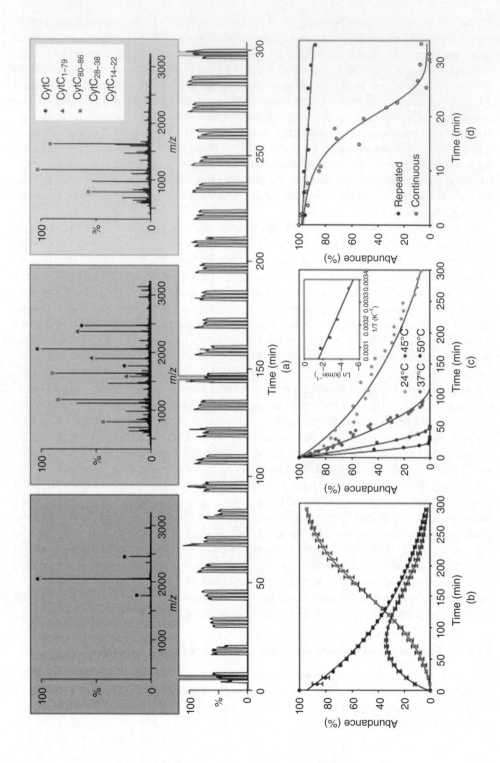

Figure 13.1   *Automated nanoESI monitoring of the tryptic digestion of cytochrome c (CytC). (a) Total-ion current chromatogram of CytC digestion monitored in triplicate over the course of 300 min. Data in the inset are spectra obtained at the beginning, middle, and end of the time course. At the beginning, the predominant species corresponds to full-length CytC (circles), and at the end numerous peptides are observed, the most prominent being CytC$_{80-86}$, CytC$_{28-38}$, and CytC$_{14-22}$ (squares). Halfway through the reaction, an intermediate fragment, CytC$_{1-79}$ (triangles), can also be detected. (b) Plotting the relative abundances of these peptides allows the quantitative monitoring of the digestion reaction. Error bars correspond to three standard deviations from the mean. The amount of CytC decreases exponentially, enabling the determination of first-order rate constants. (c) Monitoring this reaction, specifically the disappearance of intact CytC, at different temperatures (darker markers correspond to higher temperatures while lighter markers correspond to lower temperatures) demonstrates how the reaction velocity increases at higher temperatures. From the Arrhenius plot (inset), the activation energy and pre-exponential factor can be determined. (d) In protein-destabilizing solution conditions, a different reaction profile for the disappearance of intact CytC is determined when the solution is sampled continuously versus repeatedly. This result likely is due to pH effects in the emitter associated with prolonged electrospraying and highlights the benefits of the repeated sampling method advanced here [18]. Reprinted from Chemistry & Biology, 15, Painter, A.J., Jaya, N., Basha, E., Vierling, E., Robinson, C.V., Benesch, L.P., Real-time Monitoring of Protein Complexes Reveals Their Quaternary Organization and Dynamics, 246–253. Copyright (2008), with permission from Elsevier. See colour plate section for colour figure*

Michaelis–Menten kinetics is used to describe the progress of enzymatic reactions, which can be summarized by the following simplified equation [30]:

$$E + S \underset{}{\overset{K_S}{\rightleftharpoons}} ES \xrightarrow{k_{cat}} E + P \tag{13.1}$$

where E is for enzyme, S is for substrate, ES is for the enzyme–substrate complex, and P refers to the reaction product. During the first few milliseconds of the reaction, the intermediate (ES in Equation 13.1) is formed. This stage of the enzymatic process is referred to as pre-steady state (see also Section 13.1.2). The Michaelis constant (K_M) is the dissociation constant of the enzyme–substrate complex. Thus, it describes the affinity of a substrate to the active site of an enzyme. It is an important feature of almost all the enzymes which follow Michaelis–Menten kinetics [31]. The K_M values are defined by the ratio [30]:

$$K_M = \frac{[ES]}{[E][S]} \tag{13.2}$$

The initial reaction velocity (v) can be obtained from:

$$v = \frac{[E]_0[S]k_{cat}}{K_M + [S]} \tag{13.3}$$

or

$$v = \frac{v_{max}[S]}{K_M + [S]} \tag{13.4}$$

where v_{max} is the reaction velocity limit observed at high concentrations of substrate. The Michaelis constant is normally calculated based on a series of measurements of initial reaction velocities in the presence of substrate at varied concentrations.

(1) Injection of 1st set of subplugs of SM and B

(2) Injection of 2nd, 3rd, and 4th sets of subplugs of SM and Enzymes

(3) In-capillary mixing and incubation

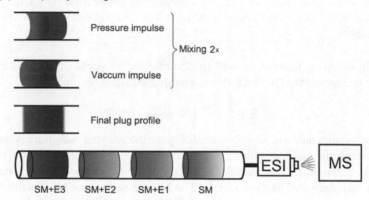

(4) CE separation and MS profiling

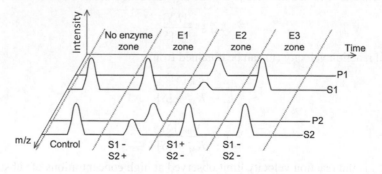

Figure 13.2 *Schematic representation of MINISEP-MS assay. In this assay, capillary electrophoresis (CE) is interfaced with an electrospray ionization mass spectrometer. (1) Sub-plugs of substrate mixture (SM) and buffer (B) are injected into the capillary. (2) Sub-plugs of SM and various enzymes (E) are injected, resulting in plugs separated by a running buffer. (3) In-capillary mixing is performed by applying two pressure and vacuum impulses. The mixing step is followed by incubation to allow enzymatic reactions to occur. (4) CE separation and MS detection of substrates (S) and products (P) is performed. Substrates and products co-migrate in zones corresponding to each plug. The "+" and "−" symbols indicate the presence or absence of reactivity with the corresponding enzyme, respectively [23]. Reprinted with permission from Mironov, G.G., St-Jacques, A.D., Mungham, A., Eason, M.G., Chica, R.A., Berezovski, M.V. (2013) Bioanalysis for Biocatalysis: Multiplexed Capillary Electrophoresis–Mass Spectrometry Assay for Aminotransferase Substrate Discovery and Specificity Profiling. J. Am. Chem. Soc. 135: 13728–13736. Copyright (2013) American Chemical Society*

Stopped-flow mixing is considered to be a gold standard methodology for reaction kinetics studies. It is also a viable option for mass spectrometric studies of enzyme mechanisms [3, 32]. Kolakowski *et al.* [33] demonstrated hyphenation of a stopped-flow mixing instrument with ESI-MS detection. The reactants were monitored within a few seconds after the merger of the substrate solutions. The model reaction used in that work was the hydrolysis of acetylcholine to choline. Pseudo-first-order kinetic profiles were characterized, and compared with reference data [33]. However, the general impression is that continuous flow methods are more suited for on-line MS, especially ESI-MS. They also offer great flexibility when it comes to adjusting incubation times, and observing early stages of the reactions. In the pioneering work, Lee *et al.* [34] utilized a simple ESI-MS method with a thermostated reservoir to record temporal profiles of various enzymatic reactions. The reaction monitoring was conducted over a few minutes. Using this method, it was possible to determine the Michaelis–Menten constant of the enzymatic hydrolysis of *O*-nitrophenyl β-D-galactopyranoside by lactase.

Enzyme reaction intermediates can be characterized, in sub-second timescale, using the so-called *pulsed flow* method [35]. It employs a direct on-line interface between a rapid-mixing device and a ESI-MS system. It circumvents chemical quenching. By way of this strategy, it was possible to detect the intermediate of a reaction catalyzed by 5-enolpyruvoyl-shikimate-3-phosphate synthase [35]. The time-resolved ESI-MS method was also implemented in measurements of pre-steady-state kinetics of an enzymatic reaction involving *Bacillus circulans* xylanase [36]. The pre-steady-state kinetic parameters for the formation of the covalent intermediate in the mutant xylanase were determined. The MS results were in agreement with those obtained by stopped-flow ultraviolet–visible spectroscopy. In a later work, hydrolysis of *p*-nitrophenyl acetate by chymotrypsin was used as a model system [27]. The chymotrypsin-catalyzed hydrolysis follows the mechanism [27]:

$$E + S \underset{}{\overset{K_d}{\rightleftharpoons}} ES \overset{k_2}{\underset{P_1}{\searrow}} EP_2 \overset{k_3}{\longrightarrow} E + P_2$$

$$(13.5)$$

Under steady-state conditions, $[EP_2]$, $[E]$, and $[ES]$ are approximately constant while $[P_1]$ and $[P_2]$ increase over time. While the Michaelis–Menten kinetics describes behavior of the *steady-state* system, using a special reaction chamber with adjustable volume [37] (cf. Figure 4.3), and by tuning the experimental conditions (on-line addition of solvent prior to ESI), it was possible to monitor the *pre-steady-state* accumulation of acetylated chymotrypsin [27]. In other work, van den Heuvel *et al.* [38] presented a method, based on ESI-MS, for the monitoring of DNA hydrolysis by DNase enzyme. It allows the detection of DNA fragments and intact non-covalent protein–DNA complexes in a single experiment. Li *et al.* [39] integrated the rapid-mixing approach with ESI-TOF-MS to monitor enzymatic reactions at very short times (6–7 ms) with no chemical quenching. A similar method was implemented in the kinetic study of a multimeric enzyme [40]. On the other hand, Yu *et al.* [41] investigated intermediates of GlcNAc-6-*O*-sulfotransferase by ESI-FT-ICR-MS. They found that relative binding constants obtained in the solution- and the gas-phase measurements were similar suggesting that the binding domain is preserved during desolvation. By means of ESI-MS, kinetic parameters of the reaction catalyzed by Stf0 sulfotransferase could be determined [42]. The results pointed to the occurrence of the rapid equilibrium random sequential Bi-Bi process (two substrates converted to two products). The above studies are interesting because they combine the information on the enzymatic activity with the identification of modified protein catalyst species. Biocatalysis with a metal-dependent enzyme, in the presence of Cd^{2+}, could also be revealed by TRMS [43].

While the largest number of kinetic measurements are conducted using ESI, Wiseman *et al.* [44] implemented electrosonic spray ionization (ESSI) to monitor dynamic changes related to the formation of enzyme–substrate and enzyme–substrate–inhibitor complexes. The advantage of ESSI is that it preserves solution protein and protein complex structures and tags each protein with a charge state that is characteristic of its conformation. It is relatively straightforward to switch from ESI to ESSI or extractive electrospray ionization (EESI) by implementing a few inexpensive commercially available connectors and tubing. Most biochemists have less choice when it comes to the mass analyzers since these are the most expensive components of MS systems. However, the selection of the mass analyzer is important for many applications related to biocatalysis. When conducting kinetic characterization of enzymes it is desirable to collect quantitative datasets. This goal can be achieved using triple quadrupole (QqQ) mass spectrometers operated in the multiple reaction monitoring (MRM) mode (see Chapters 3 and 8) [7, 45].

13.1.4   Application of Microfluidic Systems

Microfluidic systems can be used to measure kinetic parameters of enzymatic reactions with sub-millisecond resolution [46]. As outlined in Chapter 7, MS is occasionally coupled with digital microfluidic chips to enable investigation of fast biochemical phenomena, for example pre-steady-state enzyme kinetics. In one related report, Nichols and Gardeniers [47] disclosed a digital microfluidic device that facilitates the investigation of pre-steady-state reaction kinetics (Figure 13.3a). It is based on rapid quenching and MALDI-TOF-MS detection. The results obtained for protein tyrosine phosphatase agree with reference data (Figure 13.3b; [48]) proving the practicality of this methodology. In other related work, droplets containing enzyme and substrate were combined on the microchip (Figure 13.4) [49]. Small amounts of the reactants were adsorbed on the channel wall, and analyzed directly by desorption/ionization on silicon (DIOS)-MS. This microreaction system enables the determination of the initial reaction velocity. Notably, DIOS does not require the application of an organic matrix to enhance ionization – possibly

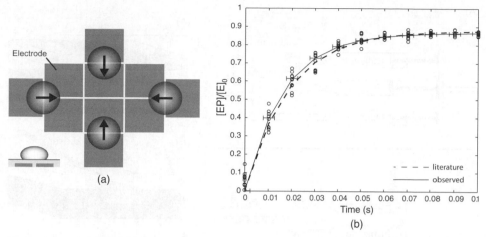

(a)

(b)

Figure 13.3   *(a) Simplified top-down and cross-sectional (inset) view of one experimental unit on the chip. Droplets are loaded robotically or manually at each of the four outermost positions shown. During loading, all the central electrodes are set to negative DC voltage and all the outer electrodes are set to positive DC voltage, to facilitate easier, more accurate droplet placement. This also allows smaller volumes to be accurately dispensed, as capillary forces in the pipette tip can be overcome. The droplets are then sequentially combined using AC voltage (enzyme with substrate; then quench; then matrix). The cross section shows a droplet over an electrode gap and a hydrophobic dielectric. Wires are not shown. During analysis, charging is negligible if the system is grounded, since small areas of the wires are in direct (uninsulated) contact with the matrix. (b) Literature data [48] and observed data. Greater scatter is observed at earlier time points. The horizontal error bars represent two standard deviations from the mean mixing time for these droplet volumes and actuation frequencies, as determined in separate experiments. E, unphosphorylated protein tyrosine phosphatase; EP, phosphorylated protein tyrosine phosphatase [47]. Adapted with permission from Nichols, K.P., Gardeniers, H.J.G.E. (2007) A Digital Microfluidic System for the Investigation of Pre-Steady-State Enzyme Kinetics Using Rapid Quenching with MALDI-TOF Mass Spectrometry. Anal. Chem. 79: 8699–8704. Copyright (2007) American Chemical Society*

reducing heterogeneity of sample deposits. However, DIOS-MS has not been in common use over the past few years, which suggests that these prototype microchips may not become popular in biochemical analysis. Although microfluidic systems are not easily accessible for every biochemist, the use of microchips in studies of enzymatic processes is justified. Many enzymes are available in small quantities or are very expensive while microchips (used in conjunction with MS) enable characterization of these biocatalysts using minute volumes of these precious samples.

13.1.5   Biochemical Waves

Mass spectrometry has been implemented as an on-line detection tool to monitor the transmission of chemical signals due to natural processes such as diffusion and convection as well as a bienzymatic autocatalytic process [6]. Using a mass spectrometer as the detector, it was found that an enzyme-accelerated chemical wave propagates faster than a chemical wave propelled by other processes. The two enzymes involved in the process (pyruvate

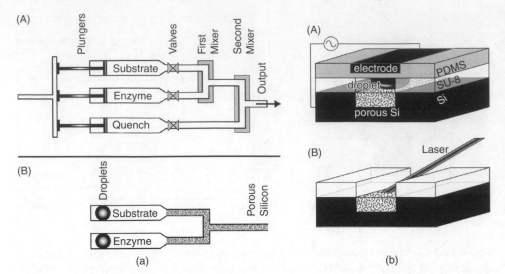

Figure 13.4 *(a) Comparison of the bench-scale stopped flow system for kinetic analysis (A) with the microfluidic analogue (B). The microfluidic system is simpler, allows faster mixing, and permits analysis of the reaction at the point where mixing initiates. The microfluidic system is analogous to a bench-scale capillary tubing-based system, but with a small percentage of the reaction products depositing on the capillary walls, and capable of being analyzed in place after the experiment using mass spectrometry. (b) Cross section of porous Si kinetic analysis chip during enzymatic reaction: as the droplet travels down the channel depositing reaction product on the channel walls (A); and during DIOS-MS analysis (B). The chip consists of a patterned porous silicon "floor," with SU-8 walls, a removable polydimethylsiloxane (PDMS) lid, and an electrode embedded in the PDMS. After the droplet has traveled the length of the channel, the lid is removed, and the bottom half of the chip is placed directly into a mass spectrometer [49]. Adapted with permission from Nichols, K.P., Azoz, S., Gardeniers, H.J.G.E. (2008) Enzyme Kinetics by Directly Imaging a Porous Silicon Microfluidic Reactor Using Desorption/Ionization on Silicon Mass Spectrometry. Anal. Chem. 80: 8314–8319. Copyright (2008) American Chemical Society*

kinase and adenylate kinase) work co-operatively – catalyzing production of adenosine diphosphate (ADP) and adenosine triphosphate (ATP). They induce formation of a front propagating along a high-aspect-ratio drift cell towards the ion source of an ion trap mass spectrometer (Figure 13.5). Isotopically labeled ^{13}C-ATP was used as the trigger of the accelerated chemical wave. With this substrate, one could easily distinguish between the two chemical waves (passive and accelerated) in a single experiment, reducing the bias due to the possible experimental instabilities [6].

13.2   Time-resolved Mass Spectrometry in Systems and Synthetic Biology

Isotopically labeled substrates are often utilized in analytical protocols related to microbial metabolomics (e.g., [50, 51]). The collected data sets enable the incorporation of the isotopic labels into the metabolites of interest to be tracked in the time domain

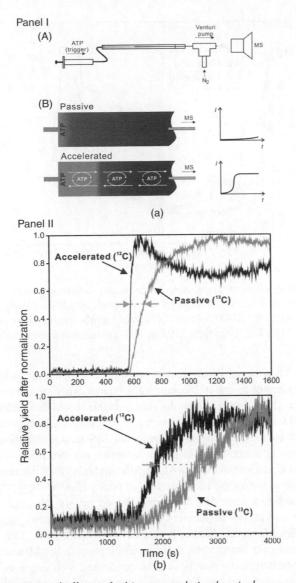

Figure 13.5 *Spatiotemporal effects of a bioautocatalytic chemical wave revealed by time-resolved mass spectrometry. (a) Investigation of a chemical wave due to "passive" transduction and a bienzymatic amplification system. (A) Experimental setup incorporating a horizontal drift cell and mass spectrometer. (B) Schematic representation of chemical wave propagation in the drift cell due to the passive and the enzyme-accelerated transduction. (b) Transduction of labeled and unlabeled ATP along the drift cell. Concentration of the $^{13}C_{10}$-ATP trigger : 10^{-2} M (A) and 5×10^{-3} M (B). Exponential smoothing with a time constant of 4.1 s has been applied, and followed by normalization (scaling to the maximal value). The dashed line denotes the time lapse between half-maxima of the normalized curves (0.5 level) corresponding to the passive and accelerated chemical transduction: 93 and 740 s in the case of the 10^{-2} and 5×10^{-3} M trigger solutions, respectively [6]. Adapted from Ting, H., Urban, P.L. (2014) Spatiotemporal Effects of a Bioautocatalytic Chemical Wave revealed by Time-resolved Mass Spectrometry. RSC Adv. 4: 2103-2108 with permission from the Royal Society of Chemistry*

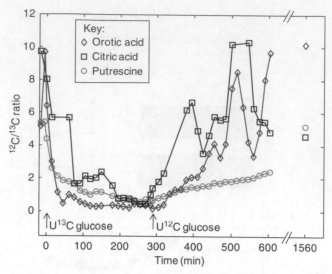

Figure 13.6 *Example of dynamic labeling of three metabolites in glucose-limited chemostat cultures of Escherichia coli [51]. Reprinted from Trends in Microbiology, 19, Winder, C.L., Dunn, W.B., Goodacre, R., TARDIS-based Microbial Metabolomics: Time and Relative Differences in Systems, 315–322. Copyright (2011), with permission from Elsevier*

(*cf.* Figure 13.6). The obtained information is useful for metabolic pathway elucidation, and to follow the fates of atoms in biochemical systems. This area of research is referred to as *fluxomics* (cf. [52]). In the metabolic flux analysis, biological cells are often grown on media enriched with a particular isotope (e.g., ^{13}C-glucose). Incorporation of the heavy carbon atoms is then followed by monitoring the MS peaks of various isotopologues. The extracts from cells labeled with heavy isotopes are normally analyzed by liquid chromatography (LC)-MS. However, MALDI-MS methods (with 9-aminoacridine matrix) have also been used to follow the incorporation of heavy isotopes into primary metabolites [53, 54]. The so-called *kinetic MS imaging* – based on soft desorption/ionization MS combined with *in-vivo* metabolic labeling of tissue with deuterium – can produce images which bear kinetic information about biological processes [55]. The results allow for correlation of metabolic activity of specific lipids found in different tumor regions. Imaging incorporation of heavy isotopes can be conducted with single-cell resolution [54]. However, the temporal resolution of such studies is rather low – typically, in the order of hours.

Real-time monitoring of enzymatic reactions is also useful for fundamental studies in synthetic biology. Bujara *et al.* [56] implemented ESI-QqQ-MS in the monitoring of a multi-enzyme reaction chain (Figure 13.7). This method may enable optimization of complex metabolic reaction networks for chemical synthesis. The on-line sampling system encompassed sampling of the reaction mixture via membrane, reduction of flow, dilution, and ionization.

Development of artificial biochemical circuits based on nucleic acids can help researchers understand the mechanisms of complex biological systems, and design strategies for various requirements [57]. Operation of such circuits can be verified by conducting MS

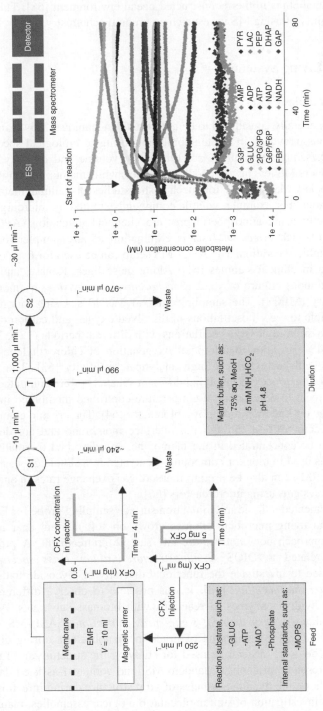

Figure 13.7 *Setup for mass spectrometry-based quantitative real-time analysis. The outlet flow of a continuous stirred enzyme membrane reactor (EMR) is continuously analyzed by multiple reaction monitoring in an electrospray ionization triple quadrupole mass spectrometer after flow reduction (S1), dilution (T) and another flow reduction (S2) [56]. Reprinted by permission from Macmillan Publishers Ltd: Nature Chemical Biology [Bujara, M., Schümperli, M., Pellaux, R., Heinemann, M., Panke, S. (2011) Optimization of a Blueprint for In Vitro Glycolysis by Metabolic Real-Time Analysis. Nature Chem. Biol. 7: 271–271]. Copyright (2011)*

analysis of reaction mixtures. MS-based methodology has also been used to track catalytic interconversions of metabolites in the reconstructed ocean environment [58]. This effort exemplifies the possibility of using MS in research on prebiotic chemistry.

13.3   Monitoring Living Systems

13.3.1   Microbial Samples

Mass spectrometry is generally considered to be a destructive technique. This feature can be regarded as a serious drawback in the temporal monitoring of biological specimens. In fact, some matrices/samples/specimens are limited by volume and quantity. Direct chemical sampling renders MS less invasive to living objects while maintaining high speed of analysis [59]. It makes MS suitable for mechanistic studies of cellular metabolism and cell–host interactions. For example, volatile compounds emitted by microorganisms can be analyzed directly using atmospheric pressure chemical ionization (APCI)-MS [60]. The advantage of such *direct analysis* is elimination of any sample treatment, including chromatographic separation. The APCI-MS setup can be used for fast detection of pathogenic bacteria in clinical samples [61]. Along these lines, Roussel and Lloyd [62] monitored a continuous culture of yeast (*Saccharomyces cerevisiae*) by membrane inlet mass spectrometry (MIMS). The sampling interval was 12 s. Using this method, the researchers were able to record oscillations of dissolved oxygen and carbon dioxide. Interestingly, it was possible to observe oscillations with different periods (13 h, 36 min, and 4 min) [62]. MIMS also allowed microbial oxygenation of chloroethylenes to be monitored over hours with an extremely high analytical sensitivity [63]. The kinetic descriptors (v_{max}/K_M) of the processes involving different substrates were computed based on the collected data [63]. In other work, a temperature-controlled membrane interface was implemented in the on-line monitoring of a bioreactor [64]. The organic compounds produced in the bioreactor diffuse through the interface membrane and are detected. Alternatively, they can be concentrated in the membrane, and desorbed thermally [64]. A similar approach was used to monitor compounds generated in a simulated wastewater treatment plant [65]. MIMS can also be used to measure gas exchange rates in anaerobic and aerobic biological systems using microreactors [66].

Hsu *et al.* [67] implemented a liquid microjunction surface sampling probe for ESI-MS to probe metabolites in living microbial colonies grown on soft nutrient agar in Petri dishes. This on-line approach does not require any sample pretreatment. A prototype ionization technique related to DIOS – *nanostructure-initiator mass spectrometry* (NIMS) – was also used to investigate the functional diversity of glycoside hydrolases secreted by *Clostridium thermocellum* [68]. It was possible to observe differences in rates and yields of individual enzymes in reactions with biomass substrates. Another novel ionization technique – *easy ambient sonic-spray ionization* (EASI)-MS – was successfully implemented in the monitoring of lipids in cyanobacteria during different growth phases [69]. Moreover, MS can be used for studying the structure and function of viruses [70]. Time-resolved proteolysis, tandem MS, and computer-assisted database searching can shed light on the dynamic changes of viral structure during infection [70]. Since ESI-MS enables investigation of supramolecular biological assemblies, maturation of native viral capsids could be monitored by means of this technique [71].

13.3.2   Plant and Animal Samples

Apart from monitoring microbial systems, several atmospheric pressure interfaces have been applied to probe the chemical composition of specimens obtained from higher organisms. TRMS using a membrane inlet interface has been instrumental in the investigation of photosynthetic water oxidation [72–75]. It showed that two water molecules yielding O_2 are rapidly exchangeable in the O_2 evolving complex of photosystem II [74]. The process of water binding to the O_2 evolving complex in photosystem II was characterized kinetically using TRMS [76]. The technique further enabled the discovery that Ca^{2+} interacts with the binding of one of the two substrate water molecules [77].

By combining microdialysis with nanoESI-MS, low-volume, low-concentration releases of small proteins in a three-dimensional neural cell culture system could be detected [78]. While microdialysis removes interferents, it also has an effect on the temporal resolution of the device (~ 1 min). Alternatively, a microdialysis probe can be coupled with paper-spray ionization MS which enables the monitoring of glucose in a cell culture medium [79].

Probe electrospray ionization (PESI) and TOF-MS were used for direct profiling of phytochemicals in different parts of a fresh tulip bulb [80], which emphasized the possibility of conducting *in-vivo* MS analysis of less sensitive biological matrices such as plant tissues. Recently, Pan *et al.* [81] demonstrated *single-probe MS* which can conduct metabolomic analysis of individual living cells in real time. The diameter of this probe is < 10 μm which makes the device compatible with eukaryotic cells. Atmospheric pressure ion sources are particularly suitable for analysis of live biological specimens. Cellular metabolism does not need to be quenched before analysis. For instance, in *laser ablation electrospray ionization* (LAESI)-MS, cells are irradiated by a laser beam in order to extract small amounts of cytosolic components, and to transfer them to the ESI plume [82].

In-vivo and real-time monitoring of secondary metabolites, released from living animals and plants, using *field-induced direct ionization* MS, was also presented (Figure 13.8) [83]. Although real-time monitoring of metabolite profiles was not clearly demonstrated, one may suspect that the method could potentially be adopted for this purpose. However, considering that animals are positioned in front of the mass spectrometer (in the presence of an electric field and hot gas), one can only speculate on the practicality of this analytical approach.

It is also feasible to monitor enzymatic reactions taking place inside biological tissues by bringing the specimens very close to the ion source [84]. Biotransformation of alliin to allicin by allinase in raw garlic cloves could be monitored *in vivo* by *internal* EESI-MS. The reactants were extracted by an infused solution running throughout the tissue. The extract was ionized on the edge of the specimen [84].

In an elegant study, organic compounds present in the blood of rats were monitored with a membrane probe and EI-MS [85]. The rats were exposed to volatile compounds. Curves describing the decay of these compounds over a few minutes were recorded. Interestingly, the elimination of dichloromethane from the bloodstream followed first-order kinetics with a decay constant of ~ 0.15 min^{-1} [85]. From the examples given above it is clear that there is great interest in the monitoring of fast biochemical processes (milliseconds to minutes). However, certain processes occurring *in vivo* can be studied retrospectively with very low temporal resolution (day or several days). For instance, off-line analysis of hair specimens by MALDI-MS imaging or LC-MS enabled chemical profiles to be obtained reflecting the intake of medicines over several days [86].

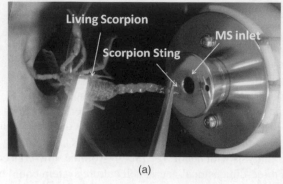

(a)

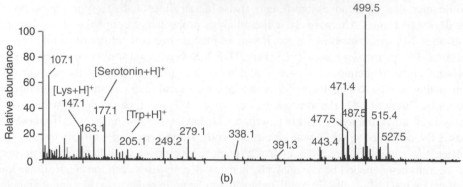

(b)

Figure 13.8 *Real-time monitoring of animal metabolites by mass spectrometry. (a) Photograph of in-vivo analysis of a living scorpion by field-induced direct ionization mass spectrometry. (b) Mass spectrum obtained by field-induced direct ionization mass spectrometry analysis of the secretion released from a living scorpion upon stimulation [83]. Reprinted by permission from Macmillan Publishers Ltd: Scientific Reports [Hu, B., Wang, L., Ye, W.-C., Yao, Z.-P. (2013) In Vivo and Real-time Monitoring of Secondary Metabolites of Living Organisms by Mass Spectrometry. Sci. Rep. 3: 2104]. Copyright (2013). See colour plate section for colour figure*

13.4   Concluding Remarks

Similar to the other areas of TRMS applications, ESI also plays an important role as a universal interface and ion source in the investigation of biochemical processes. Other ionization techniques and interfaces are used with some success in specific applications (e.g., LDI in combination with microfluidic chips and quenching). Adaptation of flow mixers for use with ESI-MS has enabled the straightforward kinetic characterization of enzymatic reactions. Biokinetic studies can also be performed using less common techniques such as liquid water beam desorption MS (a technique in which a liquid filament in a vacuum is irradiated by a focused infrared laser pulse to vaporize and analyze reactant species [87]). In fact, some novel ion sources have ingeniously been utilized to obtain profiles of metabolites of live biological specimens. Overall, the reviewed biochemical TRMS studies involve broad time ranges spanning from milliseconds (pre-steady-state kinetics) to days (metabolism in higher organisms).

References

1. Glish, G.L., Vachet, R.W. (2003) The Basics of Mass Spectrometry in the Twenty-first Century. Nature Rev. Drug Discov. 2: 140–150.
2. Fabris, D. (2005) Mass Spectrometric Approaches for the Investigation of Dynamic Processes in Condensed Phase. Mass Spectrom. Rev. 24: 30–54.
3. Northrop, D.B., Simpson, F.B. (1997) New Concepts in Bioorganic Chemistry. Beyond Enzyme Kinetics: Direct Determination of Mechanisms by Stopped-flow Mass Spectrometry. Bioorg. Med. Chem. 5: 641–644.
4. Thermo Scientific. (2010) Process Mass Spectrometry in Biotechnology, http://www.thermoscientific.com/content/dam/tfs/ATG/EPD/EPD%20Documents/Catalogs%20&%20Brochures/Online%20Process%20Analyzers/Process%20Mass%20Spectrometers/D19632~.pdf (accessed September 20, 2015).
5. Koskinen, A., Klibanov, A. (eds) (2012) Enzymatic Reactions in Organic Media. Springer, Berlin.
6. Ting, H., Urban, P.L. (2014) Spatiotemporal Effects of a Bioautocatalytic Chemical Wave Revealed by Time-resolved Mass Spectrometry. RSC Adv. 4: 2103–2108.
7. Norris, A.J., Whitelegge, J.P., Faull, K.F., Toyokuni, T. (2001) Analysis of Enzyme Kinetics Using Electrospray Ionization Mass Spectrometry and Multiple Reaction Monitoring: Fucosyltransferase V. Biochemistry 40: 3774–3779.
8. Liesener, A., Karst, U. (2005) Monitoring Enzymatic Conversions by Mass Spectrometry: a Critical Review. Anal. Bioanal. Chem. 382: 1451–1464.
9. Thomas, J.J., Shen, Z., Crowell, J.E., Finn, M.G., Siuzdak, G. (2001) Desorption/Ionization on Silicon (DIOS): A Diverse Mass Spectrometry Platform for Protein Characterization. Proc. Natl. Acad. Sci. USA 98: 4932–4937.
10. Hu, L., Jiang, G., Xu, S., Pan, C., Zou, H. (2006) Monitoring Enzyme Reaction and Screening Enzyme Inhibitor Based on MALDI-TOF-MS Platform with a Matrix of Oxidized Carbon Nanotubes. J. Am. Soc. Mass Spectrom. 17: 1616–1619.
11. Min, D.-H., Yeo, W.-S., Mrksich, M. (2004) A Method for Connecting Solution-phase Enzyme Activity Assays with Immobilized Format Analysis by Mass Spectrometry. Anal. Chem. 76: 3923–3929.
12. Smith, L.A., Caprioli, R.M. (1983) Following Enzyme Catalysis in Real-time Inside a Fast Atom Bombardment Mass Spectrometer. Biol. Mass Spectrom. 10: 98–102.
13. Smith, L.A., Caprioli, R.M. (1984) Enzyme Reaction Rates Determined by Fast Atom Bombardment Mass Spectrometry. Biol. Mass Spectrom. 11: 392–395.
14. Heidmann, M., Fonrobert, P., Przybylski, M., Platt, K.L. Seidel, A., Oesch, F. (1988) Conjugation Reactions of Polyaromatic Quinones to Mono- and Bisglutathionyl Adducts: Direct Analysis by Fast Atom Bombardment Mass Spectrometry. Biomed. Environ. Mass Spectrom. 15: 329–332.
15. Anderson, K.S. (2005) Detection of Novel Enzyme Intermediates in PEP-utilizing Enzymes. Arch. Biochem. Biophys. 433: 47–58.
16. Kelleher, N.L., Hicks, L.M. (2005) Contemporary Mass Spectrometry for the Direct Detection of Enzyme Intermediates. Curr. Opin. Chem. Biol. 9: 424–430.
17. Gross, J.W., Frey, P.A. (2002) Rapid Mix-quench MALDI-TOF Mass Spectrometry for Analysis of Enzymatic Systems. Methods Enzymol. 354: 27–49.

18. Painter, A.J., Jaya, N., Basha, E., Vierling, E., Robinson, C.V., Benesch, L.P. (2008) Real-time Monitoring of Protein Complexes Reveals Their Quaternary Organization and Dynamics. Chem. Biol. 15: 246–253.

19. Clarke, D.J., Stokes, A.A., Langridge-Smith, P., Mackay, C.L. (2010) Online Quench-flow Electrospray Ionization Fourier Transform Ion Cyclotron Resonance Mass Spectrometry for Elucidating Kinetic and Chemical Enzymatic Reaction Mechanisms. Anal. Chem. 82: 1897–1904.

20. Houston, C.T., Taylor, W.P., Widlanski, T.S., Reilly, J.P. (2000) Investigation of Enzyme Kinetics Using Quench-flow Techniques with MALDI-TOF Mass Spectrometry. Anal. Chem. 72: 3311–3319.

21. Gross, J.W., Hegeman, A.D., Vestling, M.M., Frey, P.A. (2000) Characterization of Enzymatic Processes by Rapid Mix-quench Mass Spectrometry: The Case of dTDP-Glucose 4,6-Dehydratase. Biochemistry 39: 13633–13640.

22. Pi, N., Leary, J.A. (2004) Determination of Enzyme/Substrate Specificity Constants Using a Multiple Substrate ESI-MS Assay. J. Am. Soc. Mass Spectrom. 15: 233–243.

23. Mironov, G.G., St Jacques, A.D., Mungham, A., Eason, M.G., Chica, R.A., Berezovski, M.V. (2013) Bioanalysis for Biocatalysis: Multiplexed Capillary Electrophoresis-Mass Spectrometry Assay for Aminotransferase Substrate Discovery and Specificity Profiling. J. Am. Chem. Soc. 135: 13728–13736.

24. Sun, S., Slaney, T.R., Kennedy, R.T. (2012) Label Free Screening of Enzyme Inhibitors at Femtomole Scale Using Segmented Flow Electrospray Ionization Mass Spectrometry. Anal. Chem. 84: 5794–5800.

25. Urban, P.L., Goodall, D.M., Bruce, N.C. (2006) Enzymatic Microreactors in Chemical Analysis and Kinetic Studies. Biotechnol. Adv. 24: 42–57.

26. Kim, H.Y., Pilosof, D., Dyckes, D.F., Vestal, M.L. (1984) On-line Peptide Sequencing by Enzymic Hydrolysis, High Performance Liquid Chromatography, and Thermospray Mass Spectrometry. J. Am. Chem. Soc. 106: 7304–7309.

27. Wilson, D.J., Konermann, L. (2004) Mechanistic Studies on Enzymatic Reactions by Electrospray Ionization MS Using a Capillary Mixer with Adjustable Reaction Chamber Volume for Time-resolved Measurements. Anal. Chem. 76: 2537–2543.

28. Hsieh, F.Y., Tong, X., Wachs, T., Ganem, B., Henion, J. (1995) Kinetic Monitoring of Enzymatic Reactions in Real Time by Quantitative High-performance Liquid Chromatography-Mass Spectrometry. Anal. Biochem. 229: 20–25.

29. Ge, X., Sirich, T.L., Beyer, M.K., Desaire, H., Leary, J.A. (2001) A Strategy for the Determination of Enzyme Kinetics Using Electrospray Ionization with an Ion Trap Mass Spectrometer. Anal. Chem. 73: 5078–5082.

30. Kaltashov, I.A., Eyles, S.J. (2012) Chapter 6, Kinetic Studies by Mass Spectrometry. In: Mass Spectrometry in Structural Biology and Biophysics: Architecture, Dynamics, and Interaction of Biomolecules, 2nd Edition. John Wiley & Sons, Inc., Hoboken.

31. Michaelis, L., Menten, M.L. (1913) Die Kinetik der Invertinwirkung. Biochem. Z. 49: 333–369.

32. Kolakowski, B.M., Konermann, L. (2001) From Small-molecule Reactions to Protein Folding: Studying Biochemical Kinetics by Stopped-flow Electrospray Mass Spectrometry. Anal. Biochem. 292: 107–114.

33. Kolakowski, B.M., Simmons, D.A., Konermann, L. (2000) Stopped-flow Electrospray Ionization Mass Spectrometry: A New Method for Studying Chemical Reaction Kinetics in Solution. Rapid Commun. Mass Spectrom. 14: 772–776.
34. Lee, E.D., Mück, W., Henion, J.D., Covey, T.R. (1989) Real-time Reaction Monitoring by Continuous-introduction Ion-spray Tandem Mass Spectrometry. J. Am. Chem. Soc. 111: 4600–4604.
35. Paiva, A.A., Tilton Jr R.F., Crooks, G.P., Huang, L.Q., Anderson, K.S. (1997) Detection and Identification of Transient Enzyme Intermediates Using Rapid Mixing, Pulsed-flow Electrospray Mass Spectrometry. Biochemistry 36: 15472–15476.
36. Zechel, S.L., Konermann, L., Withers, S.G., Douglas, D.J. (1998) Pre-steady State Kinetic Analysis of an Enzymatic Reaction Monitored by Time-resolved Electrospray Ionization Mass Spectrometry. Biochemistry 37: 7664–7669.
37. Wilson, D.J., Konermann, L. (2003) A Capillary Mixer with Adjustable Reaction Chamber Volume for Millisecond Time-resolved Studies by Electrospray Mass Spectrometry. Anal. Chem. 75: 6408–6414.
38. van den Heuvel, R.H.H., Gato, S., Versluis, C., Gerbaux, P., Kleanthous, C., Heck, A.J.R. (2005) Real-time Monitoring of Enzymatic DNA Hydrolysis by Electrospray Ionization Mass Spectrometry. Nucleic Acids Res. 33: e96.
39. Li, Z., Sau, A.K., Shen, S., Whitehouse, C., Baasov, T., Anderson, K.S. (2003) A Snapshot of Enzyme Catalysis Using Electrospray Ionization Mass Spectrometry. J. Am. Chem. Soc. 125: 9938–9939.
40. Li, Z., Song, F., Zhuang, Z., Dunaway-Mariano, D., Anderson, K.S. (2009) Monitoring Enzyme Catalysis in the Multimeric State: Direct Observation of *Arthrobacter* 4-Hydroxybenzoyl-coenzyme A Thioesterase Catalytic Complexes Using Time-resolved Electrospray Ionization Mass Spectrometry. Anal. Biochem. 394: 209–216.
41. Yu, Y., Kirkup, C.E., Pi, N., Leary, J.A. (2004) Characterization of Noncovalent Protein–Ligand Complexes and Associated Enzyme Intermediates of GlcNAc-6-*O*-Sulfotransferase by Electrospray Ionization FT-ICR Mass Spectrometry. J. Am. Soc. Mass Spectrom. 15: 1400–1407.
42. Pi, N., Huang, M.B., Gao, H., Mougous, J.D., Bertozzi, C.R., Leary, J.A. (2005) Kinetic Measurements and Mechanism Determination of Stf0 Sulfotransferase Using Mass Spectrometry. Anal. Biochem. 341: 94–104.
43. Roberts, A., Furdui, C., Anderson, K.S. (2010) Observation of a Chemically Labile, Noncovalent Enzyme Intermediate in the Reaction of Metal-dependent *Aquifex pyrophilus* KDO8PS by Time-resolved Mass Spectrometry. Rapid Commun. Mass Spectrom. 24: 1919–1924.
44. Wiseman, J.M., Takáts, Z., Gologan, B., Davisson V.J., Cooks, R.G. (2005) Direct Characterization of Enzyme–Substrate Complexes by Using Electrosonic Spray Ionization Mass Spectrometry. Angew. Chem. Int. Ed. 44: 913–916.
45. Norris, A.J., Whitelegge, J.P., Faull, K.F., Toyokuni, T. (2001) Kinetic Characterization of Enzyme Inhibitors Using Electrospray-Ionization Mass Spectrometry Coupled with Multiple Reaction Monitoring. Anal. Chem. 73: 6024–6029.
46. Song, H., Ismagilov, R.F. (2003) Millisecond Kinetics on a Microfluidic Chip Using Nanoliters of Reagents. J. Am. Chem. Soc. 125: 14613–14619.

47. Nichols, K.P., Gardeniers, H.J.G.E. (2007) A Digital Microfluidic System for the Investigation of Pre-steady-state Enzyme Kinetics Using Rapid Quenching with MALDI-TOF Mass Spectrometry. Anal. Chem. 79: 8699–8704.
48. Zhang, Z.Y., Palfey, B.A., Wu, L., Zhao, Y. (1995) Catalytic Function of the Conserved Hydroxyl Group in the Protein Tyrosine Phosphatase Signature Motif. Biochemistry 34: 16389–16396.
49. Nichols, K.P., Azoz, S., Gardeniers, H.J.G.E. (2008) Enzyme Kinetics by Directly Imaging a Porous Silicon Microfluidic Reactor Using Desorption/Ionization on Silicon Mass Spectrometry. Anal. Chem. 80: 8314–8319.
50. Yuan, J., Bennett, B.D., Rabinowitz, J.D. (2008) Kinetic Flux Profiling for Quantitation of Cellular Metabolic Fluxes. Nature Protoc. 3: 1328–1340.
51. Winder, C.L., Dunn, W.B., Goodacre, R. (2011) TARDIS-based Microbial Metabolomics: Time and Relative Differences in Systems. Trends Microbiol. 19: 315–322.
52. Klein, S., Heinzle, E. (2012) Isotope Labeling Experiments in Metabolomics and Fluxomics. Wiley Interdiscip. Rev. Syst. Biol. Med. 4: 261–272.
53. Urban, P.L., Schmidt, A.M., Fagerer, S.R., Amantonico, A., Ibañez, A., Jefimovs, K., Heinemann, M., Zenobi, R. (2011) Carbon-13 Labelling Strategy for Studying the ATP Metabolism in Individual Yeast Cells by Micro-arrays for Mass Spectrometry. Mol. BioSyst. 7: 2837–2840.
54. Hu, J.-B., Chen, Y.-C., Urban, P.L. (2012) On-target Labeling of Intracellular Metabolites Combined with Chemical Mapping of Individual Hyphae Revealing Cytoplasmic Relocation of Isotopologues. Anal. Chem. 84: 5110–5116.
55. Louie, K.B., Bowen, B.P., McAlhany, S., Huang, Y., Price, J.C., Mao, J., Hellerstein, M., Northen, T.R. (2013) Mass Spectrometry Imaging for *In Situ* Kinetic Histochemistry. Sci. Rep. 3: 1656.
56. Bujara, M., Schümperli, M., Pellaux, R., Heinemann, M., Panke, S. (2011) Optimization of a Blueprint for *In Vitro* Glycolysis by Metabolic Real-time Analysis. Nature Chem. Biol. 7: 271–277.
57. Nie, J., Zhao, M.-Z., Xie, W.J., Cai, L.-Y., Zhou, Y.-L., Zhang, X.-X. (2015) DNA Cross-triggered Cascading Self-amplification Artificial Biochemical Circuit. Chem. Sci. 6: 1225–1229.
58. Keller, M.A., Turchyn, A.V., Ralser, M. (2014) Non-enzymatic Glycolysis and Pentose Phosphate Pathway-like Reactions in a Plausible Archean Ocean. Mol. Syst. Biol. 10: 725.
59. Chingin, K., Liang, J., Chen, H. (2014) Direct Analysis of *In Vitro* Grown Microorganisms and Mammalian Cells by Ambient Mass Spectrometry. RSC Adv. 4: 5768–5781.
60. Liang, J., Hang, Y., Chingin, K., Hu, L., Chen, H. (2014) Rapid Differentiation of Microbial Cultures Based on the Analysis of Headspace Volatiles by Atmospheric Pressure Chemical Ionization Mass Spectrometry. RSC Adv. 4: 25326–25329.
61. Chingin, K., Liang, J., Hang, Y., Hu, L., Chen, H. (2015) Rapid Recognition of Bacteremia in Humans Using Atmospheric Pressure Chemical Ionization Mass Spectrometry of Volatiles Emitted by Blood Cultures. RSC Adv. 5: 13952–13957.
62. Roussel, M.R., Lloyd, D. (2007) Observation of a Chaotic Multioscillatory Metabolic Attractor by Real-time Monitoring of a Yeast Continuous Culture. FEBS J. 274: 1011–1018.

63. Lauritsen, F.R., Gylling, S. (1995) On-line Monitoring of Biological Reactions at Low Parts-per-trillion Levels by Membrane Inlet Mass Spectrometry. Anal. Chem. 67: 1418–1420.

64. Creaser, C.S., Gómez Lamarca, D., Freitas dos Santos, L.M., New, A.P., James, P.A. (2003) A Universal Temperature Controlled Membrane Interface for the Analysis of Volatile and Semi-volatile Organic Compounds. Analyst 128: 1150–1156.

65. Creaser, C.S., Gómez Lamarca, D., dos Santos, L.M.F., LoBiundo, G., New, A.P. (2003) On-line Biodegradation Monitoring of Nitrogen-containing Compounds by Membrane Inlet Mass Spectrometry. J. Chem. Technol. Biotechnol. 78: 1193–1200.

66. Heinzle, E., Meyer, B., Oezemre, A., Dunn, I.J. (1998) A Microreactor with On-line Mass Spectrometry for the Investigation of Biological Kinetics. In: Ehrfeld, W. (ed.) Microreaction Technology. Springer-Verlag, Berlin, pp. 267–274.

67. Hsu, C.-C., ElNaggar, M.S., Peng, Y., Fang, J., Sanchez, L.M., Mascuch, S.J., Møller, K.A., Alazzeh, E.K., Pikula, J., Quinn, R.A., Zeng, Y., Wolfe, B.E., Dutton, R.J., Gerwick, L., Zhang, L., Liu, X., Månsson, M., Dorrestein, P.C. (2013) Real-time Metabolomics on Living Microorganisms Using Ambient Electrospray Ionization Flow-probe. Anal. Chem. 85: 7014–7018.

68. Deng, K., Takasuka, T.E., Heins, R., Cheng, X., Bergeman, L.F., Shi, J., Aschenbrener, R., Deutsch, S., Singh, S., Sale, K.L., Simmons, B.A., Adams, P.D., Singh, A.K., Fox, B.G., Northen, T.R. (2014) Rapid Kinetic Characterization of Glycosyl Hydrolases Based on Oxime Derivatization and Nanostructure-initiator Mass Spectrometry (NIMS). ACS Chem. Biol. 9: 1470–1479.

69. Liu, Y., Zhang, J., Nie, H., Dong, C., Li, Z., Zheng, Z., Bai, Y., Liu, H., Zhao, J. (2014) Study on Variation of Lipids during Different Growth Phases of Living Cyanobacteria Using Easy Ambient Sonic-spray Ionization Mass Spectrometry. Anal. Chem. 86: 7096–7102.

70. Trauger, S.A., Junker, T., Siuzdak, G. (2003) Investigating Viral Proteins and Intact Viruses with Mass Spectrometry. Top. Curr. Chem. 225: 265–282.

71. Snijder, J., Rose, R.J., Veesler, D., Johnson, J.E., Heck, A.J.R. (2013) Studying 18 MDa Virus Assemblies with Native Mass Spectrometry. Angew. Chem. Int. Ed. 52: 4020–4023.

72. Messinger, J., Badger, M., Wydrzynski, T. (1995) Detection of *One* Slowly Exchanging Substrate Water Molecule in the S_3 State of Photosystem II. Proc. Natl. Acad. Sci. USA 92: 3209–3213.

73. Hillier, W., Messinger, J., Wydrzynski, T. (1998) Kinetic Determination of the Fast Exchanging Substrate Water Molecule in the S_3 State of Photosystem II. Biochemistry 37: 16908–16914.

74. Haumann, M., Junge, W. (1999) Photosynthetic Water Oxidation: a Simplex-scheme of its Partial Reactions. Biochim. Biophys. Acta 1411: 86–91.

75. Beckmann, K., Messinger, J., Badger, M.R., Wydrzynski, T., Hillier, W. (2009) On-line Mass Spectrometry: Membrane Inlet Sampling. Photosynth. Res. 102: 511–522.

76. Hillier, W., Wydrzynski, T. (2000) The Affinities for the Two Substrate Water Binding Sites in the O_2 Evolving Complex of Photosystem II Vary Independently during S-state Turnover. Biochemistry 39: 4399–4405.

77. Hendry, G., Wydrzynski, T. (2003) [18]O Isotope Exchange Measurements Reveal that Calcium Is Involved in the Binding of One Substrate-water Molecule to the Oxygen-evolving Complex in Photosystem II. Biochemistry 42: 6209–6217.
78. Olivero, D., LaPlaca, M., Kottke, P.A. (2012) Ambient Nanoelectrospray Ionization with In-line Microdialysis for Spatially Resolved Transient Biochemical Monitoring within Cell Culture Environments. Anal. Chem. 84: 2072–2075.
79. Liu, W., Wang, N., Lin, X., Ma, Y., Lin, J.-M. (2014) Interfacing Microsampling Droplets and Mass Spectrometry by Paper Spray Ionization for Online Chemical Monitoring of Cell Culture. Anal Chem. 86: 7128–7134.
80. Yu, Z., Chen, L.C., Suzuki, H., Ariyada, O., Erra-Balsells, R., Nonami, H., Hiraoka, K. (2009) Direct Profiling of Phytochemicals in Tulip Tissues and *In Vivo* Monitoring of the Change of Carbohydrate Content in Tulip Bulbs by Probe Electrospray Ionization Mass Spectrometry. J. Am. Soc. Mass Spectrom. 20: 2304–2311.
81. Pan, N., Rao, W., Kothapalli, N.R., Liu, R., Burgett, A.W.G., Yang, Z. (2014) The Single-probe: A Miniaturized Multifunctional Device for Single Cell Mass Spectrometry Analysis. Anal. Chem. 86: 9376–9380.
82. Nemes, P., Vertes, A. (2007) Laser Ablation Electrospray Ionization for Atmospheric Pressure, In Vivo, and Imaging Mass Spectrometry. Anal. Chem. 79: 8098–8106.
83. Hu, B., Wang, L., Ye, W.-C., Yao, Z.-P. (2013) *In Vivo* and Real-time Monitoring of Secondary Metabolites of Living Organisms by Mass Spectrometry. Sci. Rep. 3: 2104.
84. Zhang, H., Chingin, K., Zhu, L., Chen, H. (2015) Molecular Characterization of Ongoing Enzymatic Reactions in Raw Garlic Cloves Using Extractive Electrospray Ionization Mass Spectrometry. Anal. Chem. 87: 2878–2883.
85. Brodbelt, J.S., Cooks, R.G., Tou, J.C., Kallos, G.J., Dryzga, M.D. (1987) In Vivo Mass Spectrometric Determination of Organic Compounds in Blood with a Membrane Probe. Anal. Chem. 59: 454–458.
86. Poetzsch, M., Steuer, A.E., Roemmelt, A.T., Baumgartner, M.R., Kraemer, T. (2014) Single Hair Analysis of Small Molecules Using MALDI-Triple Quadrupole MS Imaging and LC-MS/MS: Investigations on Opportunities and Pitfalls. Anal. Chem. 86: 11758–11765.
87. Charvat, A., Bögehold, A., Abel, B. (2006) Time-resolved Micro Liquid Desorption Mass Spectrometry: Mechanism, Features, and Kinetic Applications. Austr. J. Chem. 59: 81–103.

14

Final Remarks

Most standard chemical procedures enable preparation, handling, and analysis of static samples. Thus, in many of the past studies, the relationship between concentration and time has not been considered to a great extent. Over the past few years, scientists have become interested in dynamic processes, which lead to the formation of temporal or spatial gradients. Mass spectrometry (MS) is a prime tool for chemical characterization of various matrices. Therefore, it is appealing to extend the current mass spectrometric toolkit to enable analysis of temporal properties of dynamic samples without losing chemical information.

14.1 Current Progress

For chemists, the mass spectrum is an imprint of a sample at the point when it was introduced to the mass spectrometer. The chemical composition of an unstable matrix is "volatile". Such dynamic matrices/samples undergo continuous changes. Mass spectrometry allows chemists to record and preserve information on the chemical composition of dynamic samples with extraordinary accuracy and objectivity. For more than half a century, the ability of mass spectrometers to collect data with temporal resolution has been utilized in hyphenated systems incorporating reaction chambers or separation columns. Thus, the technology of transferring separation effluents to an ion source has radically improved. This progress has led to superior sensitivities of the hyphenated systems [gas chromatography (GC)-MS, liquid chromatography (LC)-MS, capillary electrophoresis (CE)-MS]. The range of existing and potential applications of time-resolved mass spectrometry (TRMS) is very broad. Various mass spectrometric approaches have been adopted in order to attain satisfactory temporal resolutions, and to match the requirements of diverse applications. While MS cannot equalize the performance of optical detection systems when it comes to the recording speed, the temporal resolutions achieved to date (milliseconds and less) emphasize the potential of MS in studies of highly dynamic phenomena.

Time-Resolved Mass Spectrometry: From Concept to Applications, First Edition.
Pawel Lukasz Urban, Yu-Chie Chen and Yi-Sheng Wang.
© 2016 John Wiley & Sons, Ltd. Published 2016 by John Wiley & Sons, Ltd.

Just as MS has almost become a "universal" tool for characterization of chemical species (from small molecules to large biomolecular complexes), during the past four decades, TRMS has grown into a multi-disciplinary field. The areas of applications of TRMS include elucidation of reaction kinetics (including slow and fast as well as inorganic, organic, electrochemical and biochemical reactions), identification of reaction intermediates and pathways, chemical process monitoring, environmental monitoring, screening catalysts, investigation of protein folding, and various fundamental processes in chemistry, physics, and biology. Condensed and gas-phase phenomena are investigated using diverse TRMS methods. Overall, TRMS supports developments in various sub-fields of fundamental and applied sciences as well as industrial applications.

Various mass spectrometric techniques facilitate TRMS analyses. Interfaces and ion sources are critical for sample introduction to a mass spectrometer, especially in the case of real-time monitoring of dynamic samples. Temporal resolution of mass analyzers also plays a role when millisecond-timescale processes are recorded over time. However, there exist TRMS methods in which steady or quasi-steady samples are generated. In those cases, the role of MS hardware is diminished while the sample processing steps (e.g., fast quenching, development of product concentration gradients) are of utmost importance.

14.2   Instrumentation

In terms of the hardware, TRMS methods described in this book use most common types of ion sources and analyzers. Electrospray ionization (ESI), electron ionization (EI), atmospheric pressure chemical ionization (APCI), or photoionization systems, and their modified versions, are all widely used in TRMS measurements. The newly developed atmospheric pressure ionization schemes such as desorption electrospray ionization (DESI) and Venturi easy ambient sonic-spray ionization (V-EASI) have already found applications in this area. Mass analyzers constitute the biggest and the most costly part of MS hardware. Few laboratories can afford purchasing different types of mass spectrometers for use in diverse applications. Therefore, the choice of mass spectrometer for TRMS is not always dictated by the optimum specifications of the instrument but its availability. Fortunately, many real-time measurements can be conducted using different mass analyzers equipped with atmospheric pressure inlets – with better or worse results. For example, triple quadrupole mass spectrometers excel at quantitative capabilities; however, in many cases, popular ion trap (IT)-MS instruments can be used instead. On the other hand, applications of TRMS in fundamental studies often require a particular type of instrument (e.g., Fourier transform ion cyclotron resonance mass spectrometer for photodissociation studies on trapped ions).

One should note that the field of MS is loaded with terms which – in some cases – overlap with one another. This overlapping terminology may certainly be confusing for early-stage researchers. For example, the differences between certain concepts may seem to be subtle [e.g., fused droplet ESI *vs.* extractive electrospray ionization (EESI); electrospray-assisted laser desorption/ionization (ELDI) *vs.* laser ablation electrospray ionization (LAESI); desorption electrospray ionization (DESI) *vs.* easy ambient sonic-spray ionization (EASI); liquid microjunction surface sampling (LMJ-SSP) *vs.* nanospray desorption electrospray ionization (nanoDESI)]. On the other hand, some dissimilar concepts have similar names (e.g., DESI and nanoDESI). When discussing various time-resolved approaches in this book, we tried to follow the terminology used in the original reports. Hopefully, the next few years will bring consolidation of the terminology covering the approaches disclosed over the past two decades, and better classification of the existing mass spectrometric techniques. The readers are encouraged to verify the suitability of this nomenclature for their sub-field and specific applications, and search for updates in the most recent review articles.

Smart fluid handling strategies were proposed to accommodate the requirements of specific measurements, such as following early stages of biochemical reactions, protein folding events, and fast electrochemical processes. To this end, microfluidics and microreaction approaches serve TRMS well, providing ingenious solutions for the experiments which contribute new knowledge to chemical science. Fluid dynamics can help to characterize the mixing process in the early stage of reactions tracked by TRMS. Especially in the case of fast reactions, it is vital to discern the substrate solution merging time (duration), mixing time, incubation time, and reaction time (see Chapter 4). Estimations of the incubation times in fast TRMS methods are mostly based on observations. In order to characterize these processes based on first principles it would be desirable to carry out numerical simulations that could take into account the properties of fluids as well as the geometry of the interfaces. Certainly, collaborations across disciplines (including mass spectrometrists, chemical and mechanical engineers) can bring an added value to the understanding of the processes that occur in many of the systems designed for TRMS studies.

Nowadays, it is desirable to work towards new applications of TRMS in order to demonstrate its usefulness in other areas of chemical science to enable new discoveries. For example, it would be appealing to set up analytical procedures which would allow monitoring of protein complexes in tissues and cells, and at different locations and time points within single cells [1]. Successful implementation of robotic systems with MS has already been presented [2–4]. Robots speed up sample preparation, and enable prompt delivery of fast-changing samples to the MS interfaces. They can perform simple operations on the analyzed sample: movement, addition of reagent, mixing, MS analysis. However, accuracy and speed of sample handling conducted by robotic MS systems is nowadays limited. We believe that, following further developments, robotic systems will enhance operation of modern TRMS platforms. TRMS also provides new insights on the dynamics of sample treatment procedures used in chemistry. It may assist characterization of new methods for fast bias-free processing of dynamic (unstable) samples. The forecasted wide-spread use of TRMS should also encourage new fundamental developments in MS technology.

14.3   Software

During the past two decades, software for data acquisition and analysis in MS has become more user-friendly. This trend can be largely attributed to the developments of computer technology, and better understanding of software ergonomics by the developers. Students and technicians can now quickly get trained in the operation of sophisticated high-end mass spectrometers. The low-level settings are often "hidden" to make the operation of the software easier and to reduce the possibility of accidental misconfiguration. On the other hand, expert users can still access some of these settings in the software to fine-tune the instruments for special tasks. Mass spectrometry software packages often incorporate sections facilitating collection of data while hyphenating MS systems with separation systems (chromatography, electrophoresis, or ion mobility). Thus, the analysis workflows can readily be streamlined for high-throughput and high-performance operations. Nevertheless, most software packages have not yet been customized for time-resolved analyses of dynamic samples such as reaction mixtures. One can expect that the new versions will enable real-time processing of the ion traces (e.g., dividing the analyte ion currents by the internal standard ion currents) and kinetic analysis (fitting pre-processed ion currents with kinetic equations to calculate reaction rates). For the time-being, these operations are often carried out using separate programs or scripts.

14.4   Limitations

There are still numerous problems to solve in order to warrant wide-spread use of TRMS. For example, many fast reactions are conducted in solvents that are incompatible with the ionization techniques used in MS. Reaction mixtures cannot always be directly "pumped" to the mass spectrometer using conventional ion sources and interfaces. Systems for solvent exchange need to be developed and made available. However, there always exists the risk that on-line sample treatment may influence sample composition and relative concentrations of reactants (e.g., in the case of chemical equilibria). Moreover, it is hard to verify the presence of such possible artifacts. There exist only few model dynamic systems that may be used as reference in the validation of newly developed TRMS methods.

Ion suppression is a serious issue in TRMS. It can cause confounding effects in the recorded mass spectra. The unstable sample matrix can produce changes in the ion-time records which are not due to changes in the original samples. Effort needs to be made to mitigate the influence of ion suppression effects in temporal measurements conducted by MS. In all cases, one has to verify the existence of such effects, and find out how and to what extent they can bias the experimental observations. Yet another challenge is that some TRMS setups have little control over the timing (duration) of reactant incubation, and limited accuracy of the set time points. This limitation can affect the accuracy of the information derived from the kinetic measurements conducted by MS.

References

1. Sharon, M. (2010) How Far Can We Go with Structural Mass Spectrometry of Protein Complexes? J. Am. Soc. Mass Spectrom. 21: 487–500.
2. Painter, A.J., Jaya, N., Basha, E., Vierling, E., Robinson, C.V., Benesch, L.P. (2008) Real-time Monitoring of Protein Complexes Reveals Their Quaternary Organization and Dynamics. Chem. Biol. 15: 246–253.
3. Bennett, R.V., Morzan, E.M., Huckaby, J.O., Monge, M.E., Christensen, H.I., Fernández, F.M. (2014) Robotic Plasma Probe Ionization Mass Spectrometry (RoPPI-MS) of Non-planar Surfaces. Analyst 139: 2658–2662.
4. Chiu, S.-H., Urban, P.L. (2015) Robotics-assisted Mass Spectrometry Assay Platform Enabled by Open-source Electronics. Biosens. Bioelectron. 64: 260–268.

Index

Time-Resolved Mass Spectrometry: From Concept to Applications, First Edition.
Pawel Lukasz Urban, Yu-Chie Chen and Yi-Sheng Wang.
© 2016 John Wiley & Sons, Ltd. Published 2016 by John Wiley & Sons, Ltd.